Introductory
ALGEBRA

Patricia Hooper
Department of Mathematics and Computer Science
Western Kentucky University

Linda Pulsinelli
Department of Mathematics and Computer Science
Western Kentucky University

Introductory ALGEBRA

MACMILLAN PUBLISHING CO., INC.
NEW YORK

Collier Macmillan Publishers
London

We gratefully acknowledge permission to reprint the table of Square Roots that appears on the inside of the back cover:

From *Intermediate Algebra,* Second Edition, by C. L. Johnston and Alden T. Willis. © 1979 by Wadsworth Publishing Company, Inc. Reprinted by permission of Wadsworth Publishing Company, Belmont, California 94002.

Macmillan Publishing Co., Inc.
866 Third Avenue, New York, New York 10022

Collier Macmillan Canada, Inc.

Library of Congress Cataloging in Publication Data

Hooper, Patricia (Patricia I.)
    Introductory algebra.

    Includes index.
    1. Algebra. I. Pulsinelli, Linda. II. Title.
QA154.2.H66   512.9   82-15214
ISBN 0-02-357100-4   AACR2

Printing: 2 3 4 5 6 7 8   Year: 3 4 5 6 7 8 9

ISBN 0-02-357100-4

Preface

To the Students

This introductory algebra text was written with you in mind at all times. Having taught basic algebra to college students for many semesters, we were anxious to write a book that you would find interesting and readable. We did not pretend to ourselves that you would be wildly enthusiastic about taking this course. On the other hand, we knew from experience that if the principles of basic algebra were presented clearly and logically with plenty of opportunity for you to practice each new skill, you would find that you could indeed succeed in learning the fundamentals of algebra and be ready for your next mathematics course.

In presenting each new concept, we have tried to relate it to an already familiar idea. Many examples are worked in detail for your reference and the book is sprinkled with **Trial Runs,** which check to see if you have understood each new skill. Each section concludes with an **Exercise Set** to provide plenty of practice with the new concepts presented in that lesson. Answers to the odd-numbered exercises are included at the end of the book. Each chapter contains a short summary and a set of **Review Exercises** designed to tie together the material discussed in that chapter.

We hope that you find our explanations clear and readable and that your experience with this course is a successful one. Algebra is indeed the key to understanding all higher-level mathematics, and the skills learned in this course will help you succeed in all future mathematics courses. Our success as teachers and authors depends on your success as students. We have attempted to write this book as though we were speaking to each of you in person. The rest is up to you.

To the Instructor

One of the most difficult challenges in teaching an introductory algebra course for college students is maintaining student interest and enthusiasm. Students in such a course rarely visualize themselves as good mathematics students. Often, they have not taken more than general mathematics or consumer mathematics courses at the high school level. Even if they *have* taken some algebra, they have not been particularly successful with it, or in the case of returning students, they have retained very little. In almost all cases, they are frightened by their first college mathematics course.

Our first goal was to structure this book with the idea in mind that if a student achieves success quickly, he or she will be motivated to continue. To realize that objective, we have attempted to break down the course content into a sequence of fundamental skills. Imbedded within each section are several short sets of **Trial Run** exercises that allow the student to see whether he or she has followed the preceding explanation of a new manipulative technique. The answers for these exercises appear immediately following each Trial Run.

Each section ends with a lengthy **Exercise Set** designed to reinforce the new skills. Every odd-numbered exercise corresponds exactly to the following even-numbered exercise, allowing for flexibility in making assignments. The answers to the odd-numbered exercises are included at the end of the book, and the even-numbered answers appear in the *Instructor's Resource Manual*. Following each Exercise Set in the book is a short **Checkup** consisting of no more than 10 exercises to evaluate the student's mastery of the new skills. Each checkup exercise is keyed to a slightly different concept to enable the teacher to diagnose the student's weaknesses. To help correct those weaknesses, **Supplementary Exercise Sets** are included in the *Instructor's Resource Manual*. Each Supplementary Set contains exercises that are comparable to those in the book. Every Supplementary Exercise Set is followed by a **Quiz** that is comparable to the corresponding Checkup in the book. Since the pages of the manual are perforated, they can be easily removed for duplication if the instructor so desires. The answers to the Supplementary Exercises and Quizzes are included only in the *Instructor's Resource Manual*, along with six comparable versions of each **Chapter Test.** Each chapter also includes a set of **Review Exercises** with answers in the back of the book.

The vocabulary level of a book of this type seems crucial to its success with students, and we have made a concerted attempt to word our explanations in simple, straightforward language. From our experience with basic algebra students, we have discovered that every necessary skill can be explained in simple terms understandable to almost every reader. We have assumed very little in the way of mathematical backgrounds, but we have found that in most cases a student will reach the desired generalizations after several well-chosen examples have been presented. This book does not attempt to teach basic number facts, but it does try to refresh students on those arithmetic skills which are in need of review and which have direct bearing upon mastery of an algebraic skill.

The general approach of the book is to start with a fundamental idea with which the students are familiar and proceed to a related skill in the most straightforward, intuitive way possible. New concepts are presented using examples, and generalizations are presented in boxes as postscripts rather than as starting points. (Number properties, for instance, are mentioned by name, but they are reinforced by repetition rather than by requiring memorization of words.) By encountering the material in manageable chunks, it is possible for most students to keep up with their daily work without becoming bogged down with an overwhelming amount of new material to be mastered at one time.

Realistically speaking, all the desired content of a good high school Algebra I course cannot be covered in 3 hours per week during one semester. It is possible, however, to develop students' skills in operations with integers, polynomials, natural number exponents, factoring, linear equations, quadratic equations, and rational algebraic expressions. A bit more time would allow for exposure to inequalities, irrational numbers, and graphing straight lines. Each of these topics is treated in depth with applied problems included where relevant in a chapter section entitled "Switching from Words to Algebra."

This book is designed to be of use in several types of instructional programs.

1. *Conventional lecture course*
   The explanations in the book serve to reinforce the lecture, and the instructor may make appropriate assignments from the Exercise Sets. The Checkups may be used as self-quizzes.
2. *Modified self-paced course*
   The student can work through the book with deadlines set for completion of each lesson or chapter. Successful mastery of each skill via Checkups can precede the student's continuing to the next lesson. Chapter Tests may be given to the entire group at a prescribed time.
3. *Completely self-paced course*
   The student may work through the book at his or her own rate, using the Trial Runs and Checkups to determine when he or she is ready to attempt a Chapter Test.

Acknowledgments

The writing of this book would not have been possible without the assistance of many people, and we would like to express our appreciation to Maxine Worthington for typing the manuscript; to Bob Pulsinelli for carefully perusing every page; to our families for cooperating with our obsessive work schedules; to David Almand for assistance with the Exercise sections, and to our colleagues at Western Kentucky University for their encouragement. We also thank our reviewers—Professors Cynthia Siegel, University of Missouri—St. Louis; Matt Kaufmann, Purdue University; Ned W. Schillow, Lehigh County Community College; Michael R. Karelius, American River College; Ronald R. Young, Harrisburg Area Community College; Ara B. Sullenberger, Tarrant County Junior College—South Campus; Peter A. Lindstrom, Genesee Community College; Helen Joan Dykes, Northern Virginia Community College; Mary Scott, University of Florida; and Ernest Palmer, Grand Valley State College—for their careful scrutiny and helpful comments; Wayne Yuhasz and Susan Saltrick, our Mathematics Editors, for their thoughtful guidance; and Elaine Wetterau, our Production Supervisor, for her expertise. But most of all we must thank our Mathematics 055 students at Western Kentucky University for convincing us that despite their weak mathematical backgrounds a well-written textbook could provide the key to their success. We have attempted to write such a book and we dedicate the following pages to them.

P. I. H.
L. R. P.

Contents

1 Working with Numbers

Algebra can be thought of as the bridge between arithmetic and higher mathematics. A person with even a limited knowledge of algebra can solve many everyday problems that could otherwise be solved only by trial and error. Although we expect you to know very little (if any) algebra, we assume that your basic arithmetic facts are in good shape and that you know how to add, subtract, multiply, and divide.

In this chapter we

1. Add, subtract, multiply, and divide whole numbers and integers.
2. Observe some properties of addition and multiplication.
3. Switch from words to numbers.

1.1 Adding and Multiplying Whole Numbers

As a small child, you learned to count "one, two, three," and so forth. These counting numbers are also called **natural numbers.**

> Natural numbers: 1, 2, 3, 4, 5, . . .

The dots after the 5 mean "and so on"; in other words, this list continues indefinitely. If we include the number zero in this list, we obtain a new list of numbers called the **whole numbers.**

> Whole numbers: 0, 1, 2, 3, 4, 5, . . .

1.1-1 The Number Line

To picture the whole numbers, we shall use a **number line.** We draw a line and choose a zero point and a length to represent 1 unit. Then all points spaced 1 unit apart to the right of zero are labeled with the whole numbers in order.

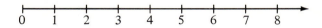

The arrow shows that this line goes on and on indefinitely so that numbers such as 24 and 793 also correspond to points.

To illustrate a whole number on the number line, we put a solid dot at the point corresponding to that number. This is called **plotting a point** on the number line.

Example 1. Plot 4 on the number line.

Solution

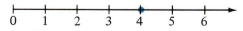

Example 2. Plot 0 on the number line.

Solution

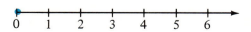

The number line also gives us a handy way of comparing numbers. If we look at a number, for example 3, we notice that all the numbers to the right of 3 are larger than 3. All numbers to the left of 3 are smaller than 3. For instance:

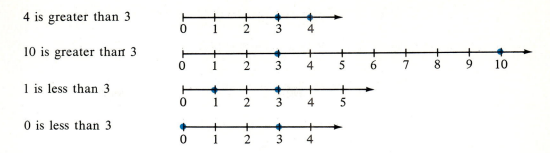

4 is greater than 3

10 is greater than 3

1 is less than 3

0 is less than 3

We may write these statements using the symbol $<$, which means "is less than" or the symbol $>$, which means "is greater than."

$$4 > 3 \text{ means "4 is greater than 3"}$$

$$0 < 3 \text{ means "0 is less than 3"}$$

$$25 > 6 \text{ means "25 is greater than 6"}$$

Such statements are called **inequalities** and we may summarize them as follows, for any numbers A and B.

> Less than: $A < B$ if A lies to the left of B on the number line.
> Greater than: $A > B$ if A lies to the right of B on the number line.

Example 3. Place a $<$ or $>$ symbol between the numbers.

$$12 \_\_\_\_\_ 3 \qquad 0 \_\_\_\_\_ 1$$
$$5 \_\_\_\_\_ 7 \qquad 110 \_\_\_\_\_ 99$$

Solution

$$12 > 3 \qquad \text{because 12 lies to the right of 3}$$
$$5 < 7 \qquad \text{because 5 lies to the left of 7}$$
$$0 < 1 \qquad \text{because 0 lies to the left of 1}$$
$$110 > 99 \qquad \text{because 110 lies to the right of 99}$$

We shall see that the number line is a useful tool in adding, subtracting, multiplying, and dividing numbers.

1.1-2 Adding Whole Numbers

Recall that when we add two numbers, the answer is called the **sum.**

Example 4. Find the sum of 3 and 2.

Solution. $3 + 2 = 5$. The sum of 3 and 2 is 5.

We might have used the number line to help find this sum. To compute $3 + 2$, we could start at 3 and move 2 more units to the right, like this:

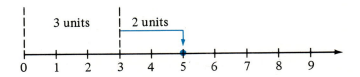

Notice that we indeed end up at 5.

Now look at the sum of 2 and 3, using the number line.

Example 5. Find the sum of 2 and 3.

Solution. We must find $2 + 3$. This time we shall start at point 2 and move 3 units to the right.

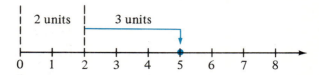

Again, we end up at 5, so $2 + 3 = 5$.

We have discovered that $3 + 2 = 5$ and $2 + 3 = 5$. Let's look at some more sums computed on the number line.

Example 6. Find $2 + 8$ and $8 + 2$.

Solution. For $2 + 8$:

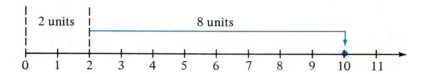

For $8 + 2$:

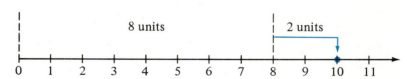

So

$$2 + 8 = 10 \qquad \text{and} \qquad 8 + 2 = 10$$

Example 7. Find $1 + 6$ and $6 + 1$.

Solution. For $1 + 6$:

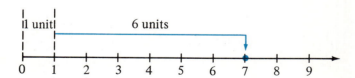

For $6 + 1$:

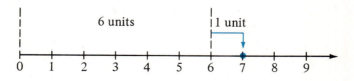

So

$$1 + 6 = 7 \qquad \text{and} \qquad 6 + 1 = 7$$

From these examples, it seems that we could safely say that

When we add two numbers, it does not matter in what order we do it. The sum will be the same.

In other words,

$$3 + 2 = 2 + 3$$
$$2 + 8 = 8 + 2$$
$$1 + 6 = 6 + 1$$
$$1 + 23 = 23 + 1$$
$$100 + 13 = 13 + 100$$
$$0 + 8 = 8 + 0$$

This law is called the **commutative property for addition.** If we use the symbols A and B to represent *any* whole numbers, we may write:

Commutative Property for Addition
$$A + B = B + A$$

Do not be disturbed by the fact that we have used letters to stand for numbers. Using letters just allows us to state a law *in general* without using specific numbers.

Suppose that we wish to add three numbers together. The human brain can only add two numbers together at once, so to compute the sum

$$5 + 2 + 8$$

perhaps you think "5 + 2 is 7, and 7 + 8 is 15. So 5 + 2 + 8 = 15." Perhaps your classmate says "2 + 8 is 10, and 5 + 10 is 15. So 5 + 2 + 8 = 15." Who is right and who is wrong? You are both right!

You added 5 and 2 first and then added the 8, whereas your classmate added 2 and 8 first and then added the 5. Let's use parentheses to show both approaches. The parentheses will say "do this first."

Your approach:

$$5 + 2 + 8 = (5 + 2) + 8$$
$$= 7 + 8$$
$$= 15$$

Classmate's approach:

$$5 + 2 + 8 = 5 + (2 + 8)$$
$$= 5 + 10$$
$$= 15$$

Let us use the number line to illustrate both approaches:
Your approach:

$$(5 + 2) + 8$$

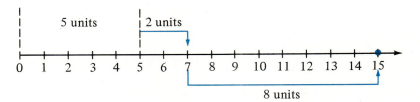

Classmate's approach:

$$5 + (2 + 8)$$

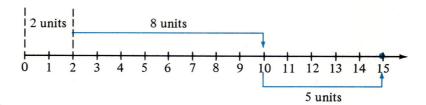

From this example we may conclude that

$$(5 + 2) + 8 = 5 + (2 + 8)$$

or, in words, we might say:

> When we are adding three numbers, it does not matter if we add the first two and then the third, *or* add the second two and then add the first. The sum will be the same.

For instance,

$$(2 + 9) + 3 = 2 + (9 + 3)$$
$$11 + 3 = 2 + 12$$
$$14 = 14$$

Example 8. Use parentheses to show two approaches to finding the sum $27 + 12 + 18$.

Solution

$$(27 + 12) + 18$$
$$27 + (12 + 18)$$

Notice that the second approach is easier to compute in this case because $12 + 18 = 30$ and $27 + 30 = 57$. You will get the same answer either way, but often one approach involves easier arithmetic than the other.

This second law is called the **associative property for addition.** Again using the letters A, B, and C to represent *any* whole numbers, we may write:

> **Associative Property for Addition**
> $$(A + B) + C = A + (B + C)$$

Notice that the commutative property deals with the addition of *two* quantities and the associative property deals with the addition of *three* quantities. To use the commutative property, we switch the order in which two quantities are added.

$$5 + 10 = 10 + 5$$

To use the associative property, we group the three quantities differently without switching their order.

$$(5 + 10) + 6 = 5 + (10 + 6)$$

1.1-3 Multiplying Whole Numbers

There are several ways to indicate that two numbers are to be multiplied. For instance, to multiply 4 times 3, we may write

$$4 \times 3$$
$$4 \cdot 3$$
$$4(3)$$
$$(4)3$$
$$(4)(3)$$

Each of these says the same thing, but we shall not use the first notation, 4×3, because of confusion that might occur when we begin to use the letter x to mean something else.

The answer to a multiplication problem is called the **product**. One way to find the product of two numbers is to realize that multiplication is just repeated addition.

$$4 \cdot 3 \quad \text{means} \quad 3 + 3 + 3 + 3$$

$$5 \cdot 2 \quad \text{means} \quad 2 + 2 + 2 + 2 + 2$$

$$6 \cdot 1 \quad \text{means} \quad 1 + 1 + 1 + 1 + 1 + 1$$

Since we learned how to add first, we know that

$$4 \cdot 3 = 12$$

$$5 \cdot 2 = 10$$

$$6 \cdot 1 = 6$$

Do you suppose that there is a **commutative property for multiplication** like the commutative property for addition?

Does $4 \cdot 3 = 3 \cdot 4$?

$$4 \cdot 3 \quad \text{means} \quad 3 + 3 + 3 + 3 = 12$$

$$3 \cdot 4 \quad \text{means} \quad 4 + 4 + 4 = 12$$

So $4 \cdot 3 = 3 \cdot 4$.
Similarly,

$$5 \cdot 2 = 2 \cdot 5$$

$$6 \cdot 1 = 1 \cdot 6$$

$$7 \cdot 12 = 12 \cdot 7$$

so we may conclude:

> It does not matter in what order we multiply two numbers. The product will be the same.

or using A and B to represent any whole numbers, we may write:

> **Commutative Property for Multiplication**
> $$A \cdot B = B \cdot A$$

How do we multiply three whole numbers? If you were asked to multiply

$$5 \cdot 2 \cdot 9$$

you might think, "5 · 2 is 10, and 10 · 9 is 90. So the product is 90." A classmate might think "2 · 9 is 18, and 5 · 18 is 90. So the product is 90." The answer is the same in either case.

Again, parentheses indicate the grouping done in the two approaches to 5 · 2 · 9. Your approach:

$$5 \cdot 2 \cdot 9 = (5 \cdot 2) \cdot 9$$
$$= \quad 10 \quad \cdot 9$$
$$= 90$$

Classmate's approach:

$$5 \cdot 2 \cdot 9 = 5 \cdot (2 \cdot 9)$$
$$= 5 \cdot \quad 18$$
$$= 90$$

As you may have guessed, the law that guarantees the same answer with either approach is called the **associative property for multiplication**.

Example 9. Illustrate the associative property on the product 3 · 9 · 4.

Solution. $(3 \cdot 9) \cdot 4 = 3 \cdot (9 \cdot 4)$.

Using A, B, and C to represent any whole numbers, we may write:

> **Associative Property for Multiplication**
>
> $$(A \cdot B) \cdot C = A \cdot (B \cdot C)$$

Trial Run   *Place the symbol $<$ or $>$ between each pair of numbers.*

1. 3 _____ 9

2. 2 _____ 0

According to the commutative property for addition:

3. $3 + 4 = 4 +$ _____

4. $8 + 7 =$ _____

According to the associative property for addition:

5. $4 + (5 + 1) =$ _____ $+ 1$

6. $(7 + 0) + 9 =$ _____

According to the commutative property for multiplication:

7. $9 \cdot 7 =$ _____ $\cdot 9$

8. $4 \cdot 5 =$ _____

According to the associative property for multiplication:

9. $5(9 \cdot 7) = $ _____ 7

10. $(2 \cdot 11)5 = $ _____

1.1-4 Using Symbols of Grouping

We have seen already how parentheses tell us to "do this first."

Example 10. Find $(3 + 2) + 16$.

Solution

$$(3 + 2) + 16$$
$$= \quad 5 \quad + 16$$
$$= 21$$

Example 11. Find $(4 \cdot 25) \cdot 7$.

Solution

$$(4 \cdot 25) \cdot 7$$
$$= \quad 100 \quad \cdot 7$$
$$= \quad 700$$

Example 12. Find $(3 + 2) \cdot 6$.

Solution

$$(3 + 2) \cdot 6$$
$$= \quad 5 \quad \cdot 6$$
$$= \quad 30$$

Example 13. Find $3 + (2 \cdot 6)$.

Solution

$$3 + (2 \cdot 6)$$
$$= 3 + \quad 12$$
$$= 15$$

Notice in Examples 12 and 13 that the placing of the parentheses changed the problem completely. Parentheses ( ), brackets [ ], and curly braces { } all give you directions about the order in which operations are to be performed. In problems containing parentheses *and* brackets, you must remember to *work inside the innermost symbols of grouping first*.

For instance, to simplify $3[2 + (4 \cdot 5)]$, we proceed as follows:

$$3[2 + (4 \cdot 5)]$$
$$= 3[2 + 20]$$
$$= 3[22]$$
$$= 66$$

Notice that we proceeded in an orderly fashion, writing down each step of calculation.

Example 14. Simplify $[2 + (8 \cdot 9) + 1] + 5$.

Solution

$$[2 + (8 \cdot 9) + 1] + 5$$
$$= [2 + 72 + 1] + 5$$
$$= \qquad 75 \qquad + 5$$
$$= 80$$

Example 15. Simplify $(3 + 2)(8 + 1)$.

Solution

$$(3 + 2) \, (8 + 1)$$
$$= \quad 5 \quad \cdot \quad 9$$
$$= 45$$

But suppose that parentheses are not used in a problem which involves several operations. There must be some accepted rule for working such problems as

$$3 + 2 \cdot 7$$

$$5 \cdot 2 + 6 \cdot 8$$

$$7 + 6 \cdot 3 + 1$$

To avoid confusion, mathematicians have agreed that

> If there are no parentheses to indicate order of operations, first perform multiplications from left to right, then perform additions from left to right.

Example 16. Simplify $3 + 2 \cdot 7$.

Solution

$$3 + 2 \cdot 7$$
$$= 3 + \quad 14$$
$$= 17$$

Example 17. Simplify $5 \cdot 2 + 6 \cdot 8$.

Solution

$$5 \cdot 2 + 6 \cdot 8$$
$$= \quad 10 \quad + \quad 48$$
$$= 58$$

Example 18. Simplify $7 + 6 \cdot 3 + 1$.

Solution

$$7 + 6 \cdot 3 + 1$$
$$= 7 + 18 + 1$$
$$= \quad 25 \quad + 1$$
$$= 26$$

We know one way to work a problem such as $3(2 + 5)$. Parentheses tell us to add first, so

$$3(2 + 5)$$
$$= 3 \cdot 7$$
$$= 21$$

Another way to work this problem is to multiply 3, the number outside the parentheses, times each of the numbers within the parentheses and then add those products. Let's see if the result is the same.

$$3(2 + 5) = 3 \cdot 2 + 3 \cdot 5$$
$$= \ 6 \ + \ 15$$
$$= 21$$

It seems that the first method is easier, but later we shall have to use the second method, so we shall note here that

> To multiply a number times the sum of numbers in parentheses, we may multiply the number outside the parentheses times each of the numbers inside the parentheses and add those products together.

This new law is the **distributive property for multiplication over addition**. Using A, B, and C to represent whole numbers, we may write:

> **Distributive Property for Multiplication over Addition**
> $$A(B + C) = A \cdot B + A \cdot C$$

Example 19. Use the distributive property for multiplication over addition to simplify $5(7 + 1)$.

Solution

$$5(7 + 1)$$
$$= 5 \cdot 7 + 5 \cdot 1$$
$$= \ 35 \ + \ 5$$
$$= 40$$

Example 20. Use the distributive property for multiplication over addition to simplify $10(17 + 19)$.

Solution

$$10(17 + 19)$$
$$= 10 \cdot 17 + 10 \cdot 19$$
$$= \ 170 \ + \ 190$$
$$= 360$$

1.1-5 Working with Zero and One

The whole number zero has some special properties that we should note here.

Example 21. Simplify $17 + 0$.

Solution. $17 + 0 = 17$.

Example 22. Simplify $0 + 4$.

Solution. $0 + 4 = 4$.

Example 23. Simplify $(6 + 0) + 9$.

$$(6 + 0) + 9$$
$$= \quad 6 \quad + 9$$
$$= 15$$

It should come as no surprise to you that

If zero is added to any number, the answer is that number.

This law is called the **addition property of zero** (or the **identity property for addition**). Using A to represent any whole number, we write:

Addition Property of Zero

$$A + 0 = A$$

What happens when we multiply any number times zero? Recalling that multiplication is repeated addition, look at $5 \cdot 0$ and $0 \cdot 3$.

$$5 \cdot 0 = 0 + 0 + 0 + 0 + 0 = 0$$
$$0 \cdot 3 = 3 \cdot 0 = 0 + 0 + 0 = 0$$

It seems, then, that

If zero is multiplied times any number, the answer is zero.

This law is called the **multiplication property of zero**, and using A to represent any whole number, we may write:

Multiplication Property of Zero

$$A \cdot 0 = 0$$

What happens when we multiply any number by 1? Look at $5 \cdot 1$ and $1 \cdot 3$.

$$5 \cdot 1 = 1 + 1 + 1 + 1 + 1 = 5$$
$$1 \cdot 3 = 3 \cdot 1 = 1 + 1 + 1 = 3$$

It seems that

If any number is multiplied times 1, the answer is that number.

This law is called the **multiplication property of 1** (or the **identity property for multiplication**). Letting A represent any whole number, we write

$$\boxed{\begin{array}{c} \textbf{Multiplication Property of 1} \\ A \cdot 1 = A \end{array}}$$

Trial Run

1. According to the distributive property for multiplication over addition:

$$8(2 + 9) = \underline{\hspace{1cm}} + \underline{\hspace{1cm}}$$

Simplify.

_____ **2.** $2(3 + 7)$

_____ **3.** $4[5 + (3 \cdot 2)]$

_____ **4.** $8 + 3 \cdot 5 + 2$

_____ **5.** $(7 + 2)(3 + 1)$

_____ **6.** $3 + (7 + 0)$

_____ **7.** $5[2 + 7 \cdot 0]$

_____ **8.** $8 \cdot 1 + 0$

_____ **9.** $3(2 \cdot 0)$

_____ **10.** $4(5 + 0)$

ANSWERS
1. $8 \cdot 2 + 8 \cdot 9$. **2.** 20. **3.** 44. **4.** 25. **5.** 36. **6.** 10. **7.** 10. **8.** 8. **9.** 0.
10. 20.

1.1-6 Switching from Words to Numbers

It is important for an algebra student to be able to change word statements into number statements. Such an ability will allow us to solve many "real-life" problems which could otherwise be solved only by trial and error.

Let's see if we can practice switching some word expressions into number expressions.

Example 24. If Henry is 18 years old now, how old will he be in 3 years?

Solution. Addition is the operation to use here. Henry's age will be $18 + 3$.

Example 25. Nancy's hourly wage is $3. If she works 32 hours this week, what will be her gross pay? (Gross pay is pay before any deductions are made.)

Solution. Since Nancy receives $3 for every hour she works, we must multiply hours worked times hourly wage. She will receive 32($3) for her week's work.

Example 26. Nancy now earns $3 per hour, but next week she will receive a raise of $1 per hour. Write an expression to represent her new hourly wage.

Solution. Nancy's new hourly wage will be $3 + $1.

Example 27. If Nancy works 30 hours next week, what will be her gross pay?

Solution. We must multiply hours worked (30) times hourly wage ($3 + $1) to obtain her week's pay of 30($3 + $1).

Example 28. Tom weighs 20 pounds more than twice his son's weight. If Tom's son weighs 73 pounds, write an expression for Tom's weight.

Solution. Tom's son weighs 73 pounds, so twice his son's weight is 2(73) pounds. Tom weighs 20 pounds more than that, so Tom weighs 2(73) + 20.

There are several key phrases that will help you switch from words to numbers. We shall list those phrases here for you to translate. The answers are given below; try not to look at them until you have attempted the switch yourself.

Words	*Numbers*
1. the sum of 7 and 9	_____
2. 6 more than 25	_____
3. 6 increased by 25	_____
4. twice as much as 7	_____
5. 14 doubled	_____
6. 5 times as large as 10	_____

Now you may check your answers.

1. $7 + 9$ **2.** $25 + 6$ **3.** $6 + 25$ **4.** $2 \cdot 7$ **5.** $14 \cdot 2$ **6.** $5 \cdot 10$

One kind of problem that crops up often involves some sort of rate, such as miles per hour, dollars per hour, miles per gallon, revolutions per minute, and so on. Each of these rates is described as

something *per* something else

and tells us how many units of the first thing we can expect for *one* unit of the other thing. For instance:

"23 miles per gallon" means that we can expect to travel 23 miles on *one* gallon of gas

"$3 per hour" means that we can expect $3 for every *one* hour we work

"78 revolutions per minute" means that we can expect a record to spin around 78 times for every *one* minute it plays

"55 miles per hour" means that we can expect to travel 55 miles for every *one* hour we drive

How should we compute how far we could travel in 3 hours? If we travel 55 miles in 1 hour, we must multiply that distance by 3, since we would have traveled 3 times as long. So we could travel $3 \cdot 55$ or 165 miles in 3 hours.

Example 29. How many revolutions will a record make during a 4-minute song if it is spinning at 78 revolutions per minute (rpm)?

Solution. Since the record makes 78 revolutions in 1 minute, it will make $4 \cdot 78 = 312$ revolutions in 4 minutes.

Example 30. If Juan's car averages 23 miles per gallon (mpg), how far can he expect to travel on a full tank of gas? Juan's gas tank holds 20 gallons.

Solution. Since Juan's car can travel 23 miles on 1 gallon of gas, he can expect it to travel $20 \cdot 23 = 460$ miles on 20 gallons of gas.

Trial Run   *Change each word expression to a number expression.*

_____ **1.** Leroy's electric bill this month was $15 more than his bill last month. If his bill last month was $33, what is his bill this month?

_____ **2.** Jeff has 15 tapes, but Carol has 3 times as many as Jeff. How many tapes does Carol have?

_____ **3.** If Sara drives at an average speed of 50 miles per hour, how far can she drive in 3 hours?

_____ **4.** The length of a table is 7 inches longer than twice its width. If the width of the table is 15 inches, what is its length?

_____ **5.** A stamp collector purchases 18 stamps at $2 per stamp. What was the amount of the purchase?

ANSWERS
1. $33 + $15.   **2.** 3(15).   **3.** 3(50) miles.   **4.** 2(15) + 7 inches.   **5.** 18($2).

EXERCISE SET 1.1

1. Plot the points corresponding to these numbers on a number line.
   (a) 3   (b) 0   (c) 11   (d) 1

2. Plot the points corresponding to these numbers on a number line.
   (a) 5   (b) 1   (c) 12   (d) 0

3. Tell what number corresponds to each letter on the number line.

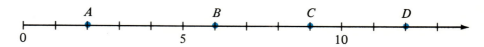

4. Tell what number corresponds to each letter on the number line.

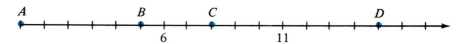

In Exercises 5–20, perform the operation.

_____ 5. $3 + 2 + 1$ _____ 6. $10 + 6 + 7$

_____ 7. $2 \cdot 3 \cdot 8$ _____ 8. $5 \cdot 6 \cdot 4$

_____ 9. $14 + (8 + 2)$ _____ 10. $(6 + 9) + 5$

_____ 11. $8(5 \cdot 4)$ _____ 12. $(7 \cdot 2) \cdot 2$

_____ 13. $2(6 + 1)$ _____ 14. $5(3 + 8)$

_____ 15. $(6 \cdot 9) + 11$ _____ 16. $7 + (3 \cdot 10)$

_____ 17. $6 \cdot 7 + 5$ _____ 18. $4 + 9 \cdot 8$

_____ 19. $3 \cdot 5 + 7 \cdot 7$ _____ 20. $6 \cdot 12 + 8 \cdot 2$

In Exercises 21–30, simplify.

_____ 21. $2(8 + 3 \cdot 2)$ _____ 22. $(10 + 5 \cdot 3)2$

_____ 23. $(8 + 7)(3 + 1)$ _____ 24. $(7 + 2)(3 + 3)$

_____ 25. $3[1 + 2(3 + 5)]$ _____ 26. $[3(1 + 4) + 2]6$

_____ 27. $4 + [3 + 4(1 + 1)]$ _____ 28. $[6 + 2(1 + 3)] + 7$

_____ 29. $5[0 + 2(1 + 0)]$ _____ 30. $6[5(0 + 3) + 0]$

According to the commutative property for addition:

31. $7 + 2 = 2 +$ _____   32. $6 + 9 = 9 +$ _____

33. $0 + 5 = 5 +$ _____   34. $117 + 0 = 0 +$ _____

35. $6 + 20 =$ _____   36. $83 + 7 =$ _____

According to the associative property for addition:

37. $(6 + 3) + 1 = 6 +$ _____

38. $(8 + 1) + 10 = 8 +$ _____

39. $7 + (2 + 5) =$ _____ $+ 5$

40. $5 + (8 + 9) =$ _____ $+ 9$

41. $(1 + 3) + 2 =$ _____

42. $13 + (2 + 4) =$ _____

According to the commutative property for multiplication:

43. $9 \cdot 8 = 8$ _____

44. $63 \cdot 10 = 10$ _____

45. $0 \cdot 5 = 5$ _____

46. $19 \cdot 0 = 0$ _____

47. $8 \cdot 13 =$ _____

48. $12 \cdot 5 =$ _____

According to the associative property for multiplication:

49. $(3 \cdot 4)2 = 3$ _____

50. $(6 \cdot 5)7 = 6$ _____

51. $9(8 \cdot 6) =$ _____ 6

52. $10(3 \cdot 4) =$ _____ 4

53. $7(6 \cdot 5) =$ _____

54. $(3 \cdot 9)2 =$ _____

According to the distributive property for multiplication over addition:

55. $3(7 + 2) = 3 \cdot 7 +$ _____

56. $5(8 + 3) = 5 \cdot 8 +$ _____

57. $9(6 + 1) =$ _____ $+ 9 \cdot 1$

58. $10(4 + 2) =$ _____ $+ 10 \cdot 2$

59. $7(3 + 5) =$ _____ $+$ _____

60. $6(8 + 5) =$ _____ $+$ _____

In Exercises 61–65, change each word expression to a number expression. Do not simplify.

_____ **61.** If one calculator costs $18, how much will 8 calculators cost?

_____ **62.** Darlene scored 67 on her first math test. If she scored 14 points higher on the second test, what score did she receive on the second test?

_____ **63.** If Tony can walk at a rate of 4 miles per hour, how far can he walk in 5 hours?

_____ **64.** Pat drove in town for 2 hours at 35 miles per hour and on the highway for 5 hours at 55 miles per hour. How far did she drive?

_____ **65.** Each month, Carl's wife earns $100 more than twice Carl's salary. If Carl's monthly salary is $600, what is his wife's monthly salary?

Name _____   Date _____

CHECKUP 1.1

Simplify Problems 1–6.

_____ 1. 5 + (3 + 4)

_____ 2. 4(8 · 3)

_____ 3. 7(5 + 1)

_____ 4. 6 · 4 + 3

_____ 5. (6 + 2)(3 + 1)

_____ 6. 4[2 + 3(1 + 7)]

_____ 7. According to the commutative property for addition,
5 + 13 = _____.

_____ 8. According to the associative property for multiplication,
(3 · 5) · 2 = _____.

_____ 9. According to the distributive property of multiplication over addition,
3(7 + 5) = _____.

_____ 10. Lanette weighs 8 pounds more than her sister. If her sister's weight is
110, write a number expression for Lanette's weight.

1.2 Subtracting and Dividing Whole Numbers

1.2-1 Subtracting Whole Numbers

To subtract whole numbers, we may again make use of the number line. When we *added* 7 + 3, recall that we started at point 7 and moved 3 units to the *right,* ending up at point 10. The sum of 7 and 3 was 10. If we wish to *subtract* 3 from 7, it seems logical that we should again start at point 7 and move 3 units to the *left.*

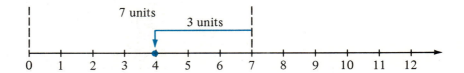

The answer to the subtraction problem 7 − 3 is 4. You may recall that the answer to a subtraction problem is called the **difference**.

Example 1. Use a number line to find the difference when 9 is subtracted from 10.

Solution. We must find 10 − 9, so we start at 10 and move 9 units to the left.

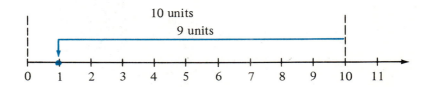

Since we end up at 1, we know that 10 − 9 = 1.

Is there a commutative property for subtraction? Does it matter in which order two numbers are subtracted? Is it true that 7 − 4 = 4 − 7? Suppose that you had $7 and spent $4; your financial condition could be described as "$7 − $4 = $3." But suppose that you had $4 and wished to spend $7; your financial condition could be described as "$4 − $7," and this certainly would *not* leave you with $3. Since these answers are different, we conclude that

$$7 - 4 \quad \text{does } not \text{ equal} \quad 4 - 7$$

and we write this as

$$7 - 4 \neq 4 - 7$$

So for subtraction, the order in which two numbers are subtracted does matter.

> There is *no* commutative property for subtraction.

Let us check on the associative property for subtraction involving three numbers. If the associative property works, then (9 − 5) − 3 should give the same answer as 9 − (5 − 3). Let's compute each difference, paying attention to the parentheses.

$$(9 - 5) - 3 = 4 - 3 = 1$$
$$9 - (5 - 3) = 9 - 2 = 7$$

Since 1 ≠ 7, we can see that

There is *no* associative property for subtraction.

As before, parentheses tell us what to do first. Let's try some examples involving the operations of addition, subtraction, and multiplication.

Example 2. Simplify $26 + (15 - 4)$.

    Solution

$$26 + (15 - 4)$$
$$= 26 + \quad 11$$
$$= 37$$

Example 3. Simplify $(7 - 1) + (9 - 3)$.

    Solution

$$(7 - 1) + (9 - 3)$$
$$= \quad 6 \quad + \quad 6$$
$$= \quad 12$$

Example 4. Simplify $(19 - 3) - (27 - 20)$.

    Solution

$$(19 - 3) - (27 - 20)$$
$$= \quad 16 \quad - \quad 7$$
$$= \quad 9$$

Example 5. Simplify $3(15 - 8)$.

    Solution

$$3(15 - 8)$$
$$= 3(7)$$
$$= 21$$

Example 5 may make you wonder about a distributive property for multiplication over subtraction. Recall that the distributive property for multiplication over addition allowed us to do problems like this:

$$4(8 + 2)$$
$$= 4 \cdot 8 + 4 \cdot 2$$
$$= 32 + 8$$
$$= 40$$

Let's see if we can work the example $3(15 - 8)$ this way and get the same answer of 21.

$$3(15 - 8)$$
$$= 3 \cdot 15 - 3 \cdot 8$$
$$= 45 - 24$$
$$= 21$$

Indeed, we do get the same answer because there is a **distributive property for multiplication over subtraction.** Using A, B, and C to represent any whole numbers, we may write:

Distributive Properties

$$A(B + C) = A \cdot B + A \cdot C$$
$$A(B - C) = A \cdot B - A \cdot C$$

Example 6. Simplify $5(16 - 12)$ using two methods.

Solution

$$\text{Method 1:} \quad 5(16 - 12) \\ = 5(4) = 20$$

$$\text{Method 2:} \quad 5(16 - 12) \\ = 5 \cdot 16 - 5 \cdot 12 \\ = 80 - 60 = 20$$

The first method seems more direct, but later we shall be forced to use the second method, so be sure that you follow it now.

Trial Run *Simplify.*

_____ **1.** $8 - 2 - 3$

_____ **2.** $12 - (6 - 2)$

_____ **3.** $(34 - 19) - 2$

_____ **4.** $(8 + 3) - (4 + 2)$

_____ **5.** $9 + 3(5 - 2)$

_____ **6.** $(12 - 5)(9 - 8)$

ANSWERS

1. 3.   **2.** 8.   **3.** 13.   **4.** 5.   **5.** 18.   **6.** 7.

1.2-2 Dividing Whole Numbers

There are several ways to indicate the operation of division. To show 6 divided by 3, we may write any of the following:

$$6 \div 3$$

$$3\overline{)6}$$

$$6/3$$

$$\frac{6}{3}$$

and each of these would yield the answer 2. Recall that the answer to a division problem is called the **quotient**. When a division problem is written in fraction form such as $\frac{6}{3}$, we label the parts as follows:

$$\underset{\text{bar}}{\text{fraction}} \longrightarrow \frac{6 \longleftarrow \text{numerator}}{3 \longleftarrow \text{denominator}}$$

Every division problem corresponds to a multiplication problem. For instance,

$$\frac{6}{3} = 2 \qquad \text{because} \qquad 3 \cdot 2 = 6$$

$$\frac{40}{8} = 5 \qquad \text{because} \qquad 8 \cdot 5 = 40$$

$$\frac{19}{1} = 19 \qquad \text{because} \qquad 1 \cdot 19 = 19$$

This idea can be very useful when you are "stuck" on a division problem. If you know your multiplication facts, you should be able to handle division.

Is there a commutative property for division? Does it matter in what order we divide two whole numbers? Do you think that $6 \div 3$ and $3 \div 6$ will yield the same answer?

$$6 \div 3 = \frac{6}{3} = 2$$

$$3 \div 6 = \frac{3}{6} = \frac{1}{2}$$

Indeed, the answers are very different: $6 \div 3 \neq 3 \div 6$, and we conclude that

> There is *no* commutative property for division.

What about an associative property for dividing three numbers? Do you think that $(24 \div 6) \div 2$ will yield the same quotient as $24 \div (6 \div 2)$?

$$\begin{aligned} &(24 \div 6) \div 2 \\ = \quad &4 \quad \div 2 = 2 \end{aligned}$$

$$\begin{aligned} &24 \div (6 \div 2) \\ = 24 \div \quad &3 \quad = 8 \end{aligned}$$

We have discovered that $(24 \div 6) \div 2 \neq 24 \div (6 \div 2)$, and we must conclude that

> There is *no* associative property for division.

Paying attention to parentheses, let's see how we may combine addition, subtraction, multiplication, and division in simplifying some problems.

Example 7. Simplify $\dfrac{(7 + 2)}{3}$.

Solution

$$\frac{(7 + 2)}{3}$$
$$= \frac{9}{3}$$
$$= 3$$

Example 8. Simplify $\dfrac{(8 - 3)}{5}$.

Solution

$$\frac{(8 - 3)}{5}$$
$$= \frac{5}{5}$$
$$= 1$$

Example 9. $16 + \left(\dfrac{10}{2}\right)$.

Solution

$$16 + \left(\frac{10}{2}\right)$$
$$= 16 + 5$$
$$= 21$$

Example 10. $\left(\dfrac{84}{7}\right) - 9$.

Solution

$$\left(\frac{84}{7}\right) - 9$$
$$= 12 - 9$$
$$= 3$$

Example 11. $7\left[\left(\dfrac{12}{2}\right) - 5\right]$.

Solution

$$7\left[\left(\frac{12}{2}\right) - 5\right]$$
$$= 7[6 - 5]$$
$$= 7[1]$$
$$= 7$$

The fraction bar acts like parentheses in division problems, telling us that we should simplify the numerator and simplify the denominator before performing the division.

Example 12. Simplify $\dfrac{7 + 8}{2 + 3}$.

Solution

$$\frac{7 + 8}{2 + 3}$$
$$= \frac{15}{5} = 3$$

Example 13. Simplify $\dfrac{5 + 19}{2 \cdot 3}$.

Solution

$$\frac{5 + 19}{2 \cdot 3}$$
$$= \frac{24}{6} = 4$$

Earlier, we learned what to do when parentheses were left out of a problem. Now we shall extend our rule to include subtraction and division.

If there are no parentheses to indicate order of operations, first perform multiplications and/or divisions from left to right, then perform additions and/or subtractions from left to right.

Example 14. Simplify $16 - \dfrac{8}{2}$.

Solution

$$16 - \frac{8}{2}$$
$$= 16 - 4 = 12$$

Example 15. Simplify $3 + \dfrac{15 - 1}{7}$.

Solution

$$3 + \frac{15 - 1}{7}$$
$$= 3 + \frac{14}{7}$$
$$= 3 + 2 = 5$$

Example 16. Simplify $6\left(9 + \dfrac{9}{3}\right) - 2$.

Solution

$$6\left(9 + \frac{9}{3}\right) - 2$$
$$= 6(9 + 3) - 2$$
$$= 6 \cdot 12 - 2$$
$$= 72 - 2$$
$$= 70$$

Example 17. Simplify $6 + \dfrac{9 + 9}{3} \div 2$.

Solution

$$6 + \frac{9 + 9}{3} \div 2$$
$$= 6 + \frac{18}{3} \div 2$$
$$= 6 + 6 \div 2$$
$$= 6 + 3$$
$$= 9$$

Example 18. Simplify $18 \div 3 \cdot 5$.

Solution

$$18 \div 3 \cdot 5$$
$$= 6 \cdot 5$$
$$= 30$$

Trial Run   *Simplify.*

———— 1. $\dfrac{9 + 7}{4}$

———— 2. $12 \div 2 \cdot 3$

———— 3. $18 - \dfrac{6}{2}$

———— 4. $\dfrac{4 \cdot 9 - 6}{2(8 - 5)}$

———— 5. $5\left(6 + \dfrac{8}{2}\right) - 3$

———— 6. $7 + \dfrac{8 + 24}{4} \div 2$

ANSWERS
1. 4.   **2.** 18.   **3.** 15.   **4.** 5.   **5.** 47.   **6.** 11.

1.2-3   Working with Zero and One

Earlier we learned that zero and 1 had some very special characteristics in addition and multiplication problems. We noted that

$$3 + 0 = 3$$
$$3 \cdot 0 = 0$$
$$3 \cdot 1 = 3$$

and we generalized to say that if A is any whole number,

$$A + 0 = A$$
$$A \cdot 0 = 0$$
$$A \cdot 1 = A$$

Do zero and 1 possess similar properties for subtraction and division? Find the following:

$$6 - 0$$
$$17 - 0$$
$$0 - 0$$

You should agree that

$$6 - 0 = 6$$
$$17 - 0 = 17$$
$$0 - 0 = 0$$

and in fact:

> If zero is subtracted from any number, the answer is that number.

This law is called the **subtraction property of zero**, and using A to represent any whole number, we may write:

$$\boxed{\begin{array}{c} \textbf{Subtraction Property of Zero} \\[4pt] A - 0 = A \end{array}}$$

Example 19. Simplify $(6 + 3) - 0$.

Solution

$$(6 + 3) - 0$$
$$= 9 - 0 = 9$$

Example 20. Simplify $29 - 0 \cdot 8$.

Solution

$$29 - 0 \cdot 8$$
$$= 29 - 0$$
$$= 29$$

Example 21. Simplify $\dfrac{2 \cdot 9 - 0}{6}$.

Solution

$$\frac{2 \cdot 9 - 0}{6}$$
$$= \frac{18 - 0}{6}$$
$$= \frac{18}{6} = 3$$

Notice what happens when we divide any number by 1.

$$\frac{6}{1} = 6 \qquad \text{because} \qquad 1 \cdot 6 = 6$$

$$\frac{1789}{1} = 1789 \qquad \text{because} \qquad 1 \cdot 1789 = 1789$$

$$\frac{0}{1} = 0 \qquad \text{because} \qquad 1 \cdot 0 = 0$$

It seems that

$$\boxed{\text{If any number is divided by 1, the answer is that number.}}$$

This law is called the **division property of 1**, and if we let A represent any whole number, we may write

$$\boxed{\begin{array}{c} \textbf{Division Property of 1} \\[4pt] \dfrac{A}{1} = A \end{array}}$$

Let's investigate what happens when we try to divide any whole number by zero. Consider

$$\frac{6}{0}$$

What should the answer be? Do you think that

$$\frac{6}{0} = 0?$$

Remember that $\frac{6}{3} = 2$ because $3 \cdot 2 = 6$.

Do you wish to say $\frac{6}{0} = 0$ because $0 \cdot 0 = 6$? Of course not! Maybe you decide that $\frac{6}{0} = 6$, which means you think that $0 \cdot 6 = 6$, which is again not true. In fact, there is *no answer* for the problem $\frac{6}{0}$.

For the problem $\frac{0}{0}$, the answer could be any number. For instance, $\frac{0}{0} = 3$ because $0 \cdot 3 = 0$, or $\frac{0}{0} = 0$ because $0 \cdot 0 = 0$. In this case there are *too many* answers.

To avoid the problem of no answer or too many answers, we agree that

Division by zero is impossible

or

Division by zero is undefined.

because there is no way to define division by zero to give us one and only one answer. In fact, if A represents any whole number, we write:

Division by Zero

$\frac{A}{0}$ is undefined

What happens if we try to divide zero by any other whole number? What is

$$\frac{0}{6}?$$

We must agree that $\frac{0}{6} = 0$ because $6 \cdot 0 = 0$. Similarly,

$$\frac{0}{1} = 0 \qquad \text{because} \qquad 1 \cdot 0 = 0$$

$$\frac{0}{795} = 0 \qquad \text{because} \qquad 795 \cdot 0 = 0$$

But $\frac{0}{0}$ is still *undefined* because of the zero in the denominator (bottom) of the fraction.

If zero is divided by any *nonzero* number, the answer is zero.

In general, if A represents any whole number except zero, we say:

Division of Zero

$$\frac{0}{A} = 0 \qquad \text{provided that } A \neq 0$$

To avoid confusion between these two situations, let us summarize them here.

If the denominator (bottom) of a fraction is zero, the fraction's value is undefined.

example: $\dfrac{17}{0}$ is undefined

If the numerator (top) of a fraction is zero and the denominator (bottom) is *not* zero, the fraction's value is zero.

example: $\dfrac{0}{17} = 0$

Example 22. Simplify $\dfrac{7-7}{2}$.

Solution

$$\frac{7-7}{2}$$
$$= \frac{0}{2} = 0$$

Example 23. Simplify $\dfrac{3 + 5 \cdot 2}{27 - 3 \cdot 9}$.

Solution

$$\frac{3 + 5 \cdot 2}{27 - 3 \cdot 9}$$
$$= \frac{3 + 10}{27 - 27}$$
$$= \frac{13}{0} \quad \text{is undefined}$$

Example 24. Simplify $\dfrac{2 \cdot 8 - 4 \cdot 4}{6 + 11}$.

Solution

$$\frac{2 \cdot 8 - 4 \cdot 4}{6 + 11}$$

$$= \frac{16 - 16}{17}$$

$$= \frac{0}{17} = 0$$

Example 25. Simplify $\frac{29(8 - 2 \cdot 4)}{56 - 7 \cdot 8}$.

Solution

$$\frac{29(8 - 2 \cdot 4)}{56 - 7 \cdot 8}$$

$$= \frac{29(8 - 8)}{56 - 56}$$

$$= \frac{29 \cdot 0}{0}$$

$$= \frac{0}{0} \quad \text{is undefined}$$

What happens when any nonzero whole number is divided by itself?

$$\frac{7}{7} = 1 \quad \text{because} \quad 7 \cdot 1 = 7$$

$$\frac{185}{185} = 1 \quad \text{because} \quad 185 \cdot 1 = 185$$

$$\frac{1}{1} = 1 \quad \text{because} \quad 1 \cdot 1 = 1$$

It seems that

> If any nonzero number is divided by itself, the answer is 1.

Indeed, if A is any nonzero whole number, we may write

$$\frac{A}{A} = 1 \quad \text{(where } A \neq 0\text{)}$$

Keep in mind that this rule does not hold when $A = 0$ because $\frac{0}{0}$ is still undefined.

Example 26. Simplify $\frac{6 + 3}{13 - 4}$.

Solution

$$\frac{6 + 3}{13 - 4}$$

$$= \frac{9}{9} = 1$$

Example 27. Simplify $\dfrac{64 - 2 \cdot 7}{5(9 + 1)}$.

Solution

$$\dfrac{64 - 2 \cdot 7}{5(9 + 1)}$$

$$= \dfrac{64 - 14}{5(10)}$$

$$= \dfrac{50}{50} = 1$$

Example 28. Simplify $\dfrac{17(6 - 3 \cdot 2)}{5 \cdot 9 - 3 \cdot 15}$.

Solution

$$\dfrac{17(6 - 3 \cdot 2)}{5 \cdot 9 - 3 \cdot 15}$$

$$= \dfrac{17(6 - 6)}{45 - 45}$$

$$= \dfrac{17(0)}{0}$$

$$= \dfrac{0}{0} \quad \text{is undefined}$$

Trial Run *Simplify.*

_____ 1. $\dfrac{3(7 - 5)}{2 \cdot 3 - 6}$

_____ 2. $\dfrac{3 \cdot 5 - 3}{2(10 - 4)}$

_____ 3. $\dfrac{9 \cdot 2 - 3 \cdot 6}{5 + 9}$

_____ 4. $\dfrac{3 \cdot 2 - \frac{10}{5}}{\frac{9}{3} - 2 \cdot 1}$

_____ 5. $\dfrac{2 \cdot 7 - \frac{28}{2}}{15 - 4 \cdot 3}$

_____ 6. $\dfrac{9(12 - 3 \cdot 4)}{4 \cdot 6 - 8 \cdot 3}$

ANSWERS
1. Undefined.　　**2.** 1.　　**3.** 0.　　**4.** 4.　　**5.** 0.　　**6.** Undefined.

1.2-4 Switching from Words to Numbers

Let's try to switch some word expressions to number expressions using all the operations we have discussed.

Example 29. Three roommates spent $187 on groceries and $242 on rent last month. If they split expenses evenly, write an expression for each person's share.

Solution. Total expenses were $187 + $242. Splitting expenses 3 ways, each person's share will be $\dfrac{\$187 + \$242}{3}$.

Example 30. Over a period of 9 weeks, Angela gained 36 pounds. What was her average weight gain for 1 week?

Solution. We must divide the total gain of 36 pounds by the total number of 9 weeks, so the average gain is $\dfrac{36}{9}$ pounds per week.

Example 31. On a business trip, Harvey's car mileage gauge changed from 21623 to 22007. If Harvey's company pays him 25 cents per mile, what will the company pay Harvey for car expenses?

Solution. First we must compute Harvey's mileage by finding the difference between his gauge readings, which is $22007 - 21623$. Now we must multiply his payment per mile (25 cents) times the number of miles traveled. The number expression we need is $25(22007 - 21623)$ cents.

Once again, you should be on the lookout for some key words that tell you what operation to perform.

Words	*Numbers*
the difference between 12 and 9	$12 - 9$
17 less 3	$17 - 3$
24 decreased by 8	$24 - 8$
24 divided by 8	$\dfrac{24}{8}$
5 less than 12	$12 - 5$

Trial Run *Change each word expression to a number expression.*

_____ 1. A dinner for two at the Chicken King costs $6.00 and milkshakes cost $1 each. If Jim and Maria split expenses evenly, how much will each pay?

_____ 2. During 4 weeks, Vicki's living expenses were $340. What were her average expenses for 1 week?

_____ 3. The temperature at noon was 45° but 5 hours later was only 20°. What was the average number of degrees the temperature fell per hour?

_____ 4. A club has sold 150 dance tickets at $4 each. If the band must be paid $250, write an expression for the club's profit.

_____ 5. Debbie scored 10 points higher on her second history test than on her first test. If she scored 76 on the first test, what is Debbie's average on the two tests?

ANSWERS
1. $\dfrac{6}{2} + 1$. 2. $\dfrac{340}{4}$. 3. $\dfrac{45 - 20}{5}$. 4. $4(150) - 250$. 5. $\dfrac{76 + (76 + 10)}{2}$.

EXERCISE SET 1.2

1. Use a number line to illustrate each subtraction problem.
   (a) $11 - 3$ (b) $7 - 1$ (c) $13 - 6$ (d) $10 - 5$

2. Use a number line to illustrate each subtraction problem.
   (a) $12 - 5$ (b) $9 - 2$ (c) $3 - 1$ (d) $14 - 13$

In Exercises 3–34, perform the operations.

_____ 3. $6 - 2 - 1$ _____ 4. $19 - 6 - 2$

_____ 5. $10 - (8 + 1)$ _____ 6. $15 - (3 + 8)$

_____ 7. $17 - (14 - 2)$ _____ 8. $27 - (9 - 3)$

_____ 9. $(48 - 7) - 3$ _____ 10. $(34 - 11) - 9$

_____ 11. $(8 + 2) - (6 + 1)$ _____ 12. $(13 + 1) - (2 + 3)$

_____ 13. $(41 - 10) - (8 - 4)$ _____ 14. $(63 - 21) - (18 - 2)$

_____ 15. $5(9 - 3)$ _____ 16. $7(8 - 1)$

_____ 17. $6 + 2(15 - 4)$ _____ 18. $7 + 3(14 - 1)$

_____ 19. $(7 - 3)(4 - 1)$ _____ 20. $(8 - 3)(9 - 2)$

_____ 21. $\dfrac{8 + 6}{7}$ _____ 22. $\dfrac{9 + 3}{4}$

_____ 23. $\dfrac{20 - 5}{3}$ _____ 24. $\dfrac{26 - 4}{2}$

_____ 25. $\dfrac{11 + 23}{13 + 4}$ _____ 26. $\dfrac{19 + 11}{5 + 1}$

_____ 27. $\dfrac{17 - 3}{7 + 0}$ _____ 28. $\dfrac{27 + 11}{0 + 2}$

_____ 29. $\dfrac{6 + 9}{16 - 1}$ _____ 30. $\dfrac{8 + 10}{9 + 9}$

_____ 31. $\dfrac{37 - 22}{8 - 8}$ _____ 32. $\dfrac{11 - 11}{7 + 3}$

_____ 33. $\dfrac{8 - 8}{37 - 22}$ _____ 34. $\dfrac{7 + 3}{11 - 11}$

In Exercises 35–60, simplify each expression.

_____ 35. $6 + \dfrac{5}{5}$ _____ 36. $9 + \dfrac{30}{5}$

_____ 37. $10 - 6 + \dfrac{18}{2}$ _____ 38. $14 - 5 + \dfrac{27}{9}$

_____ **39.** $(10 - 6) \cdot \dfrac{18}{2}$

_____ **40.** $(14 - 5) \cdot \dfrac{27}{9}$

_____ **41.** $2\left(7 + \dfrac{10}{2}\right)$

_____ **42.** $3\left(11 + \dfrac{16}{4}\right)$

_____ **43.** $17 - \left(3 + \dfrac{14}{2}\right)$

_____ **44.** $29 - \left(6 + \dfrac{45}{3}\right)$

_____ **45.** $25 - \left(\dfrac{18}{3} - 5\right)$

_____ **46.** $37 - \left(\dfrac{25}{5} - 1\right)$

_____ **47.** $\dfrac{2(7 + 2)}{3}$

_____ **48.** $\dfrac{5(9 + 3)}{6}$

_____ **49.** $\dfrac{7(18 - 10)}{4}$

_____ **50.** $\dfrac{9(21 - 5)}{8}$

_____ **51.** $\dfrac{8 + 6 \cdot 5}{7 - \dfrac{14}{2}}$

_____ **52.** $\dfrac{3 \cdot 4 - 2}{\dfrac{100}{20} - 5}$

_____ **53.** $\dfrac{17 + \dfrac{9}{3}}{5 \cdot 5 - 5}$

_____ **54.** $\dfrac{\dfrac{44}{11} + 6}{6 \cdot 8 - 38}$

_____ **55.** $2\left[\dfrac{14}{7} + 3(8 - 2)\right]$

_____ **56.** $3\left[4(9 - 3) + \dfrac{15}{3}\right]$

_____ **57.** $7\left(8 - \dfrac{2 \cdot 8}{4}\right)$

_____ **58.** $6\left(10 - \dfrac{6 \cdot 9}{27}\right)$

_____ **59.** $11\left(2 \cdot 3 - \dfrac{13 + 5}{3}\right)$

_____ **60.** $13\left(\dfrac{28}{7} - \dfrac{22 - 6}{4}\right)$

In Exercises 61–66, change each word expression to a number expression. Do not simplify.

_____ **61.** If there are 100 centimeters in 1 meter, how many centimeters are in 45 meters?

_____ **62.** If Paul earns $600 per month, how much does he earn in 1 year?

_____ **63.** If Mark weighed 235 pounds before his diet and now weighs 190 pounds, how much weight did he lose?

_____ **64.** If Mark (see Problem 63) lost his weight over a 15-week period, what was his average weekly loss?

_____ **65.** If taxi fare is $2 for the first mile and $1 for every additional mile, what will be the fare for an 8-mile trip?

_____ **66.** Jack and Jill bought 5 records at $6 each and 3 tapes at $8 each. If they divide the cost in half, how much will each person pay?

CHECKUP 1.2

Simplify.

_____ **1.** $19 - (10 - 7)$

_____ **2.** $(35 - 15) - (12 - 7)$

_____ **3.** $(8 - 3)(15 - 7)$

_____ **4.** $\dfrac{35 - 15}{4}$

_____ **5.** $\dfrac{2(20 - 14)}{3 + 0}$

_____ **6.** $19 - \left(\dfrac{18}{2} - 3\right)$

_____ **7.** $\dfrac{14(11 - 7)}{3 - \dfrac{27}{9}}$

_____ **8.** $9\left(15 - \dfrac{3 \cdot 10}{6}\right)$

_____ **9.** $\dfrac{13 - \dfrac{26}{2}}{\dfrac{32}{4} + 1}$

_____ **10.** During a year, a farm with 3 equal-sharing owners took in $625,000 and had $125,000 in expenses. Write a number expression for each owner's share of the profit.

1.3 Adding Integers

On a winter day, have you ever watched the thermometer drop to a below-zero temperature? Have you ever seen your favorite football team lose yardage on a play? Have you ever borrowed money?

In each situation described, the list of whole numbers does not provide us with a way to make the necessary measurement. We need an expanded list to write numbers such as

3° below zero

a loss of 6 yards

a debt of $10.

We know that temperatures of 3° above zero and 3° below zero are very different. A gain of 6 yards is very different from a loss of 6 yards, and a credit of $10 is very different from a debt of $10.

On a thermometer, a reading below zero is often noted using a *negative sign,* so that 3° below zero would be written

⁻3°

Similarly, in recording a football team's loss of yardage, a negative sign would again be used. A loss of 6 yards would be recorded as

⁻6 yards

Similarly, a debt of $10 would be written as

⁻$10

1.3-1 Integers on the Number Line

On a number line we will illustrate these negative numbers by extending our line to the left, marking off each unit in a leftward direction from zero.

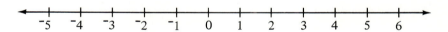

This new list of numbers, which includes all the whole numbers *and* their negative counterparts, is called the set of **integers.**

> Integers: . . . , ⁻6, ⁻5, ⁻4, ⁻3, ⁻2, ⁻1, 0, 1, 2, 3, 4, . . .

Notice that this list continues indefinitely in either direction, as shown by the arrows at both ends of the line. Points to the *right* of zero correspond to the natural numbers (also called **positive integers**) and points to the *left* of zero correspond to what we call **negative integers**.

Positive integers: 1, 2, 3, 4, 5, . . .

Negative integers: . . . , ⁻4, ⁻3, ⁻2, ⁻1.

The list of integers contains the positive integers, the negative integers, and the integer zero. Each integer corresponds to one and only one point on the number line.

Example 1. Locate ⁻3 on the number line.

Solution

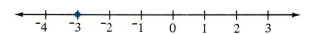

Example 2. Locate 3 on the number line.

Solution

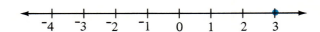

If we were to take a ruler and measure the distance between ⁻3 and 0 on our number line and then measure the distance between 0 and 3 on our number line, what would we observe? Why, those distances are the same! Each of the points, ⁻3 and 3, measures 3 *units* from zero. Their locations are on opposite sides of zero, but their distances represent the same number of units. For this reason, we say that

⁻3 is the opposite of 3

3 is the opposite of ⁻3

> When two numbers measure the same distance from zero but the two numbers are located on opposite sides of zero, we say that one number is the **opposite** of the other.

Example 3. What is the opposite of 16?

Solution. ⁻16 is the opposite of 16.

Example 4. What is the opposite of ⁻5?

Solution. 5 is the opposite of ⁻5.

Sometimes we refer to a positive integer, such as 5, as ⁺5, but the positive sign is not required. You should understand that if there is no sign in front of a number, it is considered positive. We should also agree here that

> The opposite of 0 is 0.

The number line again gives us a handy way to compare two integers. If one integer lies to the *left* of another integer, we say that the first integer is *less than* ($<$) the second integer. If one integer lies to the *right* of another integer, we say that the first integer is *greater than* ($>$) the second integer.

Example 5. Compare ⁻2 and 6.

Solution. ⁻2 lies to the left of 6, so ⁻2 $<$ 6.

Example 6. Compare ⁻5 and ⁻9.

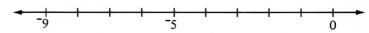

Solution. ⁻5 lies to the right of ⁻9, so ⁻5 $>$ ⁻9.

1.3-2 Performing Addition

How do we go about adding integers? First, let us recall how to add integers which are positive or zero. Remember, these are 0, 1, 2, 3, 4, . . . (the whole numbers), which we have already discussed. Let us use the number line for review.

Example 7. Use the number line to find 5 + 2.

Solution. We start at 5 and move 2 units to the right.

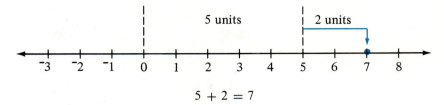

$$5 + 2 = 7$$

Example 8. Use the number line to find 3 + 1.

Solution. We start at 3 and move 1 unit to the right.

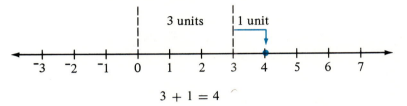

$$3 + 1 = 4$$

In each case, when we added two *positive* integers, the answer was a *positive* integer.

> To add two *positive* integers, add the units together and give the answer a *positive* sign.

What happens when we add two *negative* integers? The number line will be a help here. Remember that in locating a negative number, we must move *left* rather than right.

Example 9. Use the number line to add ⁻3 and ⁻2.

Solution. We start at ⁻3 and move 2 units to the *left*.

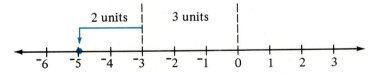

Since we ended up at ⁻5, the sum of ⁻3 and ⁻2 must be ⁻5. We write this as

$$^-3 + {}^-2 = {}^-5$$

Example 10. If a team loses 6 yards on one play, and loses 1 yard on the next play, what is the total loss?

Solution. We must find the sum of ⁻6 and ⁻1. Start at ⁻6 and move 1 unit to the *left*.

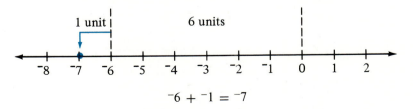

$$^-6 + {}^-1 = {}^-7$$

Example 11. If you are already in debt to your roommate for $10 and borrow another $5, what is your total debt?

Solution. We must find the sum of $^-10$ and $^-5$ dollars. Start at $^-10$ and move 5 units to the *left*.

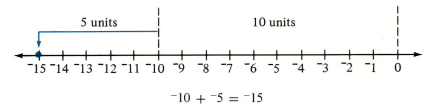

$$^-10 + {}^-5 = {}^-15$$

You are in debt $15.

Suppose that you are in debt $150 and borrow $75 more. What is your total debt? This problem is awkward to do on a number line. Can you find

$$^-150 + {}^-75$$

without a number line? Mentally, you know to start at $^-150$ and move 75 units to the *left*. Where would you end up? At $^-225$:

$$^-150 + {}^-75 = {}^-225$$

Your total debt is $225.

In each of these examples, the number of *units* in our answer was the *sum* of the units in the original numbers, but the *sign* of the answer was *negative*. We conclude:

> To add two *negative* integers, add the units together and give your answer a *negative* sign.

Example 12. Find $^-76 + {}^-13$.

Solution. $^-76 + {}^-13 = {}^-89$

Example 13. Find $(^-31 + {}^-8) + {}^-12$.

Solution

$$\begin{aligned} (^-31 + {}^-8) + {}^-12 \\ = {}^-39 + {}^-12 \\ = {}^-51 \end{aligned}$$

What happens if we try to add two numbers when one of the numbers is *positive* and one of the numbers is *negative?* The number line will give us a clue if we remember to move to the right for positive numbers and to the left for negative numbers.

Example 14. If you are in debt $10 and receive a gift of $8, what is your new financial situation?

Solution. We must find the sum of $^-10$ and 8 or $^-10 + 8$. Start at $^-10$ and move 8 units to the right.

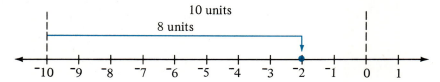

Since we end up at $^-2$, you are still in *debt* $2.

$$^-10 + 8 = {}^-2$$

Example 15. Use the number line to find ⁻6 + 13.

Solution. We start at ⁻6 and move 13 units to the *right*.

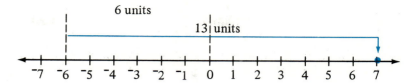

Ending up at ⁺7, we say that

$$^-6 + 13 = 7$$

Example 16. Find 9 + ⁻3.

Solution. We start at 9 and move 3 units to the left.

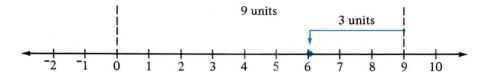

Since we end up at 6, we know that

$$9 + {}^-3 = 6$$

Example 17. Find 5 + ⁻8.

Solution. Start at 5 and move 8 units to the *left*.

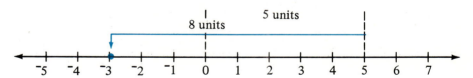

Ending up at ⁻3, we decide that

$$5 + {}^-8 = {}^-3$$

Do you see how we could add two numbers with opposite signs without using the number line? In each example, the *units* in our answer represented the *difference* between our original units. The answer will take the *sign* of the original number having *more* units. Remember, the number of units for an integer is that integer's distance from zero on the number line.

> To add a positive integer and a negative integer, find the difference between their units and give your answer the sign of the original integer having more units.

Example 18. Simplify ⁻11 + 26.

Solution. Our answer will be positive. The difference between 11 units and 26 units is 15 units.

$$^-11 + 26 = 15$$

Example 19. Simplify $^-10 + (^-17 + 2)$.

Solution. Parentheses still say "do this first."

$$^-10 + (^-17 + 2)$$
$$= ^-10 + ^-15$$
$$= ^-25$$

Example 20. Simplify $(5 + ^-16) + 3$.

Solution

$$(5 + ^-16) + 3$$
$$= ^-11 + 3$$
$$= ^-8$$

Example 21. Simplify $^-9 + 9$.

Solution. The signs are opposite, so we find the difference between their units. The difference between 9 units and 9 units is 0 units.

$$^-9 + 9 = 0$$

From this example we notice that the sum of a number and its opposite will always be zero.

You should recall that addition of whole numbers was commutative and associative. We discovered that

$$3 + 8 = 8 + 3 \qquad \text{(commutative property)}$$

$$(7 + 6) + 1 = 7 + (6 + 1) \qquad \text{(associative property)}$$

Do you think that addition of *integers* is commutative?

Example 22. Compare $^-2 + ^-6$ and $^-6 + ^-2$.

Solution

$$^-2 + ^-6 = ^-8$$

and

$$^-6 + ^-2 = ^-8$$

So we find that $^-2 + ^-6 = ^-6 + ^-2$.

Example 23. Compare $^-15 + 8$ and $8 + ^-15$.

Solution

$$^-15 + 8 = ^-7$$

and

$$8 + ^-15 = ^-7$$

So we find that $^-15 + 8 = 8 + ^-15$.

Since it does not matter in which order we add two integers, we must conclude that addition of integers is commutative. Letting A and B stand for any integers, we write:

Commutative Property for Addition

$$A + B = B + A$$

Now check the associative property for addition of integers.

Example 24. Compare $(^-7 + ^-3) + ^-5$ and $^-7 + (^-3 + ^-5)$.

Solution

$$(^-7 + {}^-3) + {}^-5 \quad \text{and} \quad {}^-7 + ({}^-3 + {}^-5)$$
$$= {}^-10 + {}^-5 \qquad\qquad = {}^-7 + {}^-8$$
$$= {}^-15 \qquad\qquad\quad\; = {}^-15$$

So we find that $(^-7 + {}^-3) + {}^-5 = {}^-7 + ({}^-3 + {}^-5)$.

Example 25. Compare $(^-12 + 2) + 9$ and $^-12 + (2 + 9)$.

Solution

$$(^-12 + 2) + 9 \quad \text{and} \quad {}^-12 + (2 + 9)$$
$$= {}^-10 + 9 \qquad\qquad = {}^-12 + 11$$
$$= {}^-1 \qquad\qquad\quad\;\; = {}^-1$$

So we find that $(^-12 + 2) + 9 = {}^-12 + (2 + 9)$.

We conclude that addition of integers is also associative. Letting A, B, and C represent any integers, we write:

Associative Property for Addition

$$(A + B) + C = A + (B + C)$$

To add several integers together when there are no parentheses, the commutative and associative properties allow us to rearrange them in any order we choose. It is often easier to add all the numbers with matching signs first and then simplify.

Example 26. Simplify $^-22 + {}^-11 + 6 + {}^-30 + 50$.

Solution

$$^-22 + {}^-11 + 6 + {}^-30 + 50$$
$$= {}^-22 + {}^-11 + {}^-30 + 6 + 50$$
$$= {}^-63 + 56$$
$$= {}^-7$$

Example 27. Simplify $9\left[\dfrac{18}{2} + ({}^-7 + 2)\right]$.

Solution

$$9\left[\frac{18}{2} + ({}^-7 + 2)\right]$$
$$= 9[9 + {}^-5]$$
$$= 9 \cdot 4$$
$$= 36$$

Example 28. Simplify $\dfrac{7({}^-6 + 6)}{6 + {}^-3}$.

Solution

$$\frac{7({}^-6 + 6)}{6 + {}^-3}$$
$$= \frac{7(0)}{3}$$
$$= \frac{0}{3}$$
$$= 0$$

Trial Run _____ **1.** Plot the points corresponding to these integers on a number line.

(a) ⁻3 (b) 2 (c) ⁻5 (d) 4

_____ **2.** Identify by number each letter on the number line.

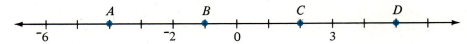

Simplify.

_____ **3.** 8 + ⁻6

_____ **4.** ⁻11 + 5

_____ **5.** ⁻7 + ⁻9

_____ **6.** ⁻3 + 3

_____ **7.** (⁻5 + ⁻6) + 10

_____ **8.** ⁻22 + ⁻7 + 3 + ⁻20 + 40

ANSWERS

1.

```
        c         a                   b       d
  <--+---●---+---+---●---+---+---+---+---●---+---●---+---+-->
    ⁻6  ⁻5  ⁻4  ⁻3  ⁻2  ⁻1   0   1   2   3   4   5   6
```

2. *A*(⁻4), *B*(⁻1), *C*(2), *D*(5).   **3.** 2.   **4.** ⁻6.   **5.** ⁻16.   **6.** 0.   **7.** ⁻1.   **8.** ⁻6.

EXERCISE SET 1.3

1. Plot the points corresponding to these integers on a number line.
   (a) ⁻7   **(b)** 4   **(c)** 0   **(d)** ⁻3

2. Plot the points corresponding to these integers on a number line.
   (a) ⁻5   **(b)** 3   **(c)** 1   **(d)** ⁻8

3. Identify by number each letter on the number line.

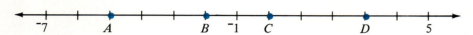

4. Identify by number each letter on the number line.

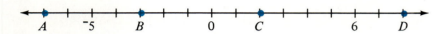

5. Use a number line to illustrate each addition problem.
   (a) 3 + 4   **(b)** ⁻2 + 6   **(c)** 6 + ⁻9   **(d)** ⁻3 + ⁻2

6. Use a number line to illustrate each addition problem.
   (a) 5 + 2   **(b)** ⁻3 + 6   **(c)** 7 + ⁻10   **(d)** ⁻1 + ⁻5

7. Compare each of the following pairs of numbers using < or >.
   _____ **(a)** 5, 12   _____ **(b)** ⁻4, ⁻10   _____ **(c)** 8, ⁻11
   _____ **(d)** ⁻3, 7

8. Compare each of the following pairs of numbers using < or >.
   _____ **(a)** 3, 9   _____ **(b)** ⁻12, ⁻2   _____ **(c)** 10, ⁻15
   _____ **(d)** ⁻2, 9

9. Compare each of the following pairs of numbers using >.
   _____ **(a)** 15, 9   _____ **(b)** ⁻8, ⁻3   _____ **(c)** 5, ⁻16
   _____ **(d)** ⁻8, 12

10. Compare each of the following pairs of numbers using >.
   _____ **(a)** 12, 5   _____ **(b)** ⁻11, ⁻4   _____ **(c)** 7, ⁻13
   _____ **(d)** ⁻9, 3

Simplify each expression.

_____ **11.** 4 + 18   _____ **12.** 12 + 5

_____ **13.** ⁻5 + ⁻7   _____ **14.** ⁻9 + ⁻12

_____ **15.** 15 + ⁻3   _____ **16.** 19 + ⁻4

_____ **17.** 3 + ⁻9   _____ **18.** 2 + ⁻15

_____ **19.** ⁻11 + 4   _____ **20.** ⁻12 + 7

_____ **21.** ⁻2 + 15   _____ **22.** ⁻3 + 22

_____ **23.** 14 + (8 + ⁻2)   _____ **24.** 20 + (10 + ⁻3)

_____ **25.** $(9 + {}^-7) + 6$

_____ **26.** $(15 + {}^-8) + 7$

_____ **27.** ${}^-15 + {}^-3 + 1$

_____ **28.** ${}^-41 + {}^-9 + 1$

_____ **29.** $({}^-12 + {}^-7) + 5$

_____ **30.** $({}^-13 + {}^-5) + 12$

_____ **31.** $2(8 + {}^-3)$

_____ **32.** $3(13 + {}^-10)$

_____ **33.** $(4 \cdot 3) + {}^-17$

_____ **34.** $(5 \cdot 6) + {}^-20$

_____ **35.** ${}^-12 + (5 \cdot 0)$

_____ **36.** ${}^-13 + (8 \cdot 0)$

_____ **37.** $2({}^-8 + 5 \cdot 4)$

_____ **38.** $3({}^-9 + 3 \cdot 6)$

_____ **39.** $(9 + {}^-3)({}^-7 + 10)$

_____ **40.** $(12 + {}^-8)({}^-9 + 16)$

_____ **41.** $3[2(8 + 5) + {}^-25]$

_____ **42.** $5[3(2 + 1) + {}^-7]$

_____ **43.** $5[{}^-11 + 6(7 + {}^-3)]$

_____ **44.** $4[{}^-10 + 3(9 + {}^-5)]$

_____ **45.** ${}^-12 + [9 + 3({}^-5 + 7)]$

_____ **46.** ${}^-13 + [10 + 2({}^-3 + 8)]$

_____ **47.** $\dfrac{24 + {}^-3}{{}^-8 + 15}$

_____ **48.** $\dfrac{33 + {}^-7}{{}^-2 + 15}$

_____ **49.** $\dfrac{{}^-3 + 30}{9 + 0}$

_____ **50.** $\dfrac{{}^-4 + 20}{8 + 0}$

_____ **51.** $\dfrac{{}^-7 + 16}{{}^-4 + 4}$

_____ **52.** $\dfrac{{}^-9 + 13}{{}^-5 + 5}$

_____ **53.** ${}^-8 + {}^-9 + \dfrac{20}{2}$

_____ **54.** ${}^-10 + {}^-3 + \dfrac{25}{5}$

_____ **55.** $\dfrac{5(7 + {}^-7)}{{}^-4 + 9}$

_____ **56.** $\dfrac{6(8 + {}^-8)}{{}^-9 + 18}$

_____ **57.** ${}^-18 + \left(6 + \dfrac{16}{4}\right)$

_____ **58.** ${}^-23 + \left(9 + \dfrac{15}{3}\right)$

_____ **59.** $2[4(9 + {}^-3) + {}^-15]$

_____ **60.** $3[2(8 + {}^-3) + {}^-7]$

_____ **61.** $4\left[\dfrac{14}{7} + 2(8 + {}^-5)\right]$

_____ **62.** $2\left[\dfrac{18}{9} + 3(7 + {}^-1)\right]$

_____ **63.** $10\left(2 \cdot 4 + \dfrac{13 + {}^-5}{4}\right)$

_____ **64.** $11\left(3 \cdot 1 + \dfrac{17 + {}^-5}{12}\right)$

_____ **65.** $6\left({}^-9 + \dfrac{2 \cdot 15}{3}\right)$

_____ **66.** $5\left({}^-11 + \dfrac{2 \cdot 18}{3}\right)$

CHECKUP 1.3

Simplify.

_____ 1. $^-3 + {}^-7$

_____ 2. $^-9 + 4$

_____ 3. $^-2 + 10$

_____ 4. $(^-3 + {}^-5) + 7$

_____ 5. $^-8 + {}^-9 + 5$

_____ 6. $3(^-3 + 8)$

_____ 7. $(7 + {}^-3)(^-4 + 10)$

_____ 8. $2 + [^-14 + 3(8 + {}^-7)]$

_____ 9. $\dfrac{^-9 + 15}{14 + {}^-8}$

_____ 10. $5\left(3 \cdot 4 + \dfrac{14 + {}^-2}{6}\right)$

Example 8. Simplify $(7 - 10) - (8 - 15)$.

Solution

$$(7 - 10) - (8 - 15)$$
$$= (7 + {}^-10) - (8 + {}^-15)$$
$$= {}^-3 - {}^-7$$
$$= {}^-3 + 7$$
$$= 4$$

Example 9. Simplify $5 - [8 - (10 - 23)]$.

Solution

$$5 - [8 - (10 - 23)]$$
$$= 5 - [8 - (10 + {}^-23)]$$
$$= 5 - [8 - {}^-13]$$
$$= 5 - [8 + 13]$$
$$= 5 - 21$$
$$= 5 + {}^-21$$
$$= {}^-16$$

If you are impatient and try to hurry through these problems, skipping steps along the way, you will find yourself making mistakes. It is important to get into the habit of dealing carefully with symbols of grouping.

Example 10. Simplify $9 - [(8 - 13) - (10 - 16) - 4]$.

Solution

$$9 - [(8 - 13) - (10 - 16) - 4]$$
$$= 9 - [(8 + {}^-13) - (10 + {}^-16) - 4]$$
$$= 9 - [{}^-5 - {}^-6 - 4]$$
$$= 9 - [{}^-5 + 6 + {}^-4]$$
$$= 9 - [{}^-9 + 6]$$
$$= 9 - {}^-3$$
$$= 9 + 3$$
$$= 12$$

Trial Run   *Simplify.*

——————— **1.** $7 - 12$

——————— **2.** ${}^-15 - 8$

——————— **3.** ${}^-9 - {}^-17$

——————— **4.** ${}^-10 - 10$

——————— **5.** $9 - (3 + 12)$

——————— **6.** $(12 - 7) - 9$

——————— **7.** $(18 - 3) - (5 + 8)$

——————— **8.** $(3 - 7) - (9 - 12)$

_____ **9.** $7 - [10 - (9 - 21)]$

_____ **10.** $13 - [(9 - 12) - (8 - 13) - 1]$

ANSWERS
1. ⁻5.   **2.** ⁻23.   **3.** 8.   **4.** ⁻20.   **5.** ⁻6.   **6.** ⁻4.   **7.** 2.   **8.** ⁻1.   **9.** ⁻15.
10. 12.

Name _____ Date _____

EXERCISE SET 1.4

Write each expression using addition and simplify.

_____ **1.** $12 - 9$

_____ **2.** $23 - 8$

_____ **3.** $11 - 27$

_____ **4.** $9 - 18$

_____ **5.** $^-15 - 7$

_____ **6.** $^-19 - 8$

_____ **7.** $12 - ^-3$

_____ **8.** $17 - ^-5$

_____ **9.** $^-9 - ^-7$

_____ **10.** $^-11 - ^-8$

Simplify each expression.

_____ **11.** $15 + (8 - 13)$

_____ **12.** $17 + (9 - 15)$

_____ **13.** $21 - (3 - 7)$

_____ **14.** $11 - (5 - 12)$

_____ **15.** $(8 - 7) - 12$

_____ **16.** $(13 - 10) - 20$

_____ **17.** $(7 - 14) - 2$

_____ **18.** $(8 - 16) - 5$

_____ **19.** $(5 - 4) - ^-2$

_____ **20.** $(10 - 3) - ^-4$

_____ **21.** $^-7 - (10 - 5)$

_____ **22.** $^-13 - (12 - 6)$

_____ **23.** $(^-4 - 2) - 4$

_____ **24.** $(^-8 - 4) - 9$

_____ **25.** $(^-5 - ^-2) - 3$

_____ **26.** $(^-9 - ^-4) - 5$

_____ **27.** $(^-10 - 5) - ^-15$

_____ **28.** $(^-12 - 6) - ^-18$

_____ **29.** $^-13 - (^-3 - 11)$

_____ **30.** $^-12 - (^-5 - 8)$

_____ **31.** $3(^-2 - ^-7)$

_____ **32.** $4(^-3 - ^-8)$

_____ **33.** $3 \cdot 5 - 17$

_____ **34.** $2 \cdot 6 - 21$

_____ **35.** $^-6 - (8 \cdot 0)$

_____ **36.** $^-9 - (6 \cdot 0)$

_____ **37.** $2(7 \cdot 4 - 8)$

_____ **38.** $5(6 \cdot 5 - 10)$

_____ **39.** $(12 - 3)(5 - {}^-2)$

_____ **40.** $(17 - 5)(2 - {}^-1)$

_____ **41.** $3[2(5 - 3) - {}^-1]$

_____ **42.** $5[3(6 - 4) - {}^-5]$

_____ **43.** $6[{}^-9 + 6(7 - 3)]$

_____ **44.** $4[{}^-11 + 7(8 - 4)]$

_____ **45.** ${}^-9 + [10 + 3(1 - {}^-2)]$

_____ **46.** ${}^-13 + [7 + 4(2 - {}^-3)]$

_____ **47.** $\dfrac{25 - 3}{4 - {}^-7}$

_____ **48.** $\dfrac{33 - 9}{6 - {}^-6}$

_____ **49.** $\dfrac{3 - {}^-15}{9 - 0}$

_____ **50.** $\dfrac{4 - {}^-17}{7 - 0}$

_____ **51.** $\dfrac{{}^-3 - {}^-12}{5 - 5}$

_____ **52.** $\dfrac{17 - {}^-2}{7 - 7}$

_____ **53.** ${}^-8 - 9 - \dfrac{16}{4}$

_____ **54.** ${}^-10 - 7 - \dfrac{20}{10}$

_____ **55.** $\dfrac{4(8 - 8)}{{}^-4 - {}^-9}$

_____ **56.** $\dfrac{3(11 - 11)}{{}^-5 - {}^-10}$

_____ **57.** ${}^-13 - \left(8 - \dfrac{10}{2}\right)$

_____ **58.** ${}^-17 - \left(9 - \dfrac{15}{3}\right)$

_____ **59.** $2[3(10 - 7) - 5]$

_____ **60.** $3[4(9 - 7) - 1]$

_____ **61.** $5\left[\dfrac{12}{6} + 3(2 - {}^-2)\right]$

_____ **62.** $6\left[\dfrac{10}{5} + 7(3 - {}^-1)\right]$

_____ **63.** $10\left(3 \cdot 3 - \dfrac{12 - {}^-6}{3}\right)$

_____ **64.** $9\left(4 \cdot 1 - \dfrac{14 - {}^-12}{13}\right)$

_____ **65.** $5\left(7 - \dfrac{2 \cdot 9}{3}\right)$

_____ **66.** $4\left(8 - \dfrac{2 \cdot 14}{4}\right)$

Name _____ Date _____

CHECKUP 1.4

Simplify.

_____ 1. $10 - 28$

_____ 2. $8 - {}^-13$

_____ 3. ${}^-17 - 5$

_____ 4. ${}^-19 - {}^-9$

_____ 5. ${}^-7 - (11 - {}^-5)$

_____ 6. $4({}^-8 - {}^-16)$

_____ 7. $({}^-5 - {}^-11)(9 - {}^-2)$

_____ 8. $5(6 \cdot 3 - {}^-2)$

_____ 9. $\dfrac{{}^-5 - {}^-30}{18 - 13}$

_____ 10. $3[4 \cdot 7 - (6 - {}^-2)]$

1.5 Multiplying and Dividing Integers

When multiplying whole numbers, we observed that multiplication was a way of performing repeated addition.

$$4 \cdot 3 = 3 + 3 + 3 + 3 = 12$$
$$5 \cdot 1 = 1 + 1 + 1 + 1 + 1 = 5$$

1.5-1 Multiplying Integers

Whenever we multiply two positive integers together, we shall be repeatedly adding positive numbers, so the answer must be positive.

> To multiply two positive integers, multiply the units and give the answer a positive sign.

Look at the product of a positive integer and a negative integer, again using repeated addition.

$$5(^-3) = {}^-3 + {}^-3 + {}^-3 + {}^-3 + {}^-3 = {}^-15$$
$$4(^-1) = {}^-1 + {}^-1 + {}^-1 + {}^-1 = {}^-4$$

In each case, we were repeatedly adding negative numbers, so the answer must be negative.

> To find the product of a positive integer and a negative integer, multiply the units and give the answer a negative sign.

Example 1. Find $3(^-6)$.

Solution. $3(^-6) = {}^-18$.

Example 2. Find $^-4 \cdot 11$.

Solution. $^-4 \cdot 11 = {}^-44$.

What happens when we multiply two negative integers? Let us look at the pattern that occurs when we consider some multiples of $^-3$.

$$4(^-3) = {}^-12$$
$$3(^-3) = {}^-9$$
$$2(^-3) = {}^-6$$
$$1(^-3) = {}^-3$$
$$0(^-3) = 0$$

As the multipliers of $^-3$ on the left are *decreasing* by 1, what is happening to the answers? You should see that they are *increasing* by 3. To continue this pattern, we must let the multipliers of $^-3$ continue to *decrease* by 1 and look at

$$^-1(^-3)$$
$$^-2(^-3)$$
$$^-3(^-3)$$
$$^-4(^-3)$$
$$^-5(^-3)$$

At the same time, the answers must continue to *increase* by 3, giving us

$$^-1(^-3) = 3$$
$$^-2(^-3) = 6$$
$$^-3(^-3) = 9$$
$$^-4(^-3) = 12$$
$$^-5(^-3) = 15$$

Observe that the product of two negative integers turns out to be a positive integer.

> To multiply two negative integers, multiply the units and give the answer a positive sign.

Example 3. Find $(^-6)(^-8)$.

Solution. $(^-6)(^-8) = 48$.

Example 4. Find $(^-8)(^-6)$.

Solution. $(^-8)(^-6) = 48$.

These two examples illustrate the fact that as with multiplication of whole numbers, multiplication of integers is **commutative**. The order in which two integers are multiplied does not matter. Letting A and B represent any integers, we write:

> **Commutative Property for Multiplication**
> $$A \cdot B = B \cdot A$$

1.5-2 Working with Symbols of Grouping in Multiplication

Symbols of grouping still give us directions regarding the order of operations.

Example 5. Simplify $9(^-2) + 7$.

Solution

$$9(^-2) + 7$$
$$= ^-18 + 7$$
$$= ^-11$$

Example 6. Simplify $^-5(3) + (^-2)(^-1)$.

Solution

$$^-5(3) + (^-2)(^-1)$$
$$= ^-15 + 2$$
$$= ^-13$$

Example 3. Simplify $^-2(6 - 19)$.

Solution

$$^-2(6 - 19)$$
$$= ^-2(6 + ^-19)$$
$$= ^-2(^-13)$$
$$= 26$$

Example 8. Simplify $^-3[^-5 - (10 - 17)]$.

Solution

$$^-3[^-5 - (10 - 17)]$$
$$= {}^-3[^-5 - (10 + {}^-17)]$$
$$= {}^-3[^-5 - {}^-7]$$
$$= {}^-3[^-5 + 7]$$
$$= {}^-3[2]$$
$$= {}^-6$$

Example 9. Simplify $[(^-3)(^-2)](^-7)$.

Solution

$$[(^-3)(^-2)](^-7)$$
$$= [6](^-7)$$
$$= {}^-42$$

Example 10. Simplify $(^-3)[(^-2)(^-7)]$.

Solution

$$(^-3)[(^-2)(^-7)]$$
$$= (^-3)[14]$$
$$= {}^-42$$

Examples 9 and 10 illustrate the fact that multiplication of integers is *associative*. In multiplication the way in which numbers are grouped does not matter. Letting A, B, and C represent integers, we write:

> **Associative Property for Multiplication**
>
> $$(A \cdot B) \cdot C = A \cdot (B \cdot C)$$

To multiply several integers together when there are no brackets, the associative property allows us to group them as we choose.

Example 11. Find $3(^-2)(^-1)(^-5)$.

Solution

$$3(^-2)(^-1)(^-5)$$
$$= {}^-6(^-1)(^-5)$$
$$= 6(^-5)$$
$$= {}^-30$$

Example 12. Find $16(^-10)(0)(3)$.

Solution

$$16(^-10)(0)(3)$$
$$= {}^-160(0)(3)$$
$$= 0(3)$$
$$= 0$$

Trial Run   *Simplify.*

_____ **1.** $4(^-10)$

_____ **2.** $^-7(^-8)$

_____ **3.** $^-6(12)$

_____ **4.** $4(^-3) + 5$

_____ **5.** $^-8(7) + (^-3)(^-5)$

_____ **6.** $^-5(8 - 17)$

_____ **7.** $^-2[^-7 - (13 - 16)]$

_____ **8.** $(^-7)(^-8)(0)(^-2)$

ANSWERS
1. $^-40$.   **2.** 56.   **3.** $^-72$.   **4.** $^-7$.   **5.** $^-41$.   **6.** 45.   **7.** 8.   **8.** 0.

1.5-3 Dividing Integers

Remember that every division statement corresponds to a multiplication statement. We said that

$$\frac{6}{3} = 2 \quad \text{because} \quad 3 \cdot 2 = 6$$

$$\frac{0}{9} = 0 \quad \text{because} \quad 9 \cdot 0 = 0$$

To work a division problem, we may look at the corresponding multiplication problem. For instance, writing $\frac{24}{8} = ?$ because $8 \cdot ? = 24$ helps us figure out what ? must be. We know that $8 \cdot 3 = 24$, so we decide that

$$\frac{24}{8} = 3$$

Example 13. Find $\frac{123}{3}$ and write the corresponding multiplication statement.

Solution. $\frac{123}{3} = 41$ because $3 \cdot 41 = 123$.

Example 14. Find $\frac{99}{9}$ and write the corresponding multiplication statement.

Solution. $\frac{99}{9} = 11$ because $9 \cdot 11 = 99$.

In each example, we found that the quotient of two positive integers was a positive integer.

> To divide a positive integer by a positive integer, divide the units and give your answer a positive sign.

Consider the quotients

$$\frac{^-36}{^-9} \quad \text{and} \quad \frac{^-18}{^-6}$$

To find the first quotient, we could say that

$$\frac{^-36}{^-9} = ? \quad \text{because} \quad ^-9 \cdot ? = ^-36$$

From our experience with multiplication, we know that $^-9(4) = ^-36$, so ? must be 4.

$$\frac{^-36}{^-9} = 4 \qquad \text{because} \qquad ^-9(4) = ^-36$$

To find the second quotient, we could say that

$$\frac{^-18}{^-6} = ? \qquad \text{because} \qquad ^-6 \cdot ? = ^-18$$

Again using the multiplication rules, we know that $^-6(3) = ^-18$, so ? must be 3.

$$\frac{^-18}{^-6} = 3 \qquad \text{because} \qquad ^-6(3) = ^-18$$

In each case, we found that the quotient of two negative integers was a positive integer.

> To divide a negative integer by a negative integer, divide the units and give the answer a positive sign.

Example 15. Find $\dfrac{^-75}{^-3}$ and write the corresponding multiplication statement.

Solution. $\dfrac{^-75}{^-3} = 25$ because $^-3(25) = ^-75$.

Example 16. Find $\dfrac{^-6}{^-6}$ and write the corresponding multiplication statement.

Solution. $\dfrac{^-6}{^-6} = 1$ because $^-6(1) = ^-6$.

Look at the quotients

$$\frac{^-35}{7} \qquad \text{and} \qquad \frac{39}{^-3}$$

To find the first quotient, we could say that

$$\frac{^-35}{7} = ? \qquad \text{because} \qquad 7 \cdot ? = ^-35$$

Multiplication facts help us see that $7(^-5) = ^-35$, so ? must be $^-5$.

$$\frac{^-35}{7} = ^-5 \qquad \text{because} \qquad 7(^-5) = ^-35$$

To find the second quotient, we could say that

$$\frac{39}{^-3} = ? \qquad \text{because} \qquad ^-3 \cdot ? = 39$$

From our experience with multiplication, we know that $^-3(^-13) = 39$, so ? must be $^-13$.

$$\frac{39}{^-3} = ^-13 \qquad \text{because} \qquad ^-3(^-13) = 39$$

In both examples, we found that the quotient of a positive integer and a negative integer was a negative integer.

> To divide a positive integer by a negative integer (or a negative integer by a positive integer), divide the units and give the answer a negative sign.

Some people prefer to learn the rules for multiplying and dividing integers in a slightly different form.

> When multiplying or dividing two integers:
>
> **1.** If the signs are the same, the answer will be positive.
> **2.** If the signs are different, the answer will be negative.

Example 17. Find $\dfrac{^-16}{^-2}$.

Solution. $\dfrac{^-16}{^-2} = 8.$

Example 18. Find $\dfrac{^-29}{1}$.

Solution. $\dfrac{^-29}{1} = {}^-29.$

Example 19. Find $\dfrac{17}{^-17}$.

Solution. $\dfrac{17}{^-17} = {}^-1.$

Remember, the rules for working with zero in a division problem continue to apply here.

Example 20. Find $\dfrac{0}{^-13}$.

Solution. $\dfrac{0}{^-13} = 0$ because $^-13(0) = 0.$

Example 21. Find $\dfrac{^-37}{0}$.

Solution. $\dfrac{^-37}{0}$ is undefined.

1.5-4 Working with Symbols of Grouping in Division

We may combine all the integer operations that we have learned by using symbols of grouping.

Example 22. Simplify $\dfrac{^-3 + 9}{^-2}$.

Solution

$$\frac{^-3 + 9}{^-2}$$

$$= \frac{6}{^-2} = {}^-3$$

Example 23. Simplify $\dfrac{2(^-5) - 8}{6}$.

Solution

$$\dfrac{2(^-5) - 8}{6}$$

$$= \dfrac{^-10 - 8}{6}$$

$$= \dfrac{^-10 + {}^-8}{6}$$

$$= \dfrac{^-18}{6} = {}^-3$$

Example 24. Simplify $\dfrac{2(^-5) + 3(^-7)}{^-31}$.

Solution

$$\dfrac{2(^-5) + 3(^-7)}{^-31}$$

$$= \dfrac{^-10 + {}^-21}{^-31}$$

$$= \dfrac{^-31}{^-31} = 1.$$

Example 25. Simplify $\dfrac{3(^-1) - \dfrac{24}{^-8}}{7}$.

Solution

$$\dfrac{3(^-1) - \dfrac{24}{^-8}}{7}$$

$$= \dfrac{^-3 - {}^-3}{7}$$

$$= \dfrac{^-3 + 3}{7}$$

$$= \dfrac{0}{7} = 0$$

Example 26. Simplify $\dfrac{5[18 + 2(^-3)]}{^-6}$.

Solution

$$\dfrac{5[18 + 2(^-3)]}{^-6}$$

$$= \dfrac{5[18 + {}^-6]}{^-6}$$

$$= \dfrac{5[12]}{^-6}$$

$$= \dfrac{60}{^-6} = {}^-10$$

Trial Run *Simplify.*

_____ 1. $\dfrac{^-16}{2}$

_____ 2. $\dfrac{27}{^-3}$

_____ 3. $\dfrac{^-24}{^-8}$

_____ 4. $\dfrac{0}{^-6}$

_____ 5. $\dfrac{^-9}{0}$

_____ 6. $\dfrac{^-4 + 10}{^-3}$

_____ 7. $\dfrac{3(^-6) + 8}{5}$

_____ 8. $\dfrac{6(^-1) - \frac{21}{^-7}}{3}$

_____ 9. $\dfrac{^-3[2(^-4) - {}^-2]}{^-11 + 2}$

ANSWERS

1. $^-8.$ **2.** $^-9.$ **3.** 3. **4.** 0. **5.** Undefined. **6.** $^-2.$ **7.** $^-2.$ **8.** $^-1.$ **9.** $^-2.$

1.5-5 ## Switching from Words to Numbers

Let's look at some word statements which involve integers and switch them to number statements.

Example 27. Yesterday, Amy's checking account showed a balance of $67. Today she deposited $100 and wrote checks for $32, $17, and $125. What is the condition of her account now?

Solution. Amy's transactions can be written as

$$67 + 100 + {}^-32 + {}^-17 + {}^-125 = 167 + {}^-174 = {}^-7$$

Amy is overdrawn $7.

Example 28. Last week, the daily temperatures in Anytown were $^-4°, {}^-3°, 0°, 8°, 11°, {}^-2°,$ and $^-3°$. What was the average temperature last week?

Solution. To find the average temperature, we must add the temperatures and divide by 7 (the number of days).

$$\frac{^-4° + {}^-3° + 0° + 8° + 11° + {}^-2° + {}^-3°}{7}$$

$$= \frac{^-12° + 19°}{7.}$$

$$= \frac{7°}{7} = 1°$$

Example 29. The estate of wealthy Uncle Donald is to be divided equally among his 5 nieces. If the estate includes a house worth $150,000 and 3 automobiles worth $11,000 each and debts of $9000 and $29,000, how much will each niece inherit?

Solution. The values of the house and automobile are positive numbers, but the debts represent negative numbers. The estate is worth

$$150{,}000 + 3(11{,}000) + {}^{-}9000 + {}^{-}29{,}000$$

Dividing this 5 ways, we have

$$\frac{150{,}000 + 33{,}000 + {}^{-}9000 + {}^{-}29{,}000}{5}$$

$$= \frac{183{,}000 + {}^{-}38{,}000}{5}$$

$$= \frac{145{,}000}{5} = \$29{,}000 \text{ for each niece}$$

Trial Run *Change each word expression to a number expression, and evaluate.*

_____ 1. A child's movie ticket costs half as much as an adult's ticket, which costs $4. A senior citizen's movie ticket costs $1 less than an adult's ticket. If 25 children's tickets, 60 adult's tickets, and 15 senior citizen's tickets were sold, write an expression for the amount received at the ticket window.

_____ 2. At the local Bingo game, Betty lost $5 per week for 3 weeks, won $7 per week for 2 weeks and lost $2 per week for 4 weeks. Write an expression for her average gain or loss per week.

_____ 3. If a balloon rises from the ground at 3 feet per second for 20 seconds and then falls at 2 feet per second for 10 seconds, what is its final height?

_____ 4. Henry and his roommate split rent and food expenses equally. This month, they spent $225 on rent and $115 on food. If Henry started the month with $250, write an expression for the money he has left after paying his share of the expenses.

_____ 5. Profits or losses are split equally among the 5 owners of the Pizza Shack. During the past week, the Shack took in $1873. Supplies cost $920, 8 workers were paid $150 each, and advertising cost $53. Write an expression for each owner's share of profits or losses.

ANSWERS

1. $25\left(\dfrac{4}{2}\right) + 60(4) + 15(4-1) = \$335.$ 2. $\dfrac{3({}^{-}5) + 2(7) + 4({}^{-}2)}{9} = \${}^{-}1.$

3. $20(3) + 10({}^{-}2) = 40$ feet. 4. $250 - \left(\dfrac{225 + 115}{2}\right) = \$80.$

5. $\dfrac{1873 + {}^{-}920 + 8({}^{-}150) + {}^{-}53}{5} = \${}^{-}60.$

EXERCISE SET 1.5

Simplify.

_____ **1.** 5($^-$6) _____ **2.** 3($^-$4)

_____ **3.** $^-$6($^-$7) _____ **4.** $^-$4($^-$8)

_____ **5.** $^-$3(8) _____ **6.** $^-$7(9)

_____ **7.** $^-$10(3)($^-$5) _____ **8.** $^-$5(7)($^-$3)

_____ **9.** 4($^-$3)(0)(9) _____ **10.** 6($^-$5)(0)(11)

_____ **11.** $^-$9(3 − 8) _____ **12.** $^-$10(5 − 12)

_____ **13.** 6($^-$10) − 20 _____ **14.** 8($^-$9) − 15

_____ **15.** 23 − 7($^-$2) _____ **16.** 29 − 3($^-$5)

_____ **17.** $\dfrac{24}{^-6}$ _____ **18.** $\dfrac{20}{^-4}$

_____ **19.** $\dfrac{^-16}{8}$ _____ **20.** $\dfrac{^-15}{5}$

_____ **21.** $\dfrac{^-35}{^-7}$ _____ **22.** $\dfrac{^-48}{^-6}$

_____ **23.** $\dfrac{11(^-4)}{22}$ _____ **24.** $\dfrac{12(^-3)}{4}$

_____ **25.** $\dfrac{15}{^-3} + \dfrac{^-12}{6}$ _____ **26.** $\dfrac{21}{^-7} + \dfrac{^-28}{4}$

_____ **27.** $\dfrac{^-20}{5} - \dfrac{^-30}{^-6}$ _____ **28.** $\dfrac{^-45}{9} - \dfrac{^-55}{^-11}$

_____ **29.** $2\left(\dfrac{^-18}{3}\right)$ _____ **30.** $5\left(\dfrac{^-54}{9}\right)$

_____ **31.** $\dfrac{^-20 + 15}{7 - 7}$ _____ **32.** $\dfrac{^-35 + 20}{5 - 5}$

_____ **33.** $\dfrac{^-40 + 15}{16 - {^-9}}$ _____ **34.** $\dfrac{^-17 + 10}{4 - {^-3}}$

_____ **35.** $\dfrac{3(^-8) + (^-2)(^-4)}{4(^-4)}$ _____ **36.** $\dfrac{5(^-9) + (^-3)(^-2)}{3(^-13)}$

_____ **37.** $\dfrac{9(^-1) + \frac{14}{^-2}}{4}$ _____ **38.** $\dfrac{3(^-9) + \frac{5}{^-1}}{8}$

_____ **39.** $\dfrac{7[18 + 5(^-2)]}{^-8}$ _____ **40.** $\dfrac{9[21 + 7(^-2)]}{^-21}$

———— 41. $\dfrac{-4[3(-2) - -7]}{-13 + 11}$

———— 42. $\dfrac{-5[6(-3) - -20]}{-17 + 12}$

———— 43. $\dfrac{20(-4)}{10} + \dfrac{(21 - 6)}{-3}$

———— 44. $\dfrac{36(-2)}{12} + \dfrac{27 - 9}{-6}$

———— 45. $\dfrac{-9(7 - 1) - 8(-3)}{-5(-8 + 6)}$

———— 46. $\dfrac{-8(10 - 3) - 9(-4)}{2(-9 + 4)}$

———— 47. $-3\left[\dfrac{-14}{7} + 3(7 - 12)\right]$

———— 48. $-5\left[\dfrac{-15}{3} + 2(9 - 12)\right]$

———— 49. $10\left[-3(6) + \dfrac{13 + -5}{-4}\right]$

———— 50. $5\left[-5(7) + \dfrac{14 + -5}{-3}\right]$

———— 51. $\dfrac{\dfrac{-15}{5} + \dfrac{0}{7}}{\dfrac{25}{-5} + 2}$

———— 52. $\dfrac{\dfrac{-18}{6} + \dfrac{0}{9}}{\dfrac{-30}{5} + 3}$

———— 53. $4\left(-9 + \dfrac{7 + -3}{6 - 8}\right) - 3$

———— 54. $5\left(-10 + \dfrac{8 + -3}{15 - 20}\right) - 6$

In Exercises 55–60, write each word expression as a number expression. Then simplify.

———— 55. Miranda borrowed $5 from her brother twice last month. This month she paid him $7 and borrowed $3. Where does Miranda's account with her brother stand now?

———— 56. In one series of plays, State University's football team completed two 7-yard passes and ran for 5 yards; the quarterback was sacked twice, for a 3-yard loss each time. What was the team's total gain or loss?

———— 57. During one week in January the temperature was $-3°F$ on two days, $4°F$ on two days, $0°F$ on two days, and $5°F$ one day. What was the average temperature during the week?

———— 58. Last week Jo Ellen worked 2 fewer hours than she worked this week. Next week she plans to work 5 hours more than she worked this week. If Jo Ellen worked 10 hours this week, how many hours per week will she be averaging?

———— 59. Carlos expects to harvest 10 tomatoes from each Big Boy plant in his garden and 11 tomatoes from each Rutgers plant. He plans to plant twice as many Rutgers plants as Big Boy plants. If he plants 16 Big Boy plants, how many tomatoes should he expect?

———— 60. A dozen medium eggs cost 9 cents less than a dozen large eggs. If a chef buys three times as many dozens of medium eggs as large eggs and he purchases 2 dozen large eggs at 89 cents each, how much will the eggs cost?

CHECKUP 1.5

Simplify.

_____ **1.** $5(^-6)$

_____ **2.** $^-6(^-8)$

_____ **3.** $\dfrac{^-35}{7}$

_____ **4.** $\dfrac{^-50}{^-5}$

_____ **5.** $^-5(3 - 9)$

_____ **6.** $\dfrac{15(^-4)}{6}$

_____ **7.** $\dfrac{5[13 + 3(^-1)]}{^-25}$

_____ **8.** $\dfrac{^-3[2(^-1) - ^-8]}{^-10 + 19}$

_____ **9.** $^-5\left[\dfrac{^-16}{4} + 3(8 - 10)\right]$

_____ **10.** At 6 P.M. the temperature was 16°. It dropped 3° each hour for the next 7 hours. Write an expression for the temperature at 1 A.M. and simplify.

Summary

In this chapter we discussed three important lists of numbers.

Natural numbers: 1, 2, 3, 4, . . .

Whole numbers: 0, 1, 2, 3, 4, . . .

Integers: . . . , ⁻3, ⁻2, ⁻1, 0, 1, 2, 3, 4, . . .

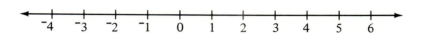

The number line gave us a handy way to picture these numbers.

We discovered how to add, subtract, multiply, and divide such numbers, using symbols of grouping to give us directions about the order in which to perform several operations.

In addition and multiplication we found that it did not matter in which order we added or multiplied two numbers (the commutative properties). Similarly, in addition and multiplication of three numbers, it did not matter how we grouped those numbers (the associative properties).

We noted that zero and 1 had some special properties that we shall summarize here, letting A represent any integer:

1. $A + 0 = A$
2. $A - 0 = A$
3. $A - A = 0$
4. $A \cdot 0 = 0$
5. $A \cdot 1 = A$
6. $\dfrac{A}{0}$ is undefined
7. $\dfrac{0}{A} = 0$, where A is not 0
8. $\dfrac{A}{A} = 1$, where A is not 0

We developed the following rules for adding integers:

1. To add two positive integers, add the units and give the answer a positive sign.
2. To add two negative integers, add the units and give the answer a negative sign.
3. To add a positive and a negative integer, take the difference between the units and give the answer the sign of the original integer having more units.

We developed the following definition for subtraction:

To subtract two integers, take the opposite of the integer being subtracted, and add.

We developed the following rules for multiplying and dividing integers:

1. The product (or quotient) of two positive integers is positive.
2. The product (or quotient) of two negative integers is positive.
3. The product (or quotient) of a positive integer and a negative integer is negative.

Name _____ **Date** _____

REVIEW EXERCISES

SECTION 1.1

1. Plot the point corresponding to each number on a number line.
   (a) -3　　**(b)** 0　　**(c)** 5　　**(d)** -6

2. Identify by number each letter on the number line.

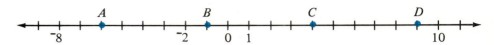

3. According to the commutative property for addition, $8 + 3 = 3 +$ _____.

4. According to the associative property for multiplication, $(3 \cdot 2)4 = 3 \cdot$ _____.

5. According to the commutative property for multiplication, $^-2 \cdot 5 = 5 \cdot$ _____.

6. According to the associative property for addition, $(3 + 0) + {}^-1 =$ _____.

7. According to the distributive property of multiplication over addition,
   $2(3 + 5) = 2 \cdot 3 +$ _____.

8. According to the distributive property of multiplication over addition,
   $2(8 + {}^-3) =$ _____.

Simplify.

SECTION 1.2

_____ 9. $3(5 + 2 \cdot 7)$

_____ 10. $(7 + 2)(8 + 0)$

_____ 11. $5[2(4 + 1) + 3]$

_____ 12. $7 + [2 + 3(4 + 1)]$

_____ 13. $17 + 2(9 - 3)$

_____ 14. $(10 - 8) \cdot \dfrac{9}{3}$

_____ 15. $\dfrac{8(10 - 6)}{9 - 5}$

_____ 16. $7\left(\dfrac{15}{3} + \dfrac{5 \cdot 0}{2}\right)$

_____ 17. $7\left(5 \cdot 1 - \dfrac{21 - 5}{8}\right)$

_____ 18. $^-6 + {}^-12$

SECTIONS 1.3–1.5

_____ 19. $8 + {}^-5$

_____ 20. $(2 \cdot 3) + {}^-11$

_____ 21. $(7 + {}^-2)({}^-5 + 1)$

_____ 22. $^-3 + \left({}^-5 + \dfrac{24}{3}\right)$

_____ 23. $32 - {}^-6$

_____ 24. $^-9 - 7$

_____ 25. $(^-3 - 9) - 11$

_____ 26. $7[4(8 - 5) - {}^-3]$

_____ 27. $\dfrac{^-9 - {}^-3}{8 - 10}$

_____ 28. $^-2(7 - 15)$

_____ 29. $\dfrac{18}{-2} - \dfrac{^-15}{3}$

_____ 30. $\dfrac{3(^-4) + (^-6)(^-2)}{^-7 - 2}$

_____ 31. $\dfrac{^-3(8 - 2) - 2(^-1)}{^-7 + (1 - 10)}$

_____ 32. $\dfrac{\dfrac{^-24}{3} + \dfrac{0}{2}}{\dfrac{10}{5} + \dfrac{8}{4}}$

_____ **33.** Julie bought 5 tapes at $7 each and 3 records at $5 each. Write an expression for the total amount she spent at Music Mart, and simplify.

_____ **34.** If Shanda worked 8 hours at $4 per hour at the Snack Shop but had $6 deducted for the food she ate during that time, write an expression for what she earned that day, and simplify.

_____ **35.** At the track, Eddie bet $2 on each of the 10 races that day. Twice he won $4 and once he won $15. Write an expression representing Eddie's total winnings (or losses) for the day, and simplify.

2 Working with Variables

In algebra we must work with **constants**, such as the integers we have already studied, but we must also work with symbols that stand for numbers. Such symbols are called **variables**; they are often represented by letters.

In this chapter we learn how to

1. Find the value of expressions containing constants and/or variables.
2. Switch from word statements to variable statements.
3. Add and subtract constants and/or variables.
4. Multiply constants times variable expressions.

2.1 Algebraic Expressions

Combinations of constants (numbers) and/or variables (letters that stand for numbers) involving the operations of addition, subtraction, multiplication, and division are called **algebraic expressions**. As in arithmetic, we use the familiar symbols to show what operation is to be performed.

$$5 + 3 \text{ means add 5 and 3}$$

$$4 + x \text{ means add 4 and the variable } x$$

$$11 - 3 \text{ means subtract 3 from 11}$$

$$y - 9 \text{ means subtract 9 from the variable } y$$

$$16 \div 2 \text{ or } \frac{16}{2} \text{ means divide 16 by 2}$$

$$a \div 13 \text{ or } \frac{a}{13} \text{ means divide the variable } a \text{ by 13}$$

$$3 \cdot 5 \text{ or } 3(5) \text{ means multiply 3 times 5}$$

$$5 \cdot x \text{ or } 5(x) \text{ or } 5x \text{ means multiply 5 times } x$$

In the last multiplication example, notice that the symbols for multiplication may be left out. From now on we agree that a number written next to a variable will always mean *multiply*.

$$6y \text{ means 6 times } y$$

$$^-5x \text{ means } ^-5 \text{ times } x$$

To make subtraction possible, remember that we agreed earlier that

$$9 \text{ minus } 3 = 9 \text{ plus the opposite of 3}$$

$$9 - 3 = 9 + {}^-3$$

When dealing with variables, we make the same definition for subtraction, so that

$$2 \text{ minus } 3x = 2 \text{ plus the opposite of } 3x$$

But what is the opposite of $3x$? We shall agree that the opposite of $3x$ is $^-(3x)$, which can also be written ^-3x. Therefore, we write

$$2 - 3x = 2 + {}^-3x$$

Practicing a little, we note that

$$\text{the opposite of } 16x \text{ is } ^-16x$$

$$\text{the opposite of } ^-8x \text{ is } 8x$$

We shall continue to write subtraction problems as addition problems using the opposite of the quantity being subtracted:

$$13 - 7x = 13 + {}^-7x$$

$$8 - {}^-4x = 8 + 4x$$

Example 1. Write an algebraic expression that means "add 9 to the variable x."

Solution. $9 + x$ or $x + 9$.

Example 2. Write an algebraic expression that represents "the product of 7 and y."

Solution. $7y$.

Example 3. Write an algebraic expression that represents "32 less than x."

Solution. $x - 32$ or $x + {}^-32$.

Example 4. Write an algebraic expression that means "divide 26 by a."

Solution. $\dfrac{26}{a}$ or $26 \div a$.

Example 5. Write an algebraic expression that means "twice the sum of y and 5."

Solution. First we must add y and 5, giving us $y + 5$. Then we multiply that sum by 2, using parentheses to show what is done first.

$$2(y + 5)$$

Example 6. Write an algebraic expression that means "subtract the variable x from 11, then divide by 4."

Solution. First we must subtract x from 11, giving us $11 - x$ (or $11 + {}^-x$); then we divide that difference by 4.

$$\frac{11 - x}{4} \qquad \text{or} \qquad \frac{11 + {}^-x}{4}$$

2.1-1 Finding the Value of Algebraic Expressions

In every algebraic expression containing variables such as

$$9 + x \qquad 7y \qquad x - 32$$

$$\frac{26}{a} \qquad \frac{2 + y}{5} \qquad 4(11 - x)$$

the number value of the expression depends on what number the variable stands for. For instance, to find the value of $9 + x$, we must know what number x represents.

$$\text{if } x = 2, \quad \text{then} \quad 9 + x = 9 + 2 = 11$$

$$\text{if } x = {}^-4, \quad \text{then} \quad 9 + x = 9 + {}^-4 = 5$$

$$\text{if } x = {}^-15, \quad \text{then} \quad 9 + x = 9 + {}^-15 = {}^-6$$

This process is called **evaluating** an expression because we let the variable take on some particular value and then find the value of the entire expression. To evaluate an algebraic expression for different values of the variable, we *substitute* (or "plug in") the value we are given, then perform the necessary arithmetic by methods learned earlier.

Example 7. Evaluate $7y$ when y is 3; when y is $^-2$; when y is 0.

Solution

$$\text{if } y = 3, \quad \text{then} \quad 7y = 7 \cdot 3 = 21$$
$$\text{if } y = ^-2, \quad \text{then} \quad 7y = 7(^-2) = ^-14$$
$$\text{if } y = 0, \quad \text{then} \quad 7y = 7(0) = 0$$

Example 8. Evaluate $x - 32$ when x is 41; when x is 12; when x is $^-6$.

Solution

$$\text{if } x = 41, \quad \text{then} \quad x - 32 = 41 - 32 = 41 + {}^-32 = 9$$
$$\text{if } x = 12, \quad \text{then} \quad x - 32 = 12 - 32 = 12 + {}^-32 = ^-20$$
$$\text{if } x = ^-6, \quad \text{then} \quad x - 32 = ^-6 - 32 = ^-6 + {}^-32 = ^-38$$

Example 9. Evaluate $\dfrac{26}{a}$ when a is 13; when a is $^-2$; when a is 0.

Solution

$$\text{if } a = 13, \quad \text{then} \quad \frac{26}{a} = \frac{26}{13} = 2$$
$$\text{if } a = ^-2, \quad \text{then} \quad \frac{26}{a} = \frac{26}{^-2} = ^-13$$
$$\text{if } a = 0, \quad \text{then} \quad \frac{26}{a} = \frac{26}{0} \text{ is undefined}$$

Example 10. Evaluate $\dfrac{2 + y}{5}$ when y is 8; when y is $^-2$; when y is $^-27$.

Solution

$$\text{if } y = 8, \quad \text{then} \quad \frac{2 + y}{5} = \frac{2 + 8}{5} = \frac{10}{5} = 2$$
$$\text{if } y = ^-2, \quad \text{then} \quad \frac{2 + y}{5} = \frac{2 + {}^-2}{5} = \frac{0}{5} = 0$$
$$\text{if } y = ^-27, \quad \text{then} \quad \frac{2 + y}{5} = \frac{2 + {}^-27}{5} = \frac{^-25}{5} = ^-5$$

Example 11. Evaluate $4(11 - x)$ when x is 6; when x is $^-1$; when x is 11.

Solution

$$\text{if } x = 6, \quad \text{then} \quad 4(11 - x) = 4(11 - 6) = 4(11 + {}^-6) = 4 \cdot 5 = 20$$
$$\text{if } x = ^-1, \quad \text{then} \quad 4(11 - x) = 4(11 - {}^-1) = 4(11 + 1) = 4 \cdot 12 = 48$$
$$\text{if } x = 11, \quad \text{then} \quad 4(11 - 11) = 4(0) = 0$$

Trial Run *Write an algebraic expression that means:*

_____ **1.** Add 7 to the variable x.

_____ **2.** Multiply $^-5$ times the variable y.

_____ **3.** Subtract 8 from the variable a.

———— **4.** Subtract the variable x from 15; then multiply by $^-3$.

———— **5.** Add $^-12$ to the variable y; then divide by 10.

———— **6.** Evaluate $6x$ when $x = ^-2$; when $x = 0$; when $x = 7$.

———— **7.** Evaluate $x - 23$ when $x = 50$; when $x = 10$; when $x = ^-5$.

———— **8.** Evaluate $\dfrac{60}{y}$ when $y = 6$; when $y = ^-12$; when $y = 0$.

———— **9.** Evaluate $\dfrac{2 + a}{^-6}$ when $a = 10$; when $a = ^-14$; when $a = ^-2$.

———— **10.** Evaluate $3(13 - m)$ when $m = 3$; when $m = ^-2$; when $m = 13$.

ANSWERS

1. $x + 7$.   **2.** ^-5y.   **3.** $a - 8$.   **4.** $^-3(15 - x)$.   **5.** $\dfrac{y + ^-12}{10}$.   **6.** $^-12$; 0; 42.
7. 27; $^-13$; $^-28$.   **8.** 10; $^-5$; undefined.   **9.** $^-2$; 2; 0.   **10.** 30; 45; 0.

2.1-2 Switching from Words to Variables

Variables and algebraic expressions give us a handy way of dealing with word expressions in which the value of one or more of the quantities is not known. For instance, suppose that movie tickets cost $3 each. We can find the cost of 2 tickets or 10 tickets or 170 tickets using the number statements:

$$\text{cost} = \$3(2) \text{ for 2 tickets}$$

$$\text{cost} = \$3(10) \text{ for 10 tickets}$$

$$\text{cost} = \$3(170) \text{ for 170 tickets}$$

In fact, we could use a variable expression to represent the cost of *any number* of tickets by letting x stand for the number of tickets sold. Then

$$3 \cdot x \qquad \text{or} \qquad 3x$$

would represent the cost of x tickets. Using such a variable expression allows us to find the total cost of any number of tickets by substituting the desired number for x into the expression $3x$.

Example 12. If part-time students at a university are charged $25 per credit hour, write a variable expression describing the cost of h credit hours. Then find the cost of 3 hours; 8 hours; 11 hours.

Solution. Since 1 hour costs $25, we know that h hours will cost $25 \cdot h$ or $25h$ dollars, and

$$\text{if } h = 3 \text{ hours}, \quad \text{then} \quad 25h = 25(3) = 75 \text{ dollars}$$

$$\text{if } h = 8 \text{ hours}, \quad \text{then} \quad 25h = 25(8) = 200 \text{ dollars}$$

$$\text{if } h = 11 \text{ hours}, \quad \text{then} \quad 25h = 25(11) = 275 \text{ dollars}$$

Example 13. If there are 25,230 people in a town, write a variable expression for the number of people after a change in population of x people has occurred. Then find the total population if 2900 people move in; if 1600 people move out.

Solution. Since the current population is 25,230, the new population will be

$$25{,}230 + x$$

$$\text{if } x = 2900, \quad \text{then} \quad 25{,}230 + 2900 = 28{,}130$$

$$\text{if } x = {}^-1600, \quad \text{then} \quad 25{,}230 + {}^-1600 = 23{,}630$$

Notice that we represented the increase by ${}^+2900$ and the decrease by ${}^-1600$.

Example 14. If a gasoline distributor services 10 local gas stations with equal allotments each week, write a variable expression to describe each station's share of the distributor's total gallons, G. Then find each station's share if the distributor has 60,000 gallons; 85,000 gallons; 0 gallons.

Solution. If each station receives an equal share of the distributor's gasoline, we must divide the total number of gallons by 10. So each station's share will be

$$\frac{G}{10}$$

$$\text{if } G = 60{,}000, \quad \text{then} \quad \frac{G}{10} = \frac{60{,}000}{10} = 6000 \text{ gallons}$$

$$\text{if } G = 85{,}000, \quad \text{then} \quad \frac{G}{10} = \frac{85{,}000}{10} = 8500 \text{ gallons}$$

$$\text{if } G = 0, \quad \text{then} \quad \frac{G}{10} = \frac{0}{10} = 0 \text{ gallons}$$

Example 15. The Internal Revenue Service allows a taxpayer to deduct $1000 for every dependent claimed. For an income of $12,000, write an expression describing the remaining taxable income after a taxpayer has made the deduction for d dependents. Then find the remaining taxable income if a taxpayer claims 1 dependent; 3 dependents; 6 dependents.

Solution. First we must write an expression for the total deduction. If $1000 may be deducted for each dependent, then $1000 \cdot d$ or $1000d$ may be deducted for d dependents. Now we must subtract that deduction from the $12,000 income, giving us

$$12{,}000 - 1000d$$

$$\text{if } d = 1, \quad \text{then} \quad 12{,}000 - 1000d = 12{,}000 - (1000 \cdot 1) = 12{,}000 - 1000$$
$$= 12{,}000 + {}^-1000 = 11{,}000 \text{ dollars}$$

$$\text{If } d = 3, \quad \text{then} \quad 12{,}000 - 1000d = 12{,}000 - (1000 \cdot 3) = 12{,}000 - 3000$$
$$= 12{,}000 + {}^-3000 = 9000 \text{ dollars}$$

$$\text{if } d = 6, \quad \text{then} \quad 12{,}000 - 1000d = 12{,}000 - (1000 \cdot 6) = 12{,}000 - 6000$$
$$= 12{,}000 + {}^-6000 = 6000 \text{ dollars}$$

Trial Run

1. If Arnold drives the speed limit, 55 miles per hour, write a variable expression for the distance he could travel in h hours. Then find the distance he travels in 3 hours; in 5 hours.

2. If there are 1252 students in a high school, write a variable expression for the number of students after a change in enrollment of x students has occurred. Then find the enrollment if 142 new students enroll; if 93 students drop out.

3. The meat from a beef calf is to be evenly divided among 5 families. Write a variable expression for the number of pounds each family will receive if the total weight of the calf is T pounds. Then find each family's share if the weight is 725 pounds; 935 pounds.

4. The cost of renting a garden tiller is $25, plus $10 for each day it is used. Write a variable expression describing the cost of renting a tiller for d days. Then find the cost for 2 days; for 5 days.

ANSWERS

1. $55h$, 165, 275.   **2.** $1252 + x$, 1394, 1159.   **3.** $\dfrac{T}{5}$, 145, 187.   **4.** $25 + 10d$, 45, 75.

EXERCISE SET 2.1

Write an algebraic expression for each word statement.

_____ **1.** Add 10 to the variable y.

_____ **2.** Add $^-5$ to the variable x.

_____ **3.** Multiply $^-3$ times the variable m.

_____ **4.** Multiply 8 times the variable n.

_____ **5.** Subtract 6 from the variable a.

_____ **6.** Subtract 11 from the variable b.

_____ **7.** Divide the variable t by $^-15$.

_____ **8.** Divide the variable s by 12.

_____ **9.** The sum of the variable x and $^-9$.

_____ **10.** The sum of the variable y and 7.

_____ **11.** The product of the variable q and 32.

_____ **12.** The product of the variable m and $^-13$.

_____ **13.** The variable n decreased by 6.

_____ **14.** The variable b decreased by $^-9$.

_____ **15.** Twice the variable x, divided by 7.

_____ **16.** Five times the variable y, divided by $^-11$.

_____ **17.** 9 more than the product of 6 and the variable k.

_____ **18.** 10 more than the product of 3 and the variable m.

_____ **19.** 5 less than the quotient of the variable p divided by 7.

_____ **20.** 13 less than the quotient of the variable p divided by $^-3$.

_____ **21.** 4 times a variable x is subtracted from 9.

_____ **22.** 8 times a variable y is subtracted from 23.

_____ **23.** $^-4$ times the sum of the variable x and 12.

_____ **24.** 9 times the sum of the variable y and 8.

Evaluate.

_____ **25.** $x + 9$ when $x = 6$; $x = {}^-4$; $x = 0$

_____ **26.** $x + 13$ when $x = 4$; $x = 0$; $x = {}^-23$

_____ **27.** $y - 3$ when $y = 10$; $y = {}^-3$; $y = 3$

_____ **28.** $y - 8$ when $y = 15$; $y = {}^-8$; $y = 8$

_____ **29.** $\dfrac{m}{5}$ when $m = 30$; $m = {}^-15$; $m = 0$

_____ **30.** $\dfrac{m}{9}$ when $m = 36$; $m = {}^-18$; $m = 0$

_____ 31. $\dfrac{45}{n}$ when $n = 9$; $n = 0$; $n = {}^-5$

_____ 32. $\dfrac{54}{n}$ when $n = 6$; $n = {}^-9$; $n = 0$

_____ 33. ${}^-7a$ when $a = {}^-6$; $a = 7$; $a = 0$

_____ 34. ${}^-9a$ when $a = 10$; $a = 0$; $a = {}^-3$

_____ 35. $4(x - 5)$ when $x = {}^-3$; $x = 10$; $x = 5$

_____ 36. $5(x - 12)$ when $x = {}^-1$; $x = 12$; $x = 3$

_____ 37. $\dfrac{m}{4} + 6$ when $m = 16$; $m = 0$; $m = {}^-4$

_____ 38. $\dfrac{m}{5} + 7$ when $m = {}^-5$; $m = 0$; $m = 25$

_____ 39. $2x + 4$ when $x = {}^-4$; $x = {}^-2$; $x = 5$

_____ 40. $5x + 10$ when $x = {}^-3$; $x = {}^-2$; $x = 1$

_____ 41. $\dfrac{a - 6}{7}$ when $a = 13$; $a = {}^-1$; $a = 6$

_____ 42. $\dfrac{a - 9}{5}$ when $a = 14$; $a = {}^-1$; $a = 9$

_____ 43. ${}^-3(7 - y)$ when $y = 10$; $y = {}^-6$; $y = 7$

_____ 44. ${}^-2(9 - y)$ when $y = {}^-4$; $y = 12$; $y = 9$

In Exercises 45–50, write a variable expression for each word expression and evaluate for the given values of the variable.

_____ 45. If the Beta's are selling lottery tickets on a car at $2 per ticket, write a variable expression for the cost of y tickets. Find the cost of buying 10 tickets; 25 tickets.

_____ 46. The cost of a charter bus to the ballgame is $600, to be divided evenly among the number of students who ride the bus. Write a variable expression for cost to each passenger if x students ride the bus. Find the cost for each student if there are 60 passengers; 30 passengers.

_____ 47. Write a variable expression for the number of gallons remaining in a 20-gallon tank if x gallons have already been used. Find how many gallons remain when 8 have been used; 13 have been used.

_____ 48. Jake earns $250 per week plus $7 per hour for each hour of overtime. Write a variable expression for Jake's pay for a week when he worked x hours of overtime. Find his pay when he worked 5 hours overtime; 15 hours overtime.

_____ 49. An oil tank that contains 200 barrels of oil begins to leak. Three barrels a day are leaking from the tank. Write a variable expression for the amount of oil in the tank after n days. How many barrels does the tank contain after 6 days; after 30 days?

_____ 50. The winning candidate in an election received 50 more votes than twice the number of votes his opponent received. If his opponent received x votes, write a variable expression for the number of votes the winner received. Find the number of votes the winner received if the loser got 120 votes; 500 votes.

CHECKUP 2.1

Write an algebraic expression for each word statement.

_____ **1.** Add ⁻3 to the variable x.

_____ **2.** Subtract 11 from the variable y.

_____ **3.** The product of the variable m and ⁻6.

_____ **4.** Five more than twice the variable b.

_____ **5.** Six less than the quotient of the variable y divided by ⁻2.

_____ **6.** Evaluate $x + 9$ when $x = ⁻12$; $x = 15$; $x = ⁻9$.

_____ **7.** Evaluate $\dfrac{24}{n}$ when $n = ⁻6$; $n = 0$; $n = 8$.

_____ **8.** Evaluate $3(y - 5)$ when $y = ⁻2$; $y = 7$; $y = 5$.

_____ **9.** Evaluate $\dfrac{m - 10}{3}$ when $m = 13$; $m = ⁻2$; $m = 10$.

_____ **10.** If it takes 14 bushels of grapes to make one case of wine, write a variable expression for the number of bushels needed to make x cases of wine. Find the number of bushels for 5 cases; 12 cases.

2.2 Combining Like Terms

We turn our attention now to learning ways of making algebraic expressions as simple as possible. This process is called **simplifying** algebraic expressions. Earlier we learned to simplify expressions such as

$$3 + 2 \cdot 5 = 3 + 10 = 13$$
$$2[5 - 6(3)] = 2[5 - 18] = 2[5 + {}^-18] = 2[{}^-13] = {}^-26$$

In each case, the final version was simpler and easier to work with than the original version.

Now we would like to learn how to rewrite expressions such as

$$3x + 5x - x$$
$$(2x + 1) + (9x - 3)$$
$$6(5x - 6)$$

in simplified form.

2.2-1 Like Terms

Let's look at some algebraic expressions and learn to call their parts by their correct mathematical names.

In the expression

$$6x + 5x + 2x$$

we say that $6x$ is one term, $5x$ is another term, and $2x$ is another **term**. Notice that the terms in our expression are separated by addition signs.

A **term** is a number alone, or a product of a number and one (or more) variables. Terms are separated from each other by addition signs.

Example 1. What are the terms in the expression $x + 3$?

Solution

x is a term   (remember that x means $1x$)
3 is a term

Example 2. What are the terms in the expression $4x + 7 - 3x$?

Solution. We could write the expression

$$4x + 7 + {}^-3x$$

Then we see that

$4x$ is a term
7 is a term
^-3x is a term

Example 3. What are the terms in the expression $5x - 3y - 10$?

Solution. We write the expression as

$$5x + {}^-3y + {}^-10$$

and then we see that

$5x$ is a term
^-3y is a term
$^-10$ is a term

Terms that contain a number alone are called **constant terms.**

3 is the constant term in the expression $x + 3$
$^-10$ is the constant term in the expression $5x - 3y - 10$

Terms that contain a variable are called **variable terms**.

x is the variable term in the expression $x + 3$
$5x$ and ^-3y are variable terms in the expression $5x - 3y - 10$

The number part of a variable term is called the **numerical coefficient** for the variable in that term. Return to the expression $6x + 5x + 2x$:

in the term $6x$, 6 is the numerical coefficient of x
in the term $5x$, 5 is the numerical coefficient of x
in the term $2x$, 2 is the numerical coefficient of x

In an expression such as $x + 3$, look at the variable term, which is x. What is the numerical coefficient of x? Could we write x as some number times x? You should agree that

$$x = 1 \cdot x$$

so we say that 1 is the coefficient of the variable x in the expression $x + 3$. Similarly, in the expression $6 - x$ or $6 + {}^-x$, the numerical coefficient of x is $^-1$.

Example 4. Identify the numerical coefficient in each variable term of the expression $11x - 9x + 2$.

Solution. Writing the expression as

$$11x + {}^-9x + 2$$

we note that 2 is the constant term. Then

in the term $11x$, 11 is the numerical coefficient of x
in the term ^-9x, $^-9$ is the numerical coefficient of x

Example 5. Identify the numerical coefficient in each variable term of the expression $^-8a - b + c$.

Solution. Writing the expression as

$$^-8a + {}^-b + c$$

we see that

in the term ^-8a, $^-8$ is the numerical coefficient of a
in the term ^-b, $^-1$ is the numerical coefficient of b
in the term c, 1 is the numerical coefficient of c

Variable terms are called **like terms** if their variable parts are *exactly* the same. All constant terms are called like terms. In the expression

$$6x + 5x + 2x$$

the terms are all *like terms,* because the variable part of each term is an x. But in the expression

$$^-8a - b + c$$

the terms are all **unlike terms**, because their variable parts are all different.

Example 6. Identify the like terms in the expression $6x + 7y - 5x - 2y$.

Solution

$6x$ and ^-5x are like terms
$7y$ and ^-2y are like terms

Example 7. Identify the like terms in the expression $3x + 2 + 9x - 8$.

Solution

$3x$ and $9x$ are like terms
2 and $^-8$ are like terms

Example 8. Identify the like terms in the expression $6A + 3a - 5A - a$.

Solution

$6A$ and ^-5A are like terms
$3a$ and ^-a are like terms

Notice that A and a are two different variables; be careful to write variables exactly as they appear.

2.2-2 Adding Terms

If you were asked, without any hints, to find the sum of

$$6x + 5x + 2x$$

what would your answer be? If you think the answer is $13x$, you are absolutely correct, and you are well on the way to understanding how to simplify algebraic expressions.

The key, of course, is to look for like terms. Can we combine the following sum into a single term?

$$3x + 2y$$

Of course not. There is no way to simplify this sum into a single term because $3x$ and $2y$ are *not like terms*.

> A sum of terms can be simplified only if it contains *like* terms.

The process of adding like terms together is called **combining like terms**; but how is this done? Recall the example

$$6x + 5x + 2x = 13x$$

The terms on the left are all like terms with x as the variable part. The answer on the right is also a term with x as the variable part. The numerical coefficients on the left are 6 and 5 and 2. The numerical coefficient on the right is 13, and

$$6 + 5 + 2 = 13$$

You should see that

$$6x + 5x + 2x = (6 + 5 + 2)x = 13x$$

which is not a surprise because of the distributive property. We kept the same variable part, and the numerical coefficient in the answer was just the sum of the numerical coefficients in the original terms.

To combine like terms in the expression

$$5x + 12y + 3 + 8x - 6y - 1$$

first we notice which terms are like terms and rearrange the expression as

$$5x + 8x + 12y + {}^-6y + 3 + {}^-1$$

so that like terms are next to each other. Then we combine like terms

$$= \underbrace{5x + 8x}_{13x} + \underbrace{12y + {}^-6y}_{6y} + \underbrace{3 + {}^-1}_{2}$$

> To combine like terms, we combine the numerical coefficients and keep the same variable part.

Although it is correct to write

$$5x + 8x + 12y + {}^-6y + 3 + {}^-1$$
$$= (5 + 8)x + (12 + {}^-6)y + (3 + {}^-1)$$
$$= \quad 13x \quad + \quad 6y \quad + \quad 2$$

including the middle step with parentheses is often more confusing than helpful. It is not a required step; you should try to combine the numerical coefficients mentally.

Example 9. Combine like terms:

$$17x + x - 20x$$

Solution

$$17x + x - 20x$$
$$= 17x + 1x + {}^-20x$$
$$= {}^-2x$$

Example 10. Combine like terms:

$${}^-5x + 3y + 7z - 6x - 10y - z$$

Solution

$${}^-5x + 3y + 7z - 6x - 10y - z$$
$$= {}^-5x - 6x + 3y - 10y + 7z - z$$
$$= {}^-5x + {}^-6x + 3y + {}^-10y + 7z + {}^-1z$$
$$= {}^-11x + {}^-7y + 6z$$
$$= {}^-11x - 7y + 6z$$

Example 11. Combine like terms:

$$x - {}^-2y + 5 - 2x - 9y$$

Solution

$$x - {}^-2y + 5 - 2x - 9y$$
$$= x - 2x - {}^-2y - 9y + 5$$
$$= 1x + {}^-2x + 2y + {}^-9y + 5$$
$$= {}^-1x + {}^-7y + 5$$
$$= {}^-x - 7y + 5$$

Notice from these last examples that it is customary to write simplified expressions with only *one sign* between the different terms. When we arrived at

$${}^-11x + {}^-7y + 6z$$

we recalled the definition of subtraction and wrote the expression as

$${}^-11x - 7y + 6z$$

We did the same thing in rewriting

$${}^-x + {}^-7y + 5 \qquad \text{as} \qquad {}^-x - 7y + 5$$

Also notice that we usually do not write a numerical coefficient of 1 or ${}^-1$. Instead, we write

$$1x \qquad \text{as} \qquad x$$
$$^-1x \qquad \text{as} \qquad {}^-x$$

and we all agree that 1 or $^-1$ is "understood" to be the numerical coefficient.

It is not "wrong" to write

$$6x + {}^-5y \qquad \text{instead of} \qquad 6x - 5y$$
$$1x + 1y \qquad \text{instead of} \qquad x + y$$
$${}^-1x + {}^-1y \qquad \text{instead of} \qquad {}^-x - y$$

but the versions on the right are considered to be in better form than the versions on the left. Answers in this book will always be in the customary form.

Trial Run

_____ 1. What are the terms in the expression $3x - 2y - 7$?

_____ 2. In the expression $7m - 9n + 11$, identify the constant terms and the variable terms.

_____ 3. Identify the numerical coefficient of each variable term in the expression $^-3a + b - 9$.

_____ 4. Identify like terms in the expression $^-10x + 5 + 6x - 9$.

Combine like terms.

_____ 5. $3x - x + 12x$

_____ 6. $^-4x + 5 + 7x - 13$

_____ 7. $^-12x + 7y + 3x - 15y$

_____ 8. $3x - 4y + 9z - 2x + 4y + 5$

ANSWERS

1. $3x$, ^-2y, $^-7$.   **2.** Constant term: 11. Variable terms: $7m$, $-9n$.
3. $^-3$ is the numerical coefficient of a; 1 is the numerical coefficient of b.
4. ^-10x and $6x$ are like terms; 5 and $^-9$ are like terms.   **5.** $14x$.   **6.** $3x - 8$.
7. $^-9x - 8y$.   **8.** $x + 9z + 5$.

2.2-3   Some Addition Properties

Earlier we learned that the order in which two numbers were added did not change their sum.

$$3 + 2 = 2 + 3$$
$$5 + {}^-11 = {}^-11 + 5$$

Since variables are letters that stand for numbers, perhaps this same **commutative property** still works. Let's see if that is so.

If we were asked to compare $3x + 2x$ and $2x + 3x$, what would we discover?

$$3x + 2x = 5x$$
$$2x + 3x = 5x$$

So you conclude that

$$3x + 2x = 2x + 3x$$

Thus the _commutative property_ for addition also applies to expressions containing variable terms.

Recall, too, that the **associative property** for addition of numbers allowed us to say that

$$(3 + 2) + 6 = 3 + (2 + 6)$$

Let's compare $(3x + 2x) + 6x$ and $3x + (2x + 6x)$:

$$(3x + 2x) + 6x = 5x + 6x = 11x$$
$$3x + (2x + 6x) = 3x + 8x = 11x$$

Our conclusion is that

$$(3x + 2x) + 6x = 3x + (2x + 6x)$$

Thus we say that the *associative property* for addition applies to expressions containing variables. The way in which we group the terms does not change the answer.

As a matter of fact, we used these properties when we rearranged the terms in algebraic expressions so that the like terms would be next to each other. Now we know that what we did then was perfectly "legal."

EXERCISE SET 2.2

Identify the terms in the following expressions.

_____ **1.** $3x - 5$ _____ **2.** $4x - 7$

_____ **3.** $2a - 7b + 9$ _____ **4.** $11a - 2b - 2$

_____ **5.** $1 - 5m + 6m$ _____ **6.** $^-1 - 9m + 4n$

_____ **7.** $13x - 5y + 2z - 7$ _____ **8.** $14x - 12y + 5z - 3$

In Exercises 9–16, identify the constant terms and variable terms.

_____ **9.** $4x - 19$ _____ **10.** $7x + 23$

_____ **11.** $2x + 3y$ _____ **12.** $6x - 5y$

_____ **13.** $-4a - 7b + 5$ _____ **14.** $^-5a + 11b - 8$

_____ **15.** $1 + 7m - 5n + 14p - 4$ _____ **16.** $2 + 5m - 9n + 3p - 5$

In Exercises 17–24, identify the numerical coefficient of each variable term.

_____ **17.** $3x + 7$ _____ **18.** $5x - 13$

_____ **19.** $4x - 12y$ _____ **20.** $^-4x + 15y$

_____ **21.** $3a + 4b - 2$ _____ **22.** $8a - 5b + 9$

_____ **23.** $3m - n - 2p$ _____ **24.** $10m + n - 4p$

In Exercises 25–32, identify like terms.

_____ **25.** $8a - 3a$ _____ **26.** $7a - 5a$

_____ **27.** $3h - 2k + 5h + 6k$ _____ **28.** $2h - 5k + 6h - 8k$

_____ **29.** $5a - 6 + 3a - 9$ _____ **30.** $7a - 8 + 2a - 15$

_____ **31.** $3x - 2y + 1 + 4y - 2x - 7$ _____ **32.** $x - y + 2 - 5y + 4x + 3$

In Exercises 33–50, combine like terms.

_____ **33.** $3a - 5a$ _____ **34.** $5a - 7a$

_____ **35.** $2x - 8x + 9x$ _____ **36.** $3x - 11x + 10x$

_____ **37.** $9 - 2y - 3y$ _____ **38.** $7 - 4y - 5y$

_____ **39.** $4m - 6 + 11m + 9$ _____ **40.** $5m - 9 + 8m + 1$

_____ **41.** $3h - k + 5 + 4h + 2k - 9$ _____ **42.** $2h - 5k - 1 + 7h + 3k - 11$

_____ **43.** $4x - 3y + 13z + 7x + 8y - 6z$ _____ **44.** $2x + 5y - 6z + 3x - y - 3z$

_____ **45.** $a + 10b - 9a + 7$ _____ **46.** $2a - 11b - 6a + 3$

_____ **47.** $3m - 5n + 4p + 2m - 6n$ _____ **48.** $4m - 3n + p + 2m - 8n$

_____ **49.** $5u - 3v + w + 2v - 5u - 7w$ _____ **50.** $6u - 2v + 4w + 2v - 4u + 5w$

Name _____ Date _____

CHECKUP 2.2

_____ **1.** Identify the terms in the expression $3 - 7x + 2y$.

_____ **2.** Identify the constant terms and the variable terms in the expression $^-4a + 11 + 7a - 13$.

_____ **3.** Identify the numerical coefficient of each variable term in the expression $3x - y - 4z - 7$.

_____ **4.** Identify like terms in the expression $5x - 2y + 4 - 7x + 2y - 1$.

In Exercises 5–10, combine like terms.

_____ **5.** $3a - 7a$

_____ **6.** $x - 2x + 5x$

_____ **7.** $5 - 3y + 11y$

_____ **8.** $4m - 5 - 3m + 7$

_____ **9.** $2x + 7y - 3x - 6$

_____ **10.** $4a - 2b + c - 3b - 4a + c$

2.3 Working with Symbols of Grouping

Parentheses and brackets are even more important in handling variable expressions than in working with only constants. They provide directions for operating with expressions, and they must be treated carefully.

2.3-1 Multiplying Single Terms by Constants

An expression such as 3(5) means to multiply 3 times 5. Repeated addition helped us see that

$$3(5) = 5 + 5 + 5 = 15$$

Similarly, an expression such as $3(5x)$ means to multiply 3 times $5x$. Repeated addition can help again.

$$3(5x) = 5x + 5x + 5x = 15x$$

Similarly,

$$4(^-6x) = {}^-6x + {}^-6x + {}^-6x + {}^-6x = {}^-24x$$
$$5(^-x) = 5(^-1x) = {}^-1x + {}^-1x + {}^-1x + {}^-1x + {}^-1x = {}^-5x$$

You should see a shortcut for doing such problems (using the associative property for multiplication):

$$3(5x) = (3 \cdot 5)x = 15x$$
$$4(^-6x) = (4 \cdot {}^-6)x = {}^-24x$$
$$5(^-x) = (5 \cdot {}^-1)x = {}^-5x$$

> To multiply a constant times a variable term, multiply the constant times the numerical coefficient and keep the variable part.

When multiplying the constant times the numerical coefficient, we continue to obey the sign rules for multiplying integers.

$$3 \cdot 5 = 15 \quad \text{so} \quad 3(5x) = (3 \cdot 5)x = 15x$$
$$^-3 \cdot 5 = {}^-15 \quad \text{so} \quad {}^-3(5x) = (^-3 \cdot 5)x = {}^-15x$$
$$3 \cdot {}^-5 = {}^-15 \quad \text{so} \quad 3(^-5x) = (3 \cdot {}^-5)x = {}^-15x$$
$$^-3 \cdot {}^-5 = 15 \quad \text{so} \quad {}^-3(^-5x) = (^-3 \cdot {}^-5)x = 15x$$

Example 1. Multiply $5(6x)$.

Solution. $5(6x) = (5 \cdot 6)x = 30x$.

Example 2. Multiply $3(^-7x)$.

Solution. $3(^-7x) = (3 \cdot {}^-7)x = {}^-21x$.

Example 3. Multiply $^-2(^-8a)$.

Solution. $^-2(^-8a) = (^-2 \cdot {}^-8)a = 16a$.

Example 4. Multiply $^-10(11x)$.

Solution. $^-10(11x) = (^-10 \cdot 11)x = {}^-110x$.

Example 5. Multiply $(^-2)(3)(5x)$.

Solution. We shall use the associative property for multiplication of integers here.

$$
\begin{array}{lll}
(^-2)(3)(5x) & or & (^-2)(3)(5x) \\
= \;^-6(5x) & & = \;^-2(3\cdot5)x \\
= (^-6\cdot5)x & & = \;^-2(15)x \\
= \;^-30x & & = (^-2\cdot15)x \\
& & = \;^-30x
\end{array}
$$

Example 6. Simplify $2(5x) + 8x$.

Solution

$$
\begin{aligned}
& 2(5x) + 8x \\
=\;& (2\cdot5)x + 8x \\
=\;& 10x + 8x \\
=\;& 18x
\end{aligned}
$$

Example 7. Simplify $4x - 2(3x)$.

Solution

$$
\begin{array}{lll}
4x - 2(3x) & or & 4x - 2(3x) \\
= 4x + \;^-2(3x) & & = 4x - (2\cdot3)x \\
= 4x + (^-2\cdot3)x & & = 4x - 6x \\
= 4x + \;^-6x & & = 4x + \;^-6x \\
= \;^-2x & & = \;^-2x
\end{array}
$$

2.3-2 Dividing Single Terms by Constants

Earlier we learned how every division statement corresponds to a multiplication statement. We said that

$$\frac{12}{4} = 3 \qquad \text{because} \qquad 4\cdot3 = 12$$

and

$$\frac{^-15}{3} = \;^-5 \qquad \text{because} \qquad 3\cdot\;^-5 = \;^-15$$

The same is true for variables. We can say that

$$\frac{45x}{5} = 9x \qquad \text{because} \qquad 5(9x) = 45x$$

$$\frac{^-39x}{3} = \;^-13x \qquad \text{because} \qquad 3(^-13x) = \;^-39x$$

Notice that we could arrive at these same answers using the following shortcut:

$$\frac{45x}{5} = \left(\frac{45}{5}\right)x = 9x$$

$$\frac{^-39x}{3} = \left(\frac{^-39}{3}\right)x = \;^-13x$$

which means that we only need to know how to divide integers. From Chapter 1 you recall the rules for signs in division are the same as those for multiplication. We conclude that

> To divide a variable term by a constant, divide the numerical coefficient by the constant and keep the variable part.

Example 8. Divide $\dfrac{16x}{^-2}$.

Solution. $\dfrac{16x}{^-2} = \left(\dfrac{16}{^-2}\right)x = {}^-8x.$

Example 9. Divide $\dfrac{^-28x}{^-4}.$

Solution. $\dfrac{^-28x}{^-4} = \left(\dfrac{^-28}{^-4}\right)x = 7x.$

Example 10. Divide $\dfrac{7a}{7}.$

Solution. $\dfrac{7a}{7} = \left(\dfrac{7}{7}\right)a = 1a = a.$

Example 11. Divide $\dfrac{^-x}{^-1}.$

Solution. $\dfrac{^-x}{^-1} = \dfrac{^-1x}{^-1} = \left(\dfrac{^-1}{^-1}\right)x = 1x = x.$

Trial Run *Simplify.*

_____ **1.** $4(5x)$

_____ **2.** $^-3(^-8x)$

_____ **3.** $^-11(3x)$

_____ **4.** $(^-2)(5)(7x)$

_____ **5.** $3(^-2x) + 10x$

_____ **6.** $\dfrac{18x}{^-3}$

_____ **7.** $\dfrac{^-32x}{^-4}$

_____ **8.** $\dfrac{^-x}{^-1}$

_____ **9.** $\dfrac{^-6a}{6}$

ANSWERS
1. $20x.$ **2.** $24x.$ **3.** $^-33x.$ **4.** $^-70x.$ **5.** $4x.$ **6.** $^-6x.$ **7.** $8x.$ **8.** $x.$ **9.** $^-a.$

2.3-3 Multiplying Constants Times Sums

In Chapter 1 we learned that a problem such as $2(3 + 7)$ could be simplified using either of two approaches.

Approach 1: $\qquad\qquad\qquad\qquad \begin{aligned} &2(3 + 7) \\ &= 2(10) = 20 \end{aligned}$

Approach 2: $\qquad\qquad\qquad\qquad \begin{aligned} &2(3 + 7) \\ &= 2\cdot 3 + 2\cdot 7 \\ &= 6 + 14 = 20 \end{aligned}$

We referred to Approach 2 as an example of the **distributive property for multiplication over addition**, which we wrote as:

> **Distributive Property**
>
> $$A(B + C) = A \cdot B + A \cdot C$$

In working with variable expressions, the distributive property will be one of our most valuable tools. Let's verify that it works, using the example $3(x + 2)$. From the repeated addition point of view, we know that

$$
\begin{aligned}
3(x + 2) &= (x + 2) + (x + 2) + (x + 2) \\
&= x + 2 + x + 2 + x + 2 \\
&= x + x + x + 2 + 2 + 2 \\
&= 3x + 6
\end{aligned}
$$

If the distributive property holds here, we would want

$$
\begin{aligned}
3(x + 2) &= 3 \cdot x + 3 \cdot 2 \\
&= 3x + 6
\end{aligned}
$$

Since we arrived at the same answer either way, we can see that the distributive property does work for multiplication over addition, even when variables occur in the expression.

Example 12. Multiply $^-3(x + 4y)$.

Solution

$$
\begin{aligned}
&^-3(x + 4y) \\
&= {}^-3 \cdot x + {}^-3(4y) \\
&= {}^-3x + ({}^-3 \cdot 4)y \\
&= {}^-3x + {}^-12y = {}^-3x - 12y
\end{aligned}
$$

Example 13. Multiply $5(6x - 3y)$.

Solution

$$
\begin{aligned}
&5(6x - 3y) \\
&= 5(6x + {}^-3y) \\
&= 5(6x) + 5({}^-3y) \\
&= (5 \cdot 6)x + (5 \cdot {}^-3)y \\
&= 30x + {}^-15y \\
&= 30x - 15y
\end{aligned}
$$

Example 14. Multiply $^-4(3x - 7y)$.

Solution

$$
\begin{aligned}
&^-4(3x - 7y) \\
&= {}^-4(3x + {}^-7y) \\
&= {}^-4(3x) + {}^-4({}^-7y) \\
&= ({}^-4 \cdot 3)x + ({}^-4 \cdot {}^-7)y \\
&= {}^-12x + 28y
\end{aligned}
$$

Working with so many signs in these problems is getting to be very awkward. Perhaps it is time to observe some patterns and take some shortcuts. Let's rework our examples and use arrows to indicate our thoughts.

Example 15. Multiply $^-3(x + 4y)$.

Solution

$$-3(x + 4y) = -3(x + 4y)$$

$$= -3x - 12y$$

Notice that we are getting the same answer as before!

Example 16. Multiply $5(6x - 3y)$.

Solution

$$5(6x - 3y) = 5(6x - 3y)$$

$$= 30x - 15y$$

Example 17. Multiply $-4(3x - 7y)$.

Solution

$$-4(3x - 7y) = -4(3x - 7y)$$

$$= -12x + 28y$$

Elimination of double signs may help us avoid errors and confusion with the distributive property. Using arrows (either in your head or on paper) may be useful.

Trial Run

———— **1.** $3(x + y)$

———— **2.** $-2(x + 3y)$

———— **3.** $5(2x - 4y)$

———— **4.** $-2(3a - 2b)$

———— **5.** $3(-2m - n)$

———— **6.** $-5(-x + 6y)$

ANSWERS
1. $3x + 3y$.　　**2.** $-2x - 6y$.　　**3.** $10x - 20y$.　　**4.** $-6a + 4b$.　　**5.** $-6m - 3n$.
6. $5x - 30y$.

2.3-4 Working with Symbols of Grouping

The algebraic expression $(4x + 5y)$ can be thought of as an example of the distributive property, if we agree that there is a 1 understood to be in front of the parentheses. In other words,

$$(4x + 5y) = 1 \cdot (4x + 5y)$$

$$= 1 \cdot (4x + 5y)$$

$$= 4x + 5y$$

We see, then, that to simplify $(4x + 5y)$, we merely "drop the parentheses." How do we simplify the expression

$$^-(3x + 2y)?$$

Again we recall that ^-x meant $^-1 \cdot x$, so that we may say

$$^-(3x + 2y) = {}^-1 \cdot (3x + 2y)$$

$$= {}^-1(3x + 2y)$$

$$= {}^-3x - 2y$$

Example 18. Simplify $(7x - 6y)$.

Solution. $(7x - 6y) = 7x - 6y.$

Example 19. Simplify $^-(x + 13)$.

Solution

$$^-(x + 13)$$
$$= {}^-1(x + 13)$$
$$= {}^-x - 13$$

Example 20. Simplify $^-(7x - 2y)$.

Solution

$$^-(7x - 2y)$$

$$= {}^-1(7x - 2y)$$

$$= {}^-1(7x - 2y)$$

$$= {}^-7x + 2y$$

Example 21. Simplify $^-(^-5x - 8y + 1)$.

Solution

$$^-(^-5x - 8y + 1)$$

$$= {}^-1(^-5x - 8y + 1)$$

$$= 5x + 8y - 1$$

This ability to find the negative (or opposite) of a quantity in parentheses will help us deal with subtraction problems involving symbols of grouping. We know by the definition of subtraction that

$$6 - 9 = 6 + {}^-9 = {}^-3$$
$$5x - 2x = 5x + {}^-2x = 3x$$

Similarly, we should agree that

$$(4x + 5y) - (3x + 2y)$$
$$= (4x + 5y) - 1(3x + 2y)$$
$$= (4x + 5y) + {}^-1(3x + 2y)$$

From what we have discovered in this section, we can now finish this problem.

$$(4x + 5y) - (3x + 2y)$$
$$= (4x + 5y) - 1(3x + 2y)$$
$$= 1(4x + 5y) + {}^-1(3x + 2y)$$
$$= 4x + 5y - 3x - 2y$$
$$= 4x - 3x + 5y - 2y$$
$$= x + 3y$$

Example 22. Simplify $(8x + 7) - (2x + 3)$.

Solution

$$(8x + 7) - (2x + 3)$$
$$= (8x + 7) + {}^-1(2x + 3)$$
$$= 8x + 7 - 2x - 3$$
$$= 8x - 2x + 7 - 3$$
$$= 6x + 4$$

Example 23. Simplify $(9a - 6b) - (10a - 5b)$.

Solution

$$(9a - 6b) - (10a - 5b)$$
$$= (9a - 6b) + {}^-1(10a - 5b)$$
$$= 9a - 6b - 10a + 5b$$
$$= 9a - 10a - 6b + 5b$$
$$= {}^-a - b$$

Once again, it seems that there should be some shortcut which we could use to skip the second step in the solution. If we mentally insert a 1 in front of the second set of parentheses, we may then multiply the quantity within those parentheses by $^-1$.

$$(9a - 6b) - 1(10a - 5b)$$
$$= 9a - 6b - 10a + 5b$$
$$= 9a - 10a - 6b + 5b$$
$$= {}^-a - b$$

Notice that the result is the same.

Example 24. Simplify $(14x + 7y) - (3x + y)$.

Solution

$$(14x + 7y) - (3x + y)$$
$$= 14x + 7y - 3x - y$$
$$= 14x - 3x + 7y - y$$
$$= 11x + 6y$$

Our shortcut, which we have verified by examples, can also be used in more complicated problems. We shall work a problem the "long way," then use a shortcut to see if the results are the same.
Simplify $5(3x + 7) - 2(3x + 1)$.
Long way:

$$5(3x + 7) - 2(3x + 1)$$
$$= 5(3x + 7) + {}^-2(3x + 1)$$
$$= 15x + 35 - 6x - 2$$
$$= 15x - 6x + 35 - 2$$
$$= 9x + 33$$

Shortcut:

$$5(3x + 7) - 2(3x + 1)$$
$$= 15x + 35 - 6x - 2$$
$$= 15x - 6x + 35 - 2$$
$$= 9x + 33$$

Since the answers are identical, we shall continue to use the shorter method.

Example 25. Simplify $3(x - 5y) + 4(^-x + 3y)$.

Solution

$$3(x - 5y) + 4(^-x + 3y)$$
$$= 3x - 15y - 4x + 12y$$
$$= 3x - 4x - 15y + 12y$$
$$= ^-x - 3y$$

Example 26. Simplify $^-2(5a - 3b) - 7(a - 3b)$.

Solution

$$^-2(5a - 3b) - 7(a - 3b)$$
$$= ^-10a + 6b - 7a + 21b$$
$$= ^-10a - 7a + 6b + 21b$$
$$= ^-17a + 27b$$

In simplifying expressions containing brackets and parentheses, we shall continue the practice of working with the innermost symbols of grouping first. We shall also use our shortcuts.

Let's simplify

$$2[5x - 3(x - 1)]$$
$$= 2[5x - 3x + 3]$$
$$= 2[2x + 3]$$
$$= 4x + 6$$

Notice that we worked with the innermost parentheses first and combined like terms within the brackets before multiplying by 2.

Example 27. Simplify $3[(5x - 2y) - (3x - y)]$.

Solution

$$3[(5x - 2y) - (3x - y)]$$
$$= 3[5x - 2y - 3x + y]$$
$$= 3[5x - 3x - 2y + y]$$
$$= 3[2x - y]$$
$$= 6x - 3y$$

Example 28. Simplify $^-4[2(a - 3b) - 5(^-3a - 2b)]$.

Solution

$$^-4[2(a - 3b) - 5(^-3a - 2b)]$$
$$= ^-4[2a - 6b + 15a + 10b]$$
$$= ^-4[2a + 15a - 6b + 10b]$$
$$= ^-4[17a + 4b]$$
$$= ^-68a - 16b$$

Example 29. Simplify $2[3x - (x + 1)] - 4[2(x + 3) - 5]$.

Solution

$$2[3x - (x + 1)] - 4[2(x + 3) - 5]$$
$$= 2[3x - x - 1] - 4[2x + 6 - 5]$$
$$= 2[2x - 1] - 4[2x + 1]$$
$$= 4x - 2 - 8x - 4$$
$$= 4x - 8x - 2 - 4$$
$$= {}^-4x - 6$$

Trial Run　　*Simplify.*

_____ **1.** $^-(3x - 5y)$

_____ **2.** $^-(^-2a - 11b + 2)$

_____ **3.** $(3x + 6) - (2x + 3)$

_____ **4.** $(8a - 5b) - (^-a - 3b)$

_____ **5.** $4(2m - 3) - 2(3m + 1)$

_____ **6.** $3[2x - (x - 2)]$

_____ **7.** $2[(x - 5y) - (2x + y)]$

_____ **8.** $3[^-2(3x - 4) - (x + 1)]$

_____ **9.** $5[3x - (x + 2)] - 2[2(x - 1) - 4]$

ANSWERS
1. $^-3x + 5y$.　　**2.** $2a + 11b - 2$.　　**3.** $x + 3$.　　**4.** $9a - 2b$.　　**5.** $2m - 14$.　　**6.** $3x + 6$.
7. $^-2x - 12y$.　　**8.** $^-21x + 21$.　　**9.** $6x + 2$.

　　You now should see that even the most complicated-looking problem involving symbols of grouping can be simplified with ease if you will be careful to include every necessary step. Not even the most experienced mathematician would take any more shortcuts than you now have learned.

EXERCISE SET 2.3

Find the indicated product or quotient.

_____ 1. $2(4x)$

_____ 2. $3(7x)$

_____ 3. $3(^-5x)$

_____ 4. $4(^-2x)$

_____ 5. $^-12(2m)$

_____ 6. $^-9(3m)$

_____ 7. $^-3(^-2x)$

_____ 8. $^-2(^-4x)$

_____ 9. $(^-2)(4)(3y)$

_____ 10. $(^-3)(5)(2y)$

_____ 11. $(^-3)(2)(^-5k)$

_____ 12. $(^-4)(3)(^-3k)$

_____ 13. $\dfrac{20x}{^-4}$

_____ 14. $\dfrac{27x}{^-3}$

_____ 15. $\dfrac{^-45x}{^-9}$

_____ 16. $\dfrac{^-34x}{^-2}$

_____ 17. $\dfrac{5a}{5}$

_____ 18. $\dfrac{7a}{7}$

_____ 19. $\dfrac{^-m}{^-1}$

_____ 20. $\dfrac{m}{^-1}$

_____ 21. $\dfrac{^-10y}{2}$

_____ 22. $\dfrac{^-15y}{3}$

_____ 23. $\dfrac{^-7n}{7}$

_____ 24. $\dfrac{^-15n}{15}$

In Exercises 25–66, simplify the expressions by removing grouping symbols and combining like terms.

_____ 25. $5x + 3(2x)$

_____ 26. $4x + 3(3x)$

_____ 27. $2(5x) - 3x$

_____ 28. $3(4x) - 7x$

_____ 29. $4(a - b)$

_____ 30. $5(a - b)$

_____ 31. $^-3(x + 2y)$

_____ 32. $^-4(x + 5y)$

_____ 33. $7(3x - 2y)$

_____ 34. $5(2x - 4y)$

_____ 35. $^-4(5a - 7b)$ _____ 36. $^-6(a - 2b)$

_____ 37. $9(^-2m - 3n)$ _____ 38. $8(^-3m - 5n)$

_____ 39. $^-3(^-x + 4y)$ _____ 40. $^-7(^-x + 2y)$

_____ 41. $^-(2x - 5y)$ _____ 42. $^-(3x - 2y)$

_____ 43. $^-(^-3a - 7b + 5)$ _____ 44. $^-(^-4a - 6b + 3)$

_____ 45. $(3x + 5) - (2x + 4)$ _____ 46. $(5x + 3) - (3x + 1)$

_____ 47. $(12a + 3b) - (5a - 2b)$ _____ 48. $(9a + 4b) - (7a - b)$

_____ 49. $(^-4m - 3n) - (2m - 3n)$ _____ 50. $(^-5m - 2n) - (m - 2n)$

_____ 51. $(7x - 3y) - (^-2x + y)$ _____ 52. $(8x - 5y) - (^-3x + 2y)$

_____ 53. $3(5m - 4) - 4(m + 2)$ _____ 54. $5(2m - 1) - 2(m + 5)$

_____ 55. $^-3(x - 2y) + 4(^-x + y)$ _____ 56. $^-2(2x - y) + 3(^-x + 2y)$

_____ 57. $4[2a - (a - 3)]$ _____ 58. $5[a - (2a - 1)]$

_____ 59. $5[(3x - 2y) - (x + y)]$ _____ 60. $7[(2x - 3y) - (x + 2y)]$

_____ 61. $^-2[5(a - 2b) + 7(2a + b)]$ _____ 62. $^-3[4(2a - b) + 5(a + 3b)]$

_____ 63. $4[^-3(x - 4) - 2(x + 6)]$ _____ 64. $5[^-6(x - 1) - 3(x + 2)]$

_____ 65. $2[3x - (x + 2)] - [3(x - 1) - 5]$ _____ 66. $3[4x - (2x + 1)] - [5(x - 2) - 7]$

CHECKUP 2.3

Simplify.

_____ **1.** $4(^-8x)$

_____ **2.** $(^-2)(4)(^-5y)$

_____ **3.** $\dfrac{48x}{^-12}$

_____ **4.** $\dfrac{10a}{^-10}$

_____ **5.** $7x - 3(4x)$

_____ **6.** $5(^-2x + 3y)$

_____ **7.** $(4x - 5) - (2x + 3)$

_____ **8.** $2(3a - 6b) - 3(^-2a + b)$

_____ **9.** $^-6[4a - (3a + 1)]$

_____ **10.** $7[3(2m - n) - 2(3m - 5n)]$

Summary

In this chapter we were introduced to the foundations of algebra: algebraic expressions involving constants (known numbers) and variables (letters that stand for numbers).

Algebraic expressions contain terms (constants or numerical coefficients times variables) that are separated by addition signs. After learning to identify like terms (terms whose variable parts are exactly the same), we learned how to combine like terms by combining the numerical coefficients.

We discovered that addition of algebraic expressions is commutative and associative, and we made much use of the distributive property for multiplication over addition. After practice with the distributive property, we learned to take some legal shortcuts to make our work with signs less confusing.

REVIEW EXERCISES

SECTION 2.1

Write an algebraic expression for each word statement.

_____ **1.** Multiply 6 times the variable x.

_____ **2.** The sum of the variable y and 9.

_____ **3.** 10 more than three times the variable m.

_____ **4.** Four times the variable k is subtracted from 23.

_____ **5.** Nine times the sum of the variable x and 7.

Evaluate.

_____ **6.** $x - 9$; when $x = 13$, $x = 2$, $x = {}^-12$.

_____ **7.** $\dfrac{56}{n}$; when $n = 14$, $n = 0$, $n = {}^-7$.

_____ **8.** $3(9 - x)$; when $x = 11$, $x = 9$, $x = {}^-2$.

_____ **9.** $2x - 7$; when $x = 5$, $x = {}^-1$, $x = 9$.

_____ **10.** $\dfrac{4a - 12}{3}$; when $a = 3$, $a = {}^-6$, $a = 0$.

In Exercises 11–13, write a variable expression for each word expression and evaluate for the given values of the variable.

_____ **11.** A hiker left Horseshoe Cave walking along the trail at 4 miles per hour. Write a variable expression for how far she had traveled after h hours. What will be the total distance walked after 3 hours?

_____ **12.** Aunt Vina left her entire estate to her three nephews. If the estate is valued at x dollars, write a variable expression for each nephew's share. Find the value of each nephew's share if the estate is worth $210,000.

_____ **13.** The cost of renting a cottage at the lake is $200 for the first week and $20 for each additional day. Write a variable expression for the cost of the cabin after x additional days. Find the total amount for staying 3 days more than a week.

SECTION 2.2

In Exercises 14–17, identify the constant terms and variable terms.

_____ **14.** $3x - 12$ _____ **15.** $^-2x + 7y$

_____ **16.** $4a - 3b + 2$ _____ **17.** $3m - 2 + 4n - 9 - p$

In Exercises 18–21, identify the numerical coefficient of each variable term.

_____ **18.** $3x - 9$ _____ **19.** $7x - 2y$

_____ **20.** $^-3a + b - 1$ _____ **21.** $m - n + 2p$

In Exercises 22–25, combine like terms.

_____ **22.** $2x - 9x + 5x$ _____ **23.** $3m - 4 + 2m + 6$

_____ **24.** $2a - 13b - 9a + 3$

_____ **25.** $^-4x - 3y + 7z + 8x + 3y - 2z$

In Exercises 26–29, find the product or quotient.

_____ **26.** $4(^-5x)$ _____ **27.** $(^-2)(^-3)(4y)$

_____ **28.** $\dfrac{^-54x}{9}$ _____ **29.** $\dfrac{^-a}{^-1}$

In Exercises 30–37, simplify.

_____ **30.** $^-3(5x) + 12x$ _____ **31.** $7(4a - 5b)$

_____ **32.** $^-(7x - 2y)$ _____ **33.** $(^-7x + 2) + (3x + 5)$

_____ **34.** $(^-9m - 3n) - (7m - 3n)$ _____ **35.** $3[4a - (7a + 2)]$

_____ **36.** $5[^-3(x + y) + 7(x - y)]$

_____ **37.** $4[5x - (x - 2)] - [5(x + 1) - 4x]$

3 Solving Equations

Our work with algebraic expressions in Chapter 2 forms the basic foundation for a very important part of algebra: using equations to solve problems. **Equations** are mathematical sentences which say that one quantity represents the same number as another quantity.

In this chapter we learn how to

1. Solve equations and check the solutions.
2. Switch from word statements to equations.

Before we begin, let's say a word about a change in the notation being used. In Chapter 2 we used a raised negative sign to indicate a negative integer (such as $^-6$) and a centered negative sign to indicate subtraction (such as $10 - 6$). We learned, however, that

$$10 - 6 \qquad \text{and} \qquad 10 + {}^-6$$

represented the same number. Similarly, we discovered that

$$-(x + 3) \qquad \text{and} \qquad {}^-1(x + 3)$$

represented the same quantity.

Therefore, at this point in our study of algebra, we shall abandon the use of the raised negative sign, using the centered negative sign to represent the sign of a number or the operation of subtraction. This simplification does *not* change the meaning of any expression, but it does allow us to eliminate many unnecessary double signs. From now on, we shall write

$$-6 \text{ to mean negative six}$$
$$10 - 6 \text{ to mean ten minus six}$$

It is not expected that this switch will present any difficulty; it is merely a change in notation.

3.1 Solving Equations with One Operation

Equations may or may not contain variables. Using the symbol $=$ to mean "is equal to" or "equals," we may write

$$2 + 3 = 6 - 1$$
$$2(5 - 8) = -6$$
$$x + 3 = 3 + x$$
$$2(x + 3) = 2x + 6$$

and we know that each of these equations is *always* true. The quantity on the left-hand side of the $=$ symbol is exactly the same as the quantity on the right-hand side. These statements are called **identities**, but they are not our primary concern. Instead, we shall concentrate on statements which are sometimes, but not always, true.

3.1-1 Satisfying Equations

We turn our attention to equations such as

$$x + 3 = 5$$
$$17 - y = 11$$
$$5x = 20$$
$$\frac{x}{3} = 4$$

Do you think each of these statements is *always* true? Will the left-hand side always equal the right-hand side, no matter what value we substitute for the variable? Indeed not. You have probably already decided that

$$x + 3 = 5 \quad \text{when} \quad x = 2, \quad \text{because} \quad 2 + 3 = 5$$

$$17 - y = 11 \quad \text{when} \quad y = 6, \quad \text{because} \quad 17 - 6 = 11$$

$$5x = 20 \quad \text{when} \quad x = 4, \quad \text{because} \quad 5 \cdot 4 = 20$$

$$\frac{x}{3} = 4 \quad \text{when} \quad x = 12, \quad \text{because} \quad \frac{12}{3} = 4$$

If an equation containing variables is only true for certain values of the variables, we call it a **conditional equation**. From now on, we shall be very concerned with finding the value of a variable that makes a particular equation true. Such a value is called the **solution** of the equation. Finding the solution is called **solving** the equation.

> The **solution** for any equation is the value of the variable that makes the equation a true statement.

For instance, in the examples above, we can say that

$$x = 2 \text{ is the solution of the equation } x + 3 = 5$$

$$y = 6 \text{ is the solution of the equation } 17 - y = 11$$

$$x = 4 \text{ is the solution of the equation } 5x = 20$$

$$x = 12 \text{ is the solution of the equation } \frac{x}{3} = 4$$

We can decide whether a certain number is a solution to an equation by substituting that number into the equation to see whether a true statement results. If we substitute a number into an equation and obtain a true statement, we say that the number **satisfies** the equation and that the number is a *solution* for the equation. For instance, in the examples above, we can say that

$$x = 2 \text{ satisfies the equation } x + 3 = 5$$

$$y = 6 \text{ satisfies the equation } 17 - y = 11$$

$$x = 4 \text{ satisfies the equation } 5x = 20$$

$$x = 12 \text{ satisfies the equation } \frac{x}{3} = 4$$

Example 1. Is $x = 3$ a solution for $2x + 1 = 7$?

Solution. Substituting 3 for x, we have

$$2(3) + 1 \stackrel{?}{=} 7$$
$$6 + 1 \stackrel{?}{=} 7$$
$$7 = 7$$

so $x = 3$ is a solution for $2x + 1 = 7$.

Example 2. Is $y = -2$ a solution for $5 - 3y = -1$?

Solution. Substituting -2 for y, we have

$$5 - 3(-2) \stackrel{?}{=} -1$$
$$5 + 6 \stackrel{?}{=} -1$$
$$11 \stackrel{?}{=} -1$$
$$11 \neq -1$$

No,

so $y = -2$ is *not* a solution for $5 - 3y = -1$.

Example 3. Does $x = 4$ satisfy the equation $2(x - 5) = -2$?

Solution. Substituting 4 for x, we have

$$2(4 - 5) \stackrel{?}{=} -2$$
$$2(-1) \stackrel{?}{=} -2$$
$$-2 = -2$$

so $x = 4$ satisfies the equation $2(x - 5) = -2$.

3.1-2 Finding Solutions by Addition and Subtraction

As we have already discovered, it is sometimes possible to find a solution just by looking at an equation.

if $x + 6 = 10$, we see that $x = 4$ is the solution

if $x - 9 = 3$, we see that $x = 12$ is the solution

if $7a = 28$, we see that $a = 4$ is the solution

if $\dfrac{y}{3} = 11$, we see that $y = 33$ is the solution

We must be careful, however. Trying to solve equations by looking or by inspection is a dangerous practice. If the answer is not immediately obvious, we may become discouraged. Moreover, there are many problems that are too complicated to be solved by inspection, such as

$$9x + 37 = -145$$
$$\frac{6(8x - 36)}{5} = 120$$

We need some steps that will help us find solutions in an orderly and accurate way.

In solving equations, our goal is to find the value of the variable that makes the equation true. We would like to end up with a statement such as

$$x = 4$$
$$x = 12$$
$$a = 4$$
$$y = 33$$

In other words, we would like to have the variable (with a numerical coefficient of 1) all by itself on one side of the equation, and some number (the solution) on the other side. We wish to **isolate the variable**.

> In solving an equation, our goal is to isolate the variable.

We must concentrate, therefore, on getting rid of any numbers that are connected to the variable by addition or subtraction or multiplication or division.

In the equation

$$x + 6 = 10$$

we must get rid of the 6 being added to the x.

In the equation

$$x - 9 = 3$$

we must get rid of the 9 being subtracted from the x.

Keep in mind that equations are statements which say that one quantity is the same as another quantity. If we make any change in the value of one side of an equation, we *must* make the same change on the other side. If we do not, we shall destroy the truth of the original equation.

In the equation $x + 6 = 10$, how could we get rid of the 6? If we *subtracted* 6 from the left-hand side, we would have

$$x + 6 - 6$$
$$x + 0$$
$$x$$

and the variable x would be isolated. *But* if we subtract 6 on the left, we *must* subtract it on the right also. Here is how the problem should be worked:

$$x + 6 = 10$$
$$x + 6 - 6 = 10 - 6$$
$$x + 0 = 4$$
$$x = 4$$

Notice that our solution, $x = 4$, is the same solution as that obtained "by looking."

In the equation $x - 9 = 3$ we could get rid of the -9 by adding 9 on the left. But we must add 9 on the right at the same time. Here are the steps:

$$x - 9 = 3$$
$$x - 9 + 9 = 3 + 9$$
$$x + 0 = 12$$
$$x = 12$$

Again, this answer matches the answer we got by inspection.

We conclude that the meaning of an equation will not be changed if we add (or subtract) some quantity to (or from) one side of that equation, as long as we add (or subtract) the same quantity to (or from) the other side.

> We may add the same quantity to both sides of an equation. We may subtract the same quantity from both sides of an equation.

Remember that we can always see if our solution satisfies the original equation by substituting it back in the equation to see if it makes the original statement true. This is called **checking the solution.**

Example 4. Solve $x + 9 = 32$ and check.

Solution

$$x + 9 = 32$$
$$x + 9 - 9 = 32 - 9$$
$$x + 0 = 23$$
$$x = 23$$

CHECK:
$$x + 9 = 32$$
$$23 + 9 \overset{?}{=} 32$$
$$32 = 32$$

Example 5. Solve $x - 15 = 100$ and check.

Solution

$$x - 15 = 100$$
$$x - 15 + 15 = 100 + 15$$
$$x + 0 = 115$$
$$x = 115$$

CHECK:
$$x - 15 = 100$$
$$115 - 15 \overset{?}{=} 100$$
$$100 = 100$$

Example 6. Solve $y + 7 = -2$ and check.

Solution

$$y + 7 = -2$$
$$y + 7 - 7 = -2 - 7$$
$$y + 0 = -9$$
$$y = -9$$

CHECK:
$$y + 7 = -2$$
$$-9 + 7 \overset{?}{=} -2$$
$$-2 = -2$$

Example 7. Solve $-5 + y = -3$ and check.

Solution

$$-5 + y = -3$$
$$-5 + y + 5 = -3 + 5$$
$$-5 + 5 + y = 2$$
$$0 + y = 2$$
$$y = 2$$

CHECK:
$$-5 + y = -3$$
$$-5 + 2 \overset{?}{=} -3$$
$$-3 = -3$$

Trial Run

_____ 1. Is $x = 3$ a solution for $3x - 1 = 8$?

_____ 2. Is $y = -5$ a solution for $6 - 2y = 16$?

_____ 3. Is $m = 9$ a solution for $\dfrac{m}{3} + 2 = -7$?

_____ 4. Is $a = 0$ a solution for $2(a - 5) = -10$?

_____ 5. Is $x = -1$ a solution for $2x + 8 = x + 7$?

Solve and check.

_____ **6.** $x + 3 = 8$

_____ **7.** $6 = y - 4$

_____ **8.** $3 + m = 3$

_____ **9.** $-5 = a + 4$

_____ **10.** $-8 + x = -4$

3.1-3 Finding Solutions by Multiplication and Division

We now have methods for solving equations such as

$$x + 3 = 19$$
$$x - 7 = 32$$

by using addition and subtraction. Consider equations such as

$$3x = 15$$
$$-6x = 24$$
$$\frac{x}{2} = 9$$
$$\frac{-x}{3} = 21$$

We must decide how to *isolate the variable* in each example by getting rid of the number operating on the variable. Notice what those numbers are doing in each case.

in $3x = 15$, 3 is multiplying x

in $-6x = 24$, -6 is multiplying x

in $\frac{x}{2} = 9$, 2 is dividing x

in $\frac{-x}{3} = 21$, 3 is dividing x and -1 is multiplying x

We would like to operate on both sides of the equation in some way so that we are left with just $1x$ or x by itself.

Look at what happens in the example $3x = 15$ if we *divide* both sides by 3.

$$3x = 15$$
$$\frac{3x}{3} = \frac{15}{3}$$
$$\left(\frac{3}{3}\right)x = \frac{15}{3}$$
$$1x = 5$$
$$x = 5$$

Is this the correct solution? Checking, we see that

$$3x = 15$$
$$3(5) \overset{?}{=} 15$$
$$15 = 15$$

Could we treat the example $-6x = 24$ in the same way? Let's divide both sides by -6.

$$-6x = 24$$
$$\frac{-6x}{-6} = \frac{24}{-6}$$
$$\left(\frac{-6}{-6}\right)x = \frac{24}{-6}$$
$$1x = -4$$
$$x = -4$$

and checking this solution, we have

$$-6x = 24$$
$$-6(-4) \overset{?}{=} 24$$
$$24 = 24$$

We may divide both sides of an equation by any number (except 0).

Example 8. Solve $10x = -150$ and check.

Solution

$$10x = -150$$
$$\frac{10x}{10} = \frac{-150}{10}$$
$$\left(\frac{10}{10}\right)x = \frac{-150}{10}$$
$$x = -15$$

CHECK:
$$10x = -150$$
$$10(-15) \overset{?}{=} -150$$
$$-150 = -150$$

Example 9. Solve $-x = 13$ and check.

Solution $-x = 13$ can be written

$$-1x = 13$$
$$\frac{-1x}{-1} = \frac{13}{-1}$$
$$\left(\frac{-1}{-1}\right)x = \frac{13}{-1}$$
$$x = -13$$

CHECK:
$$-x = 13$$
$$-(-13) \overset{?}{=} 13$$
$$13 = 13$$

Example 10. Solve $-132 = 12x$ and check.

Solution

$$-132 = 12x$$

$$\frac{-132}{12} = \frac{12x}{12}$$

$$\frac{-132}{12} = \left(\frac{12}{12}\right)x$$

$$-11 = x$$

CHECK:

$$-132 = 12x$$

$$-132 \overset{?}{=} 12(-11)$$

$$-132 = -132$$

In each of these examples, we got rid of a number *multiplying* the variable by *dividing* both sides of the equation by that number. How do you suppose that we shall get rid of a number *dividing* the variable? We might try *multiplying* both sides of the equation by that number.

Let's start with the example $\frac{x}{2}$ and try multiplying both sides of the equation by 2.

$$\frac{x}{2} = 9$$

$$2\left(\frac{x}{2}\right) = 2 \cdot 9$$

$$\left(\frac{2}{2}\right)x = 2 \cdot 9$$

$$1 \cdot x = 18$$

$$x = 18$$

Does this solution satisfy the original equation?

$$\frac{x}{2} = 9$$

$$\frac{18}{2} \overset{?}{=} 9$$

$$9 = 9$$

The next example in our list was $\frac{-x}{3} = 21$, so we shall multiply both sides of the equation by 3 to get rid of the 3, which is dividing the variable term.

$$\frac{-x}{3} = 21$$

$$3\left(\frac{-x}{3}\right) = 3 \cdot 21$$

$$\left(\frac{3}{3}\right)(-x) = 63$$

$$-x = 63$$

We have gotten rid of the 3, but now we are left with $-x$ rather than x. From the last set of examples, we know we must divide both sides of our equation by -1 to isolate x.

$$-x = 63$$

$$\frac{-x}{-1} = \frac{63}{-1}$$

$$x = -63$$

Checking our solution, we see that

$$\frac{-x}{3} = 21$$

$$\frac{-(-63)}{3} \overset{?}{=} 21$$

$$\frac{63}{3} \overset{?}{=} 21$$

$$21 = 21$$

> We may multiply both sides of an equation by any number (except 0).

Example 11. Solve $\frac{a}{5} = -8$ and check.

Solution

$$\frac{a}{5} = -8$$

$$5\left(\frac{a}{5}\right) = 5(-8)$$

$$\left(\frac{5}{5}\right)a = 5(-8)$$

$$a = -40$$

CHECK:

$$\frac{a}{5} = -8$$

$$\frac{-40}{5} \overset{?}{=} -8$$

$$-8 = -8$$

Example 12. Solve $7 = \frac{-x}{2}$ and check.

Solution

$$7 = \frac{-x}{2}$$

$$2(7) = 2\left(\frac{-x}{2}\right)$$

$$2(7) = \left(\frac{2}{2}\right)(x)$$

$$14 = -x$$

$$\frac{14}{-1} = \frac{-x}{-1}$$

$$-14 = x$$

CHECK:

$$7 = \frac{-x}{2}$$

$$7 \overset{?}{=} \frac{-(-14)}{2}$$

$$7 \overset{?}{=} \frac{14}{2}$$

$$7 = 7$$

Solve and check.

———————— **1.** $4x = 24$

———————— **2.** $-5x = 25$

———————— **3.** $3x = 0$

———————— **4.** $-x = 7$

———————— **5.** $\dfrac{x}{3} = -2$

———————— **6.** $4 = \dfrac{x}{7}$

———————— **7.** $\dfrac{-x}{4} = 5$

———————— **8.** $\dfrac{x}{6} = 0$

ANSWERS
1. 6. **2.** -5. **3.** 0. **4.** -7. **5.** -6. **6.** 28. **7.** -20. **8.** 0.

Summarizing, we have discovered that in finding the solution of an equation, we are permitted to

---

1. Add the same quantity to both sides.
2. Subtract the same quantity from both sides.
3. Multiply both sides by the same quantity (not zero).
4. Divide both sides by the same quantity (not zero).

To solve equations such as

$x + 3 = 6$, we *subtract* 3 from both sides
$x - 3 = 6$, we *add* 3 to both sides
$3x = 6$, we *divide* both sides by 3
$\dfrac{x}{3} = 6$, we *multiply* both sides by 3

In each case we wish to isolate the variable by "undoing" the operation being performed on it by constants.

To undo addition, we must subtract.
To undo subtraction, we must add.
To undo multiplication, we must divide.
To undo division, we must multiply.

EXERCISE SET 3.1

In the following exercises, decide if the given value of the variable is a solution for the equation.

_____ **1.** $5x - 1 = 9$, $x = 2$ _____ **2.** $4x - 3 = 21$, $x = 6$

_____ **3.** $9 - 2y = 7$, $y = -1$ _____ **4.** $8 - 2y = 6$, $y = -1$

_____ **5.** $\dfrac{m}{2} + 7 = 13$, $m = 12$ _____ **6.** $\dfrac{m}{4} + 2 = 8$, $m = 24$

_____ **7.** $3(a - 6) = 3$, $a = 6$ _____ **8.** $4(a - 5) = 16$, $a = 5$

_____ **9.** $7x + 14 = 0$, $x = -2$ _____ **10.** $5x + 15 = 0$, $x = -3$

_____ **11.** $4x - 5 = 3x + 2$, $x = -7$ _____ **12.** $6x - 7 = 5x + 1$, $x = -8$

Solve and check.

_____ **13.** $x + 5 = 12$ _____ **14.** $x + 9 = 16$

_____ **15.** $x - 9 = 3$ _____ **16.** $x - 8 = 7$

_____ **17.** $6 + y = 4$ _____ **18.** $9 + y = 5$

_____ **19.** $y + 2 = -6$ _____ **20.** $y + 3 = -8$

_____ **21.** $m - 11 = -6$ _____ **22.** $m - 9 = -8$

_____ **23.** $-4 + a = 6$ _____ **24.** $-6 + a = 12$

_____ **25.** $12 = k + 6$ _____ **26.** $13 = k + 9$

_____ **27.** $8 + x = 8$ _____ **28.** $-9 + x = -9$

_____ **29.** $-4 = m - 9$ _____ **30.** $-1 = m - 7$

_____ **31.** $y + 12 = 4$ _____ **32.** $y + 9 = 2$

_____ **33.** $a - 7 = -14$ _____ **34.** $a - 11 = -23$

_____ **35.** $-13 = x - 13$ _____ **36.** $-15 = x - 15$

_____ **37.** $3x = 21$

_____ **38.** $4x = 28$

_____ **39.** $2x = -18$

_____ **40.** $3x = -27$

_____ **41.** $-2x = 34$

_____ **42.** $-5x = 45$

_____ **43.** $-36 = -6x$

_____ **44.** $-48 = -6x$

_____ **45.** $4x = 0$

_____ **46.** $7x = 0$

_____ **47.** $-x = 6$

_____ **48.** $-x = 9$

_____ **49.** $\dfrac{x}{2} = 6$

_____ **50.** $\dfrac{x}{4} = 11$

_____ **51.** $-3 = \dfrac{x}{7}$

_____ **52.** $-9 = \dfrac{x}{6}$

_____ **53.** $\dfrac{-x}{7} = 8$

_____ **54.** $\dfrac{-x}{8} = 9$

_____ **55.** $\dfrac{x}{7} = 0$

_____ **56.** $\dfrac{x}{9} = 0$

_____ **57.** $\dfrac{x}{5} = -8$

_____ **58.** $\dfrac{x}{11} = -4$

_____ **59.** $-9 = \dfrac{-x}{7}$

_____ **60.** $-7 = \dfrac{-x}{6}$

Name _____ Date _____

CHECKUP 3.1

_____ **1.** Is $x = -2$ a solution for $7 - 3x = 13$?

Solve.

_____ **2.** $x + 9 = 8$

_____ **3.** $x - 9 = -2$

_____ **4.** $12 = x + 8$

_____ **5.** $x + 11 = 11$

_____ **6.** $6x = 54$

_____ **7.** $42 = -7x$

_____ **8.** $4x = 0$

_____ **9.** $\frac{x}{8} = 3$

_____ **10.** $-10 = \frac{x}{6}$

3.2 Solving Equations with Several Operations

Many equations involve more than one operation. Equations such as

$$3x + 1 = 7$$
$$5x - 2 = 13$$
$$-8y - 11 = 53$$
$$\frac{x}{2} + 10 = 29$$

require more than one step in isolating the variable.

3.2-1 Undoing Two Operations

In the example $3x + 1 = 7$, we must undo the operations of addition (get rid of the 1) and multiplication (get rid of the 3). Let's start by subtracting 1 from both sides of the equation.

$$3x + 1 = 7$$
$$3x + 1 - 1 = 7 - 1$$
$$3x = 6$$

Now we can divide both sides by 3:

$$\frac{3x}{3} = \frac{6}{3}$$
$$x = 2$$

Again, we shall check this solution.

$$3x + 1 = 7$$
$$3(2) + 1 \overset{?}{=} 7$$
$$6 + 1 \overset{?}{=} 7$$
$$7 = 7$$

In the example $5x - 2 = 13$ we must undo the operations of subtraction (get rid of the 2) and multiplication (get rid of the 5). We shall start by adding 2 to both sides; then we shall divide both sides by 5.

$$5x - 2 = 13$$
$$5x - 2 + 2 = 13 + 2$$
$$5x = 15$$
$$\frac{5x}{5} = \frac{15}{5}$$
$$x = 3$$

CHECK:
$$5x - 2 = 13$$
$$5(3) - 2 \overset{?}{=} 13$$
$$15 - 2 \overset{?}{=} 13$$
$$13 = 13$$

To solve $-8y - 11 = 53$, we shall undo the subtraction by adding 11 to both sides; then we shall undo the multiplication by dividing both sides by -8.

$$-8y - 11 = 53$$
$$-8y - 11 + 11 = 53 + 11$$
$$-8y = 64$$
$$\frac{-8y}{-8} = \frac{64}{-8}$$
$$y = -8$$

CHECK:
$$-8y - 11 = 53$$
$$-8(-8) - 11 \overset{?}{=} 53$$
$$64 - 11 \overset{?}{=} 53$$
$$53 = 53$$

To solve $\frac{x}{2} + 10 = 29$, we shall undo the addition by subtracting 10 from both sides; then we shall undo the division by multiplying both sides by 2.

$$\frac{x}{2} + 10 = 29$$
$$\frac{x}{2} + 10 - 10 = 29 - 10$$
$$\frac{x}{2} = 19$$
$$2\left(\frac{x}{2}\right) = 2(19)$$
$$x = 38$$

CHECK:
$$\frac{x}{2} + 10 = 29$$
$$\frac{38}{2} + 10 \overset{?}{=} 29$$
$$19 + 10 \overset{?}{=} 29$$
$$29 = 29$$

From these examples, some general rules appear for solving equations containing more than one operation.

> If there are no parentheses in an equation, *first* perform the necessary addition or subtraction, *then* perform the necessary multiplication or division.

Example 1. Solve $-5x + 13 = 78$.

Solution

$$-5x + 13 = 78$$
$$-5x + 13 - 13 = 78 - 13$$
$$-5x = 65$$
$$\frac{-5x}{-5} = \frac{65}{-5}$$
$$x = -13$$

Example 2. Solve $-4 = 3x + 20$.

Solution

$$-4 = 3x + 20$$
$$-4 - 20 = 3x + 20 - 20$$
$$-24 = 3x$$
$$\frac{-24}{3} = \frac{3x}{3}$$
$$-8 = x$$

Example 3. Solve $3 + \frac{x}{4} = 14$.

Solution

$$3 + \frac{x}{4} = 14$$

$$3 + \frac{x}{4} - 3 = 14 - 3$$

$$\frac{x}{4} = 11$$

$$4\left(\frac{x}{4}\right) = 4(11)$$

$$x = 44$$

Example 4. Solve $-7 - \frac{x}{6} = -2$.

Solution

$$-7 - \frac{x}{6} = -2$$

$$-7 - \frac{x}{6} + 7 = -2 + 7$$

$$\frac{-x}{6} = 5$$

$$6\left(\frac{-x}{6}\right) = 6(5)$$

$$-x = 30$$

$$\frac{-x}{-1} = \frac{30}{-1}$$

$$x = -30$$

Example 5. Solve $\frac{2x}{3} - 1 = 5$ and check.

Solution

$$\frac{2x}{3} - 1 = 5$$

$$\frac{2x}{3} - 1 + 1 = 5 + 1$$

$$\frac{2x}{3} = 6$$

$$3\left(\frac{2x}{3}\right) = 3(6)$$

$$\left(\frac{3}{3}\right)(2x) = 18$$

$$2x = 18$$

$$\frac{2x}{2} = \frac{18}{2}$$

$$x = 9$$

CHECK:

$$\frac{2x}{3} - 1 = 5$$

$$\frac{2(9)}{3} - 1 \overset{?}{=} 5$$

$$\frac{18}{3} - 1 \overset{?}{=} 5$$

$$6 - 1 \overset{?}{=} 5$$

$$5 = 5$$

3.2-2 Working with Symbols of Grouping in Equations

Parentheses and other symbols of grouping in equations still provide us with important directions. As before, we shall use shortcuts whenever possible to perform the necessary operations, but we shall continue to deal with symbols of grouping first, then combine like terms, and then solve the equation.

To solve an equation such as

$$(x + 3) + (x + 5) = 18$$

we must first simplify the left-hand side:

$$x + 3 + x + 5 = 18$$

Now we combine like terms:

$$x + x + 3 + 5 = 18$$
$$2x + 8 = 18$$

Then solve by the usual methods:

$$2x + 8 - 8 = 18 - 8$$
$$2x = 10$$
$$\frac{2x}{2} = \frac{10}{2}$$
$$x = 5$$

As you can see, solving equations containing parentheses involves methods with which you are already familiar.

Example 6. Solve $2(x + 7) - 9 = 31$ and check.

Solution

$$2(x + 7) - 9 = 31$$
$$2x + 14 - 9 = 31$$
$$2x + 5 = 31$$
$$2x + 5 - 5 = 31 - 5$$
$$2x = 26$$
$$\frac{2x}{2} = \frac{26}{2}$$
$$x = 13$$

CHECK:

$$2(x + 7) - 9 = 31$$
$$2(13 + 7) - 9 \overset{?}{=} 31$$
$$2(20) - 9 \overset{?}{=} 31$$
$$40 - 9 \overset{?}{=} 31$$
$$31 = 31$$

Solving such equations is accomplished by following these steps:

1. Deal with any symbols of grouping.
2. Combine like terms.
3. Perform necessary additions and subtractions to move constant terms to one side, keeping variable terms on the other side.
4. Perform necessary multiplication and division to isolate the variable.
5. Check your solution in the original equation.

Example 7. Solve $1 = 3(2x - 7) - 5(3x + 1)$ and check.

Solution

$$1 = 3(2x - 7) - 5(3x + 1)$$
$$1 = 6x - 21 - 15x - 5$$
$$1 = -9x - 26$$
$$1 + 26 = -9x - 26 + 26$$
$$27 = -9x$$
$$\frac{27}{-9} = \frac{-9x}{-9}$$
$$-3 = x$$

CHECK:
$$1 = 3(2x - 7) - 5(3x + 1)$$
$$1 \overset{?}{=} 3[2(-3) - 7] - 5[3(-3) + 1]$$
$$1 \overset{?}{=} 3[-6 - 7] - 5[-9 + 1]$$
$$1 \overset{?}{=} 3[-13] - 5[-8]$$
$$1 \overset{?}{=} -39 + 40$$
$$1 = 1$$

Example 8. Solve $3[x + 2(x - 7)] = -60$.

Solution

$$3[x + 2(x - 7)] = -60$$
$$3[x + 2x - 14] = -60$$
$$3[3x - 14] = -60$$
$$9x - 42 = -60$$
$$9x - 42 + 42 = -60 + 42$$
$$9x = -18$$
$$\frac{9x}{9} = \frac{-18}{9}$$
$$x = -2$$

Checking the solution is left to you.

Trial Run   *Solve and check.*

_____   **1.** $2x - 4 = -12$

_____   **2.** $-3y + 9 = 27$

_____ 3. $\frac{x}{5} + 6 = 8$

_____ 4. $5 - \frac{a}{2} = -1$

_____ 5. $8 = -3m + 2$

_____ 6. $(x + 2) + (x - 3) = 9$

_____ 7. $3(x - 2) + 7 = 22$

_____ 8. $2(x - 6) - (x + 8) = -12$

_____ 9. $11 = 3(3x - 2) - 5(x - 1)$

_____ 10. $4[2x + 3(x - 2)] = -4$

ANSWERS

1. $x = -4$.   **2.** $y = -6$. $x = 10$.   **4.** $a = 12$.   **5.** $m = -2$.   **6.** $x = 5$.
7. $x = 7$.   **8.** $x = 8$.   **9.** $3 = x$.   **10.** $x = 1$.

EXERCISE SET 3.2

Solve the following equations.

_____ **1.** $3x + 5 = 26$　　　　　_____ **2.** $4x + 6 = 26$

_____ **3.** $2x - 9 = 11$　　　　　_____ **4.** $7x - 9 = 5$

_____ **5.** $5y + 3 = -22$　　　　_____ **6.** $2y + 8 = -12$

_____ **7.** $8x + 15 = 7$　　　　　_____ **8.** $6x + 23 = 11$

_____ **9.** $-4a + 9 = 45$　　　　_____ **10.** $-5a + 10 = 55$

_____ **11.** $12 + 2m = 0$　　　　_____ **12.** $18 + 3m = 0$

_____ **13.** $-x - 7 = 6$　　　　　_____ **14.** $-x - 8 = 9$

_____ **15.** $7 - 2x = 13$　　　　　_____ **16.** $6 - 5x = 41$

_____ **17.** $24 = 3x - 15$　　　　_____ **18.** $17 = 6x - 7$

_____ **19.** $4y + 25 = 25$　　　　_____ **20.** $7y + 19 = 19$

_____ **21.** $-23 = 9 - 4a$　　　　_____ **22.** $-21 = 15 - 9a$

_____ **23.** $\dfrac{x}{6} + 3 = 5$　　　　_____ **24.** $\dfrac{x}{7} + 4 = 9$

_____ **25.** $7 + \dfrac{a}{5} = -1$　　　_____ **26.** $6 + \dfrac{a}{3} = -2$

_____ **27.** $13 = \dfrac{y}{4} + 9$　　　_____ **28.** $17 = \dfrac{y}{5} + 8$

_____ **29.** $6 - \dfrac{x}{4} = 7$　　　　_____ **30.** $9 - \dfrac{x}{2} = 12$

_____ **31.** $\dfrac{2x}{5} = 4$　　　　　_____ **32.** $\dfrac{3x}{4} = 6$

_____ **33.** $5 = \dfrac{3x}{2} - 4$　　　_____ **34.** $2 = \dfrac{5x}{3} - 8$

_____ **35.** $-9 = \dfrac{x}{5} - 9$　　　_____ **36.** $-7 = \dfrac{x}{8} - 7$

_____ **37.** $8 - \dfrac{2x}{5} = 12$　　　_____ **38.** $9 - \dfrac{3x}{2} = 15$

_____ **39.** $(x + 3) + (x - 5) = 8$　　_____ **40.** $(x + 5) + (x - 6) = 9$

_____ **41.** $12 = 4(x - 7) + 8$　　_____ **42.** $12 = 3(x - 6) + 9$

_____ **43.** $2(x - 8) - (x + 3) = -22$　_____ **44.** $3(x - 4) - (2x + 7) = -26$

_____ **45.** $2 = 4(x - 5) - 2(x + 6)$　_____ **46.** $1 = 5(x - 5) - 3(x - 2)$

_____ **47.** $2[3x + 3(x - 1)] = 6$　_____ **48.** $3[4x + 3(x - 2)] = 24$

_____ **49.** $5[2(x - 4) - 2] = 0$　_____ **50.** $6[3(x - 2) + 12] = 0$

CHECKUP 3.2

Solve.

_____ **1.** $3x - 7 = 8$

_____ **2.** $7 = 8x + 15$

_____ **3.** $9 - x = 13$

_____ **4.** $-2y - 5 = 7$

_____ **5.** $\dfrac{x}{5} - 2 = 3$

_____ **6.** $8 - \dfrac{x}{3} = 9$

_____ **7.** $\dfrac{3x}{2} - 1 = 5$

_____ **8.** $(3a - 2) - (a + 5) = -7$

_____ **9.** $2 = 2(x - 3) - 4(x + 2)$

_____ **10.** $3[2x - (x + 4)] = 6$

3.3 Solving More Equations

Sometimes an equation contains variable terms on the left-hand side *and* on the right-hand side. In such a situation, how do we go about isolating the variable and then solving the equation?

3.3-1 Solving Equations with Variables on Both Sides

In an equation such as

$$3x + 1 = x + 5$$

we must find some way to get the variable terms together on one side of the equation and the constant terms together on the other side. We shall do this by addition and subtraction.

First we shall subtract x from both sides of the equation to remove the x on the right.

$$3x + 1 - x = x + 5 - x$$
$$3x - x + 1 = x - x + 5$$
$$2x + 1 = 5$$

Now subtract 1 from both sides:

$$2x + 1 - 1 = 5 - 1$$
$$2x = 4$$

and divide both sides by 2.

$$\frac{2x}{2} = \frac{4}{2}$$
$$x = 2$$

CHECK:
$$3x + 1 = x + 5$$
$$3(2) + 1 \stackrel{?}{=} 2 + 5$$
$$6 + 1 \stackrel{?}{=} 2 + 5$$
$$7 = 7$$

Let us solve and check the equation

$$7x - 3 = 33 - 2x$$

To get all the variable terms on the left-hand side, we can add $2x$ to both sides of the equation.

$$7x - 3 + 2x = 33 - 2x + 2x$$
$$7x + 2x - 3 = 33$$
$$9x - 3 = 33$$

Now we add 3 to both sides to get all constants on the right.

$$9x - 3 + 3 = 33 + 3$$
$$9x = 36$$

To isolate the variable, we divide both sides by 9.

$$\frac{9x}{9} = \frac{36}{9}$$
$$x = 4$$

CHECK:
$$7x - 3 = 33 - 2x$$
$$7(4) - 3 \stackrel{?}{=} 33 - 2(4)$$
$$28 - 3 \stackrel{?}{=} 33 - 8$$
$$25 = 25$$

Example 1. Solve $5x + 6 = 12x + 41$ and check.

Solution

$$5x + 6 = 12x + 41$$
$$5x + 6 - 12x = 12x + 41 - 12x$$
$$5x - 12x + 6 = 12x - 12x + 41$$
$$-7x + 6 = 41$$
$$-7x + 6 - 6 = 41 - 6$$
$$-7x = 35$$
$$\frac{-7x}{-7} = \frac{35}{-7}$$
$$x = -5$$

CHECK:

$$5x + 6 = 12x + 41$$
$$5(-5) + 6 \overset{?}{=} 12(-5) + 41$$
$$-25 + 6 \overset{?}{=} -60 + 41$$
$$-19 = -19$$

Example 2. Solve $5x + 6 = 12x + 41$.

Solution

$$5x + 6 = 12x + 41$$
$$5x + 6 - 5x = 12x + 41 - 5x$$
$$5x - 5x + 6 = 12x - 5x + 41$$
$$6 = 7x + 41$$
$$6 - 41 = 7x + 41 - 41$$
$$-35 = 7x$$
$$\frac{-35}{7} = \frac{7x}{7}$$
$$-5 = x$$

Notice that Examples 1 and 2 showed two methods of solving the same problem. Either method is perfectly correct. You may move the variable terms to either side of the equation. *You* must decide to which side to move the variable terms, and then move the constants to the *other* side.

Equations involving symbols of grouping are solved in the same basic way.

Example 3. Solve $2(x + 3) + 5(x + 1) = 3x - 9$.

Solution

$$2(x + 3) + 5(x + 1) = 3x - 9$$

Work with the parentheses first.

$$2x + 6 + 5x + 5 = 3x - 9$$

Combine like terms.

$$7x + 11 = 3x - 9$$

Move the variable terms to the left-hand side.

$$7x + 11 - 3x = 3x - 9 - 3x$$
$$4x + 11 = -9$$

Move the constants to the right-hand side.

$$4x + 11 - 11 = -9 - 11$$
$$4x = -20$$

Isolate the variable.

$$\frac{4x}{4} = \frac{-20}{4}$$
$$x = -5$$

> We may summarize the steps for solving these equations as follows:
>
> **1.** Work with parentheses if necessary.
> **2.** Combine like terms.
> **3.** Move all variable terms to one side.
> **4.** Move all constant terms to the other side.
> **5.** Isolate the variable to obtain the solution.
> **6.** Check your solution.

Example 4. Solve $5(3 - 2x) + 4 = 3(4 - x)$ and check.

Solution

$$5(3 - 2x) + 4 = 3(4 - x)$$
$$15 - 10x + 4 = 12 - 3x$$
$$19 - 10x = 12 - 3x$$
$$19 - 10x + 3x = 12 - 3x + 3x$$
$$19 - 7x = 12$$
$$19 - 7x - 19 = 12 - 19$$
$$-7x = -7$$
$$\frac{-7x}{-7} = \frac{-7}{-7}$$
$$x = 1$$

CHECK:
$$5(3 - 2x) + 4 = 3(4 - x)$$
$$5(3 - 2 \cdot 1) + 4 \stackrel{?}{=} 3(4 - 1)$$
$$5(3 - 2) + 4 \stackrel{?}{=} 3(3)$$
$$5 \cdot 1 + 4 \stackrel{?}{=} 3 \cdot 3$$
$$5 + 4 \stackrel{?}{=} 9$$
$$9 = 9$$

Example 5. Solve $2[x + 3(5 - x)] = 7(x + 1) - 21$.

Solution

$$2[x + 3(5 - x) = 7(x + 1) - 21$$
$$2[x + 15 - 3x] = 7x + 7 - 21$$
$$2[-2x + 15] = 7x - 14$$
$$-4x + 30 = 7x - 14$$
$$-4x + 30 + 4x = 7x - 14 + 4x$$
$$30 = 11x - 14$$
$$30 + 14 = 11x - 14 + 14$$
$$44 = 11x$$
$$\frac{44}{11} = \frac{11x}{11}$$
$$4 = x$$

Checking the solution is left to you.

Trial Run *Solve and check.*

_____ **1.** $4x - 3 = x + 6$

_____ **2.** $3x - 9 = 8x + 6$

_____ **3.** $-3x + 7 = 2x - 18$

_____ **4.** $9 + 2x = -x - 3$

_____ **5.** $2(x + 3) + 5(x - 1) = 3x - 7$

_____ **6.** $4(2x - 3) + 2 = 2(5 - x)$

_____ **7.** $3 - 2(4 - 5x) = 4(2x - 1) + 7$

_____ **8.** $2[x + 5(x - 3)] = 4(x - 3) - 10$

ANSWERS
1. $x = 3.$ **2.** $x = -3.$ **3.** $x = 5.$ **4.** $x = -4.$ **5.** $x = -2.$ **6.** $x = 2.$
7. $x = 4.$ **8.** $x = 1.$

3.3-2   Switching from Words to Equations

We learned earlier that algebraic expressions provide a nice way to switch from words to variables.

We learned in Chapter 2 that if 1 credit hour at a university costs $25, we could find the cost of h hours using the expression

$$25 \cdot h \qquad \text{or} \qquad 25h$$

Suppose that a student had saved $150 for part-time courses. How can we figure out how many hours the student could take?

We also found that if the population of a town of 25,230 changed by x people, the final population would be

$$25,230 + x$$

How can we decide how many people would have to join the population in order for it to reach 30,000?

We agreed, too, that each of 10 gas stations would receive an equal share of the distributor's total gallons G in the amount of

$$\frac{G}{10}$$

How can we determine how many gallons the distributor must receive if each of the stations would like an allotment of 7525 gallons?

Each of these questions can be answered using an equation.

In the first example, we must find h so that

$$25h = 150$$

In the second example, we must find x so that

$$25,230 + x = 30,000$$

In the third example, we must find G so that

$$\frac{G}{10} = 7525$$

It is left to you to solve these equations. Your answers should be $h = 6$, $x = 4770$, $G = 75,250$.

Problem 1

Suppose that Gary has $1760 in his savings account but wishes to save a total of $5000 by adding deposits of $40 per week from his paycheck. For how many weeks must he continue this plan?

First we must decide what is being asked for in this problem. The unknown quantity is the number of weeks. So we let

$$w = \text{number of weeks}$$

If Gary saves $40 each week, after w weeks he will have saved $40w$ dollars. We add this to his current balance of 1760 dollars and

$$1760 + 40w = \text{total dollars saved}$$

Since Gary wants to save $5000 altogether, we know that

$$1760 + 40w = 5000$$

and we are ready to solve the equation.

$$1760 + 40w = 5000$$
$$1760 + 40w - 1760 = 5000 - 1760$$
$$40w = 3240$$
$$\frac{40w}{40} = \frac{3240}{40}$$
$$w = 81 \text{ weeks}$$

Problem 2

Suppose that Beverly has 36 yards of fencing to use for enclosing a rectangular garden. If she wishes the garden to be twice as long as it is wide, what should the garden's dimensions be? Let's use a drawing to help us.

The amount of fencing needed is $l + w + l + w$

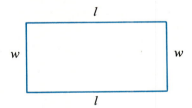

But since the garden is twice as long as it is wide, one length is actually two widths; $l = 2w$. The amount of fencing becomes $2w + w + 2w + w$ and we see that

$$2w + w + 2w + w = 36$$
$$6w = 36$$
$$\frac{6w}{6} = \frac{36}{6}$$
$$w = 6 \text{ yards}$$

The width is 6 yards and the length is $2w = 2(6) = 12$ yards.

Switching problems from words to equations should be done in an orderly way.

---

1. Write down what you want the variable to stand for.
2. Write any expressions containing the variable.
3. Write an equation from the information in the problem.
4. Solve the equation and check in the original problem.

Example 6. If Mandy earns $3 per hour, how many hours must she work in 1 week in order to earn $93 (before taxes, etc.)?

Solution

$$\text{let } h = \text{hours worked in 1 week}$$
$$3h = \text{dollars earned in 1 week}$$
$$\text{so} \quad 3h = 93$$
$$\frac{3h}{3} = \frac{93}{3}$$
$$h = 31 \text{ hours}$$

Example 7. After a Weight Worriers session, John announced: "I lost 3 pounds more than twice as much as Carla lost this month. I lost 17 pounds." How many pounds did Carla lose?

Solution

$$\text{let } p = \text{pounds Carla lost}$$
$$2p = \text{twice Carla's loss}$$
$$2p + 3 = \text{3 pounds more than twice Carla's loss}$$
$$2p + 3 = \text{John's loss}$$
$$\text{so} \quad 2p + 3 = 17$$
$$2p + 3 - 3 = 17 - 3$$
$$2p = 14$$
$$\frac{2p}{2} = \frac{14}{2}$$
$$p = 7 \text{ pounds}$$

Example 8. Charlotte now earns $6 per hour as a grounds keeper working 35 hours per week. She has been offered another job paying $280 per week for 35 hours also. How much of an increase in her hourly wage must her present employer give Charlotte to meet the better offer?

Solution

$$\text{let } x = \text{increase in hourly wage}$$
$$6 + x = \text{new hourly wage}$$
$$35(6 + x) = \text{new weekly earnings}$$
$$\text{so} \quad 35(6 + x) = 280$$
$$210 + 35x = 280$$
$$210 + 35x - 210 = 280 - 210$$
$$35x = 70$$
$$\frac{35x}{35} = \frac{70}{35}$$
$$x = 2 \text{ dollars per hour increase}$$

Example 9. Three roommates share monthly rent and food expenses equally. If each person's share of the rent is $65, how much can they spend altogether on food so that each roommate pays $150 for total monthly expenses?

Solution

$$\text{let } f = \text{total spent on food}$$
$$\frac{f}{3} = \text{each person's share of food cost}$$

$$\frac{f}{3} + 65 = \text{each person's share of total cost}$$

$$\text{so} \quad \frac{f}{3} + 65 = 150$$

$$\frac{f}{3} + 65 - 65 = 150 - 65$$

$$\frac{f}{3} = 85$$

$$3\left(\frac{f}{3}\right) = 3(85)$$

$$f = 255 \text{ dollars for food}$$

Trial Run   *Solve.*

_____ 1. Maurice buys a stereo for $500. He makes a $150 down payment and pays the balance by making payments of $25 a week. For how many weeks will he be making payments?

_____ 2. Jo Ann is buying framing for a rectangular picture that is three times as long as it is wide. The salesperson tells her that she needs 72 inches of framing. What are the dimensions of the frame?

_____ 3. Tickets to Opryland are $8 each. How many members of Pi Mu Epsilon can go if the organization has $184?

_____ 4. In physical fitness class, Ramon did 5 more than twice as many pushups as his father. If Ramon did 113 pushups, how many did his father do?

_____ 5. Ann's average in math class on 5 tests is 88. What is the total of her 5 test grades?

ANSWERS
1. 14.   2. 9 by 27 inches.   3. 23.   4. 54.   5. 440.

Name _____ Date _____

EXERCISE SET 3.3

Solve the following equations, and check.

_____ 1. $3x + 2 = x + 10$ _____ 2. $5x + 3 = 3x + 15$

_____ 3. $6x - 5 = 3x + 16$ _____ 4. $7x - 5 = 4x + 7$

_____ 5. $11x - 3 = 5x - 15$ _____ 6. $12x - 9 = 7x - 14$

_____ 7. $3x - 9 = 5 - 4x$ _____ 8. $3x - 8 = 12 - 7x$

_____ 9. $5x + 7 = 9x + 23$ _____ 10. $2x + 5 = 11x + 32$

_____ 11. $x + 3 = 5x - 9$ _____ 12. $7x + 6 = 9x - 4$

_____ 13. $-4x + 2 = x - 13$ _____ 14. $-5x + 3 = 2x - 11$

_____ 15. $3x - 8 = 5x$ _____ 16. $7x - 15 = 12x$

_____ 17. $7x + 8 = 8 - 3x$ _____ 18. $9x + 12 = 12 - 7x$

_____ 19. $13 + 4x = -x - 7$ _____ 20. $14 + 7x = -2x - 4$

_____ 21. $2(x + 1) + 3(x - 2) = 2x - 37$ _____ 22. $4(x + 1) + (x - 5) = 3x - 19$

_____ 23. $5(2x - 4) - 2 = 2(7 - x)$ _____ 24. $3(2x - 7) - 7 = 5(1 - x)$

_____ 25. $9 - 2(5 - 3x) = 5(2x - 3) - 22$ _____ 26. $7 - 3(2 - x) = 4(2x - 1) - 5$

_____ 27. $2[x + 3(x - 2)] = 5(x - 4) - 1$ _____ 28. $3[2x + (x - 5)] = 2(x - 7) + 6$

_____ 29. $3[2x - 3(2x - 4)] = 5[2(x - 4) + 2]$ _____ 30. $2[3x - (x + 4)] = 3[2(5 - x) - 6]$

_____ 31. The Waller twins' combined weight at birth was 12 pounds. If Lyle weighed 2 more pounds than Lydia, how much did each weigh?

_____ 32. If Carl finds he can save $12 a week, how many weeks will it take him to save $252?

_____ 33. A salesperson at the Specialty Shop earns a base salary of $840 per month plus a $2 bonus for each dress she sells costing the customer $100 or more. If Ann earned $914 last month, how many dresses did she sell costing $100 or more?

_____ **34.** Julia and Sara are selling tickets for a concert to benefit the Special Olympics. Sara has sold 4 times as many tickets as Julia and together they have sold 75 tickets. How many tickets has each sold?

_____ **35.** Carol is building a bookcase that will have to hold 162 books. How many shelves will the bookcase have if each shelf will hold 18 books?

_____ **36.** The area of a rectangle is the product of the length and the width. Find the width of a rectangle with a length of 15 units and an area of 195 square units.

_____ **37.** The restaurant manager tells the chairperson of the banquet committee that she will charge $200 for the use of the room plus $8 per person for the dinner. If the total cost of the banquet is $1120, how many persons will be attending?

_____ **38.** At the Bargain Barn Restaurant, the hamburger costs 10 cents less than three times the cost of the fries. If the hamburger costs 80 cents, what is the cost of the fries?

_____ **39.** The width of a room is 5 feet less than the length. If a paperhanger needs 70 feet of ceiling border, what are the dimensions of the room?

_____ **40.** The sum of six times an integer and 5 times the next consecutive integer is 49. Find the two integers.

CHECKUP 3.3

Solve.

_____ **1.** $4x - 3 = x + 6$

_____ **2.** $7x + 5 = 3x - 15$

_____ **3.** $y + 4 = 7y - 8$

_____ **4.** $-3x + 16 = 2x - 4$

_____ **5.** $16 + 3a = -a + 4$

_____ **6.** $3(3y - 1) - 9 = 5(4 + y)$

_____ **7.** If Henry drives at a rate of 55 mph, how long will it take for him to travel 495 miles?

_____ **8.** Cheri bought a shirt and a blazer. The cost of the blazer was $15 less than 3 times the cost of the shirt. If the blazer cost $45, what was the cost of the shirt?

SUMMARY

In this chapter we learned that equations are mathematical sentences which say that one quantity represents the same number as another quantity. The solution for an equation containing a variable is the value of the variable that makes the equation a true statement.

To solve equations we agreed to

1. Deal with any symbols of grouping.
2. Combine like terms.
3. Perform additions and subtractions needed to move variable terms to one side, constant terms to the other side.
4. Perform multiplication and divisions needed to isolate the variable.
5. Check the solution in the original equation.

We practiced switching from word statements to equations containing variables using an orderly approach.

1. Write down what you want the variable to stand for.
2. Write any expressions in the problem that contain the variable.
3. Write an equation from the information in the problem.
4. Solve the equation and check in the original problem.

REVIEW EXERCISES

SECTION 3.1

In the following exercises, decide if the given number is a solution for the equation.

_____ **1.** $19 - 2x = 7$, $x = 6$ _____ **2.** $\dfrac{y}{5} - 3 = -4$, $y = 5$

_____ **3.** $7(2a - 3) = 7$, $a = 1$

_____ **4.** $3m - 9 = m + 11$, $m = 10$

Solve and check.

_____ **5.** $x - 7 = 3$ _____ **6.** $5 + y = -9$

_____ **7.** $-3 = m - 7$ _____ **8.** $a + 15 = 3$

_____ **9.** $9x = 72$ _____ **10.** $2x = -36$

_____ **11.** $-4x = 28$ _____ **12.** $5x = 0$

_____ **13.** $\dfrac{x}{3} = -1$ _____ **14.** $2 = \dfrac{y}{-7}$

_____ **15.** $4x - 5 = 31$ _____ **16.** $8 - 2x = 6$

_____ **17.** $-15 = 12 + 9x$ _____ **18.** $\dfrac{x}{4} - 2 = -1$

_____ **19.** $\dfrac{-7x}{3} = 14$ _____ **20.** $9 = \dfrac{5x}{2} - 1$

SECTION 3.2

_____ **21.** $8 - \dfrac{3x}{4} = 5$ _____ **22.** $2(x - 4) - (5x + 2) = 20$

_____ **23.** $8 = -2[3(y - 4) - 7]$

SECTION 3.3

_____ **24.** $7x - 5 = 3x + 11$ _____ **25.** $4a - 7 = 3 - a$

_____ **26.** $-4x - 9 = 10 - 3x$ _____ **27.** $18 + 9x = -2x - 15$

_____ **28.** $2(x + 2) + 6 = 5(2 + x)$ _____ **29.** $3(x - 7) - (x + 1) = 0$

_____ **30.** $3[x + 2(x - 1)] = 5(x - 2) - 8$

_____ **31.** The sum of 3 times an integer and 5 times the next consecutive integer is 53. Find the two integers.

_____ **32.** The length of a rectangle is 3 times the width. If the perimeter of the rectangle is 32 feet, find its dimensions.

_____ **33.** Raymond earns $40 a day plus $7 an hour for each hour he works overtime. If last Tuesday, Raymond earned $68, how many hours of overtime did he work?

4 Operating with Polynomials

Until now, we have spent time learning how to add and subtract variable terms and multiply and divide variable terms by constants. We have not yet learned how to multiply variable expressions.

In this chapter we learn how to

1. Identify polynomials.
2. Use exponents to write products.
3. Multiply polynomials.

We have been discussing algebraic expressions which are sums of variable terms and/or constant terms. Such expressions are examples of **polynomials**. For instance,

$$2x + 1$$
$$3a + 2b - 7$$
$$x - 5y$$
$$7x$$
$$2x + 3y - 4z + 8$$

are all examples of polynomials.

Mathematicians give different names to polynomials depending on how many *terms* the polynomial contains. (Remember, terms are separated by addition or subtraction signs.)

If a polynomial contains *one* term, it is called a **monomial.**

$$7x \text{ is a monomial}$$
$$-x \text{ is a monomial}$$

If a polynomial contains *two* terms, it is called a **binomial.**

$$2x + 1 \text{ is a binomial}$$
$$x - 5y \text{ is a binomial}$$

If a polynomial contains *three* terms, it is called a **trinomial.**

$$3a + 2b - 7 \text{ is a trinomial}$$
$$6x + y - z \text{ is a trinomial}$$

If a polynomial contains *more* than three terms, it is simply referred to as a polynomial of so-many terms.

$$2x + 3y - 4z + 8 \text{ is a polynomial of four terms}$$
$$a + b - c + d + e \text{ is a polynomial of five terms}$$

An expression such as

$$xy$$

tells us to multiply x times y. There is no addition involved, so we know xy is just *one* term and we name it:

$$xy \text{ is a monomial}$$

Similarly, we see that

$$xy + x \text{ is a binomial}$$
$$xy + 2x + 3y \text{ is a trinomial}$$

If it were possible to combine any of these terms we would do so before naming the polynomial. From Chapter 2, however, we recall that we can only combine *like* terms and

Like terms must contain exactly the same variable parts.

In the expression $xy + 2x + 3y$,

the variable part of the first term is xy
the variable part of the second term is x
the variable part of the third term is y

These terms are *not* like terms so they cannot be combined, and $xy + 2x + 3y$ is indeed a trinomial.

Let us practice naming some polynomials.

Example 1. $-16x$.

Solution. $-16x$ is a monomial.

Example 2. $1 - 16x$.

Solution. $1 - 16x$ is a binomial.

Example 3. $x - y - 3$.

Solution. $x - y - 3$ is a trinomial.

Example 4. $xy + xz$.

Solution. $xy + xz$ is a binomial.

Example 5. $x + 3x$.

Solution. We must combine like terms first.

$$x + 3x = 4x, \text{ a monomial}$$

Example 6. $xy + 2xy + xz$.

Solution. We must combine like terms first.

$$xy + 2xy + xz = 3xy + xz, \text{ a binomial}$$

Trial Run *Name the following expressions.*

_____ **1.** $-3x$

_____ **2.** $-3x + 5$

_____ **3.** $3ab - 2a + b$

_____ **4.** $-2x + x - 2y$

_____ **5.** $-3m + 4n + 2m + m$

_____ **6.** $xy - 3xz + 4yz$

ANSWERS
1. Monomial.　**2.** Binomial.　**3.** Trinomial.　**4.** $-x - 2y$, binomial.
5. $4n$, monomial.　**6.** Trinomial.

4.1 Exponents

We shall now learn a shorthand way of writing products of variables or constants.

4.1-1 Using Exponents

Sometimes it is necessary to deal with products such as

$$3 \cdot 3$$
$$x \cdot x$$
$$a \cdot a \cdot a$$
$$x \cdot x \cdot y \cdot y \cdot y \cdot y$$
$$b \cdot b \cdot b \cdot b \cdot b \cdot b$$

but stringing out the factors in these multiplication problems quickly becomes very awkward. Mathematicians decided long ago to invent some notation to handle such products. They agreed to write

$3 \cdot 3 = 3^2$	(read "3 squared")
$x \cdot x = x^2$	(read "x squared")
$a \cdot a \cdot a = a^3$	(read "a cubed")
$x \cdot x \cdot y \cdot y \cdot y \cdot y = x^2 y^4$	(read "x squared times y to the fourth")
$b \cdot b \cdot b \cdot b \cdot b \cdot b = b^6$	(read "b to the sixth")

The small raised positive integer is called the **exponent** and it tells us how many times the other number, called the **base**, is to be used as a factor in a product.

3^2 says "use 3 as a factor in a product 2 times"

x^2 says "use x as a factor in a product 2 times"

a^3 says "use a as a factor in a product 3 times"

$x^2 y^4$ says "use x as a factor in a product 2 times, use y as a factor in a product 4 times, then multiply them together"

b^6 says "use b as a factor in a product 6 times"

Such expressions (sometimes called **powers**) always contain two important pieces of information: the **base** and the **exponent.**

The exponent always tells us how many times to use the base as a factor in a product.

$$\underset{\text{base}}{x}{}^{5 \leftarrow \text{exponent}} = \underbrace{x \cdot x \cdot x \cdot x \cdot x}_{\text{5 factors}}$$

$$\underset{\text{base}}{a}{}^{n \leftarrow \text{exponent}} = \underbrace{a \cdot a \cdot \ldots \cdot a}_{n \text{ factors}} \qquad \text{(where } n \text{ is a positive integer)}$$

Example 1. Evaluate 4^3.

Solution. $4^3 = 4 \cdot 4 \cdot 4 = 64$.

Example 2. Evaluate $(-2)^5$.

Solution. $(-2)^5 = (-2)(-2)(-2)(-2)(-2) = -32$.

Example 3. Use an exponent to rewrite $y \cdot y \cdot y$.

Solution. $y \cdot y \cdot y = y^3$.

Example 4. Use exponents to rewrite $x \cdot x \cdot x \cdot y \cdot y$.

Solution. $x \cdot x \cdot x \cdot y \cdot y = x^3 y^2$.

Example 5. Write a^4b without exponents.

Solution. $a^4b = a \cdot a \cdot a \cdot a \cdot b$.

Example 6. Evaluate x^2y^3 when $x = 4$, $y = 2$.

Solution

$$x^2y^3 = 4^2 \cdot 2^3$$
$$= 4 \cdot 4 \cdot 2 \cdot 2 \cdot 2 = 128$$

Example 7. Evaluate a^3b^5 when $a = -1$, $b = -2$.

Solution

$$a^3b^5 = (-1)^3(-2)^5$$
$$= (-1)(-1)(-1)(-2)(-2)(-2)(-2)(-2)$$
$$= (-1)(-32) = 32$$

Unless parentheses indicate otherwise, an exponent belongs *only* with the number to which it is attached.

Example 8. Write $2x^3$ without exponents.

Solution. $2x^3 = 2 \cdot x \cdot x \cdot x$.

Example 9. Write $(2x)^3$ without exponents.

Solution

$$(2x)^3 = (2x)(2x)(2x)$$
$$= 2 \cdot x \cdot 2 \cdot x \cdot 2 \cdot x$$
$$= 2 \cdot 2 \cdot 2 \cdot x \cdot x \cdot x$$
$$= 8 \cdot x \cdot x \cdot x$$

In Example 9, notice that we were free to regroup the factors because of the commutative and associative properties for multiplication. Notice also that these two examples represented very different numbers.

$$2x^3 = 2 \cdot x \cdot x \cdot x$$
$$(2x)^3 = 8 \cdot x \cdot x \cdot x$$

Example 10. Write $-x^4$ without exponents.

Solution. $-x^4 = -1 \cdot x \cdot x \cdot x \cdot x$.

Example 11. Write $(-x)^4$ without exponents.

Solution

$$(-x)^4 = (-1x)^4$$
$$= (-1x)(-1x)(-1x)(-1x)$$
$$= (-1)(-1)(-1)(-1)x \cdot x \cdot x \cdot x$$
$$= 1x \cdot x \cdot x \cdot x$$
$$= x \cdot x \cdot x \cdot x$$

Example 12. Evaluate $(2x^2y)^3$ when $x = -1$, $y = 4$.

Solution

$$(2x^2y)^3 = [2 \cdot (-1)^2 \cdot 4]^3$$
$$= [2(-1)(-1)4]^3$$
$$= [2 \cdot 1 \cdot 4]^3$$
$$= 8^3 = 8 \cdot 8 \cdot 8 = 512$$

As you may have already decided,

$$x^1 = x$$
$$a^1 = a$$
$$3^1 = 3$$

because the exponent of 1 tells us to use each base once as a factor in a product. An exponent of 1 is not usually written in algebra, but it is not incorrect to do so.

We have now learned how to work with the positive integers (1, 2, 3, . . .) used as exponents. We shall mention the fact that 0 can also be used as an exponent, provided that we understand that the following definition must be applied.

$$a^0 = 1 \qquad \text{provided that } a \neq 0$$

When a base other than zero is raised to the zero power, the value is 1.

So we may say that

$$2^0 = 1$$
$$5^0 = 1$$
$$x^0 = 1$$
$$(3y)^0 = 1$$
$$(-3xyz)^0 = 1$$

but

$$3x^0 = 3 \cdot x^0 = 3(1) = 3 \qquad \text{provided that } x \neq 0$$

and

$$5x^0y = 5 \cdot x^0 \cdot y = 5 \cdot 1 \cdot y = 5y \qquad \text{provided that } x \neq 0$$

Now that we have discussed the meaning of exponents, we can state a formal definition of a polynomial.

Definition of Polynomial: A polynomial in x is a sum of one or more terms of the form

$$ax^n$$

where a is a constant and n is a whole number.

It is important to note that the exponent on the variable in each term of a polynomial must be a *whole number* and that a polynomial must be expressible as a *sum* (or difference) of such terms.

Learning to recognize algebraic expressions that are *not* polynomials will help us understand what polynomials must look like. For instance,

$$\frac{1}{x} \text{ is } not \text{ a polynomial}$$

$$\sqrt{x} \text{ is } not \text{ a polynomial}$$

$$\frac{x^2 + 3x + 5}{x + 1} \text{ is } not \text{ a polynomial}$$

But

$$x^2 - 3x - 8 \text{ is a polynomial}$$
$$x^9 \text{ is a polynomial}$$
$$\frac{1}{2}x^2 + 3x \text{ is a polynomial}$$

Trial Run

_____ 1. Use exponents to rewrite 2*aaabb*.

_____ 2. Write $-x^4$ without exponents.

_____ 3. Write $(3x)^3$ without exponents.

_____ 4. Evaluate $(-2)^4$.

_____ 5. Evaluate $(-3x)^2$ when $x = 2$.

_____ 6. Evaluate $-3x^2$ when $x = 2$.

_____ 7. Evaluate $(2x^5y)^3$ when $x = -1$ and $y = 3$.

ANSWERS
1. $2a^3b^2$.   2. $-1 \cdot x \cdot x \cdot x \cdot x$.   3. $27x \cdot x \cdot x$.   4. 16.   5. 36.   6. -12.   7. -216.

4.1-2   Adding Polynomials

If you were asked to simplify an expression such as

$$(2x^2 + 7x + 3) + (x^2 + 5x + 1)$$

you should see that this is a matter of _combining like terms_. Here

$$2x^2 \text{ and } x^2 \text{ are like terms}$$
$$7x \text{ and } 5x \text{ are like terms}$$
$$3 \text{ and } 1 \text{ are like terms}$$

Our sum becomes

$$2x^2 + 7x + 3 + x^2 + 5x + 1$$
$$= 2x^2 + x^2 + 7x + 5x + 3 + 1$$
$$= 3x^2 + 12x + 4$$

because we recall that _to combine like terms, we combine the numerical coefficients and keep the variable part._

Previous work also tells us how to multiply a constant times a variable term. Remember that

$$3(2x) = (3 \cdot 2)x = 6x$$

and

$$5(-x) = (5 \cdot -1)x = -5x$$

So

$$2(9x^2) = (2 \cdot 9)x^2 = 18x^2$$

and

$$-3(4y^3) = (-3 \cdot 4)y^3 = -12y^3$$

and

$$5(-2xy) = (5 \cdot -2)xy = -10xy$$

To multiply a constant times a variable term, we multiply the constant times the numerical coefficient and keep the variable part.

The distributive property allowed us to simplify expressions such as

$$3(x + 2y) = 3x + 6y$$
$$-5(2x - y) = -10x + 5y$$
$$-(3 - x) = -3 + x$$

Similarly,

$$2(x^2 + 3x + 1) = 2x^2 + 6x + 2$$
$$-4(6 - y^2) = -24 + 4y^2$$
$$-(5a^2 - 3a - 2) = -5a^2 + 3a + 2$$

We must use all these ideas to simplify polynomial sums or differences.

Example 13. Simplify $2(x^2 - x) + 5(2x^2 - 1)$.

Solution

$$2(x^2 - x) + 5(2x^2 - 1)$$
$$= 2x^2 - 2x + 10x^2 - 5$$
$$= 12x^2 - 2x - 5$$

Example 14. Simplify $3(2a - a^2) - (a^2 - a + 3)$.

Solution

$$3(2a - a^2) - (a^2 - a + 3)$$
$$= 6a - 3a^2 - a^2 + a - 3$$
$$= -4a^2 + 7a - 3$$

Notice that it is customary to write simplified expressions with the exponents in decreasing order from left to right. This is sometimes called "writing an expression in descending powers of the variable."

Trial Run   *Simplify.*

_____ 1. $2(x + y) - 3(2x - y)$

_____ 2. $5(x^2 - 1) + 2(2x^2 - 3x)$

_____ 3. $(a^2 - 2a + 1) - (3a^2 + a + 5)$

_____ 4. $-2(x^2 - 3) + (2x^2 - 5)$

_____ 5. $(y^2 + 2y + 1) - (y^2 - 2y + 1)$

ANSWERS
1. $-4x + 5y$.   2. $9x^2 - 6x - 5$.   3. $-2a^2 - 3a - 4$.   4. 1.   5. $4y$.

EXERCISE SET 4.1

Name the following expressions.

_____ 1. $-3x^2$ _____ 2. $5y^4$

_____ 3. $7x - 4y$ _____ 4. $-3x + 5y$

_____ 5. $4ab - 2a + b$ _____ 6. $a - b + 3ab$

_____ 7. $-7m^2 + 2n^2 + 10m^2$ _____ 8. $8n - 6m + n$

_____ 9. $3a - 5b - 7a + 6b$ _____ 10. $9a - 6b - 4a + 9b$

_____ 11. $ab - ac + bc$ _____ 12. $3ab + 2ac - 5bc$

_____ 13. $4a^3 - 2a^2 + a - 5$ _____ 14. $-5a^3 - 4a^2 + 3a - 9$

Use exponents to rewrite each expression.

_____ 15. $3xxyyy$ _____ 16. $-2xxxxyy$

_____ 17. $-a \cdot a \cdot a \cdot a$ _____ 18. $-b \cdot b \cdot b \cdot b \cdot b$

_____ 19. $(-2x)(-2x)(-2x)$ _____ 20. $(3x)(3x)(3x)(3x)$

_____ 21. $3x \cdot x \cdot x - 4yyyy$ _____ 22. $x \cdot x \cdot x \cdot x + 2y \cdot y$

Write the following without exponents.

_____ 23. $8a^3$ _____ 24. $5a^4$

_____ 25. $(3a)^3$ _____ 26. $(2a)^5$

_____ 27. $-x^4$ _____ 28. $-3x^3$

_____ 29. $3x^2y^3$ _____ 30. $7x^4y^2$

In Exercises 31–40, evaluate the given expressions.

_____ 31. $(-3)^3$ _____ 32. $(-2)^5$

_____ 33. -2^4 _____ 34. -3^4

_____ 35. $(2x)^2$ when $x = -1$ _____ 36. $(-3x)^2$ when $x = 2$

_____ 37. $(xy^2)^3$ when $x = 1$ and $y = -2$ _____ 38. $(3x^5y)^2$ when $x = 1$ and $y = 2$

_____ 39. $(7xy^3)^0$ when $x = 1$ and $y = 2$ _____ 40. $(3xy^5)^0$ when $x = 2$ and $y = -1$

Simplify.

_____ 41. $-2(x + y) + 3(2x - 5y)$ _____ 42. $-3(2x - y) + 4(x + y)$

_____ 43. $4(x^2 - 3) - 5(3x - 2)$ _____ 44. $3(2x^2 - 5x) - 7(-2x + 5)$

_____ 45. $(y^2 + 3y - 5) - (2y^2 + 3y - 6)$ _____ 46. $(2y^2 - y - 4) - (y^2 - 2y + 2)$

_____ 47. $-3(a^2 + 2) + (3a^2 + 7)$ _____ 48. $(3x^2 - 6) - 3(x^2 - 2)$

_____ 49. $5(y^2 - 2y + 1) - 2(3y^2 - 5y + 4)$ _____ 50. $3(x^2 - 2x + 1) - 5(x^2 + x - 1)$

CHECKUP 4.1

_____ **1.** Name the expression $4m^2 - 2n^2 + m^2$.

_____ **2.** Name the expression $3a - 5b + 7ab$.

_____ **3.** Use exponents to write $-3 \cdot x \cdot x \cdot x \cdot x$.

_____ **4.** Use exponents to write $2a \cdot a - b \cdot b \cdot b \cdot b$.

_____ **5.** Write $(-2x)^3$ without exponents.

_____ **6.** Write $5x^3y^2$ without exponents.

_____ **7.** Evaluate $(-2)^4$.

_____ **8.** Evaluate $(-3x^2)^3$ when $x = -1$.

_____ **9.** Simplify $-2(x^2 - 5x) + 3(x - 2)$.

_____ **10.** Simplify $(2y^2 + y - 5) - (y^2 + 2y - 2)$.

4.2 Multiplying with Monomials

Let us develop some methods for multiplying monomials times monomials and finding powers of monomials.

4.2-1 Multiplying Monomials Times Monomials

Suppose that we wished to find the following products:

$$3^2 \cdot 3^4$$
$$x^3 \cdot x$$
$$y^4 \cdot y^3$$
$$a^2 \cdot a^2$$

We could rewrite each of these products using what we know about the meaning of exponents.

$$3^2 \cdot 3^4 = \underbrace{3 \cdot 3}_{2 \text{ factors}} \cdot \underbrace{3 \cdot 3 \cdot 3 \cdot 3}_{4 \text{ factors}} = 3^6$$
$$\underbrace{}_{6 \text{ factors}}$$

$$x^3 \cdot x = \underbrace{x \cdot x \cdot x}_{3 \text{ factors}} \cdot \underbrace{x}_{1 \text{ factor}} = x^4$$
$$\underbrace{}_{4 \text{ factors}}$$

$$y^4 \cdot y^3 = \underbrace{y \cdot y \cdot y \cdot y}_{4 \text{ factors}} \cdot \underbrace{y \cdot y \cdot y}_{3 \text{ factors}} = y^7$$
$$\underbrace{}_{7 \text{ factors}}$$

$$a^2 \cdot a^2 = \underbrace{a \cdot a}_{2 \text{ factors}} \cdot \underbrace{a \cdot a}_{2 \text{ factors}} = a^4$$
$$\underbrace{}_{4 \text{ factors}}$$

Do you see a rule that we could use instead of stringing out the factors? The base in our answer is the original base and the exponent in our answer is the *sum* of the original exponents.

$$3^2 \cdot 3^4 = 3^{2+4} = 3^6$$
$$x^3 \cdot x = x^3 \cdot x^1 = x^{3+1} = x^4$$
$$y^4 \cdot y^3 = y^{4+3} = y^7$$
$$a^2 \cdot a^2 = a^{2+2} = a^4$$

Notice that the bases must be the same in the original problem before we can use this rule.

$$y^4 \cdot y^3 = y^7$$

but

$$x^4 \cdot y^3 = x^4 \cdot y^3$$

To multiply powers of the *same* base, we keep the same base and *add* the exponents.

This rule is called the **First Law of Exponents** and we may state it in general, using a to represent any base and m and n to represent any whole-number exponents.

$$a^m \cdot a^n = a^{m+n}$$

Example 1. Multiply $x^6 \cdot x^9$.

   Solution. $x^6 \cdot x^9 = x^{6+9} = x^{15}$.

Example 2. Multiply $y \cdot y^3 \cdot y^5$.

   Solution. $y \cdot y^3 \cdot y^5 = y^1 \cdot y^3 \cdot y^5 = y^{1+3+5} = y^9$.

Example 3. Multiply $x^2 \cdot y$.

   Solution. $x^2 \cdot y = x^2 \cdot y$ (the bases are different).

Example 4. Multiply $x^0 \cdot x^5$.

   Solution. $x^0 \cdot x^5 = x^{0+5} = x^5$.

Example 5. Multiply $a^{11} \cdot a^0$.

   Solution. $a^{11} \cdot a^0 = a^{11+0} = a^{11}$.

Examples 4 and 5 should help you see why x^0 and a^0 were defined to be 1. From the First Law of Exponents,

$$x^0 \cdot x^5 = x^5 \quad \text{and} \quad a^{11} \cdot a^0 = a^{11}$$

but the only number that multiplies another number and leaves it unchanged is the number 1. You should agree now that

$$a^0 = 1 \qquad \text{provided that } a \neq 0$$

makes good sense.

What happens when we try to multiply monomials containing constants and one or more variables? The commutative and associative properties for multiplication will help us out here.

$$\begin{aligned}
(2x^3)(4x^2) &= 2 \cdot x^3 \cdot 4 \cdot x^2 \\
&= 2 \cdot 4 \cdot x^3 \cdot x^2 \\
&= 8x^{3+2} \\
&= 8x^5
\end{aligned}$$

$$\begin{aligned}
(-5x^8)(6x^7) &= -5 \cdot x^8 \cdot 6 \cdot x^7 \\
&= -5 \cdot 6 \cdot x^8 \cdot x^7 \\
&= -30x^{8+7} \\
&= -30x^{15}
\end{aligned}$$

Note that we found these products by multiplying the numerical coefficients and then multiplying the variable parts.

Example 6. Multiply $(4a^5)(3a^7)$.

   Solution

$$\begin{aligned}
(4a^5)(3a^7) &= 4 \cdot 3 \cdot a^5 \cdot a^7 \\
&= 12a^{5+7} = 12a^{12}
\end{aligned}$$

Example 7. Multiply $(-x^{10})(3x^9)$.

Solution

$$(-x^{10})(3x^9) = -1 \cdot 3 \cdot x^{10} \cdot x^9$$
$$= -3x^{10+9} = -3x^{19}$$

Example 8. Multiply $(3x^2)(-2x^4)(-7x)$.

Solution

$$(3x^2)(-2x^4)(-7x) = 3(-2)(-7)x^2 \cdot x^4 \cdot x$$
$$= 3(14)x^2 \cdot x^4 \cdot x^1$$
$$= 42x^{2+4+1} = 42x^7$$

If there is more than one variable in the monomials, we do the same kind of regrouping.

$$(5x^2y)(2x^3y^5) = 5 \cdot x^2 \cdot y \cdot 2 \cdot x^3 \cdot y^5$$
$$= 5 \cdot 2 \cdot x^2 \cdot x^3 \cdot y^1 \cdot y^5$$
$$= 10x^{2+3}y^{1+5}$$
$$= 10x^5y^6$$

> To multiply monomials, first multiply the numerical coefficients, then multiply the variables using the First Law of Exponents.

Example 9. Multiply $8x(3x^2y)(y^4)$.

Solution

$$8x(3x^2y)(y^4) = 8 \cdot 3 \cdot x \cdot x^2 \cdot y \cdot y^4$$
$$= 24x^3y^5$$

Example 10. Multiply $(-2xy^2z)(-9x^2y^5z^7)$.

Solution
$$(-2xy^2z)(-9x^2y^5z^7) = (-2)(-9)x \cdot x^2 \cdot y^2 \cdot y^5 \cdot z \cdot z^7$$
$$= 18x^3y^7z^8$$

Trial Run   *Multiply.*

———— **1.** $x^3 \cdot x^6$

———— **2.** $x^3 \cdot y^2$

———— **3.** $a^0 \cdot a^5$

———— **4.** $y^2 \cdot y \cdot y^3$

Multiply.

———— **5.** $(-4x^5)(7x^3)$

———— **6.** $(2a^3)(-5a^2)(6a^6)$

———— **7.** $(3x^2y^3)(2x^5y^7)$

———— **8.** $(4x^2yz^5)(-3xy^3z^7)$

ANSWERS
1. x^9.   **2.** x^3y^2.   **3.** a^5.   **4.** y^6.   **5.** $-28x^8$.   **6.** $-60a^{11}$.   **7.** $6x^7y^{10}$.
8. $-12x^3y^4z^{12}$.

4.2-2 Powers of Monomials

How would you simplify an expression such as

$$(x^2)^3?$$

We know that the exponent of 3 tells us to use the base x^2 as a factor in a product 3 times. So

$$(x^2)^3 = x^2 \cdot x^2 \cdot x^2$$
$$= x^{2+2+2}$$
$$= x^6$$

We could similarly decide that

$$(x^5)^2 = x^5 \cdot x^5 = x^{5+5} = x^{10}$$
$$(a^4)^5 = a^4 \cdot a^4 \cdot a^4 \cdot a^4 \cdot a^4 = a^{4+4+4+4+4} = a^{20}$$

Do you see a shortcut for this procedure? This is called "raising a power to a power" and it is accomplished by *multiplying* exponents.

$$(x^2)^3 = x^{2 \cdot 3} = x^6$$
$$(x^5)^2 = x^{5 \cdot 2} = x^{10}$$
$$(a^4)^5 = a^{4 \cdot 5} = a^{20}$$

> To raise a power of a base to a power, we keep the base and multiply the exponents.

This rule is called the **Second Law of Exponents** and we may state it in general.

> **Second Law of Exponents**
>
> $$(a^m)^n = a^{m \cdot n}$$
>
> where m and n are whole numbers

Example 11. Simplify $(x^7)^2$.

  Solution. $(x^7)^2 = x^{7 \cdot 2} = x^{14}$.

Example 12. $(a^9)^3$.

  Solution. $(a^9)^3 = a^{9 \cdot 3} = a^{27}$.

Suppose that we wish to simplify

$$(xy)^3$$

You should agree that this can be written as

$$(xy)^3 = (xy)(xy)(xy)$$
$$= x \cdot x \cdot x \cdot y \cdot y \cdot y$$
$$= x^3 y^3$$

Similarly,

$$(2b)^5 = (2b)(2b)(2b)(2b)(2b)$$
$$= 2 \cdot 2 \cdot 2 \cdot 2 \cdot 2 \cdot b \cdot b \cdot b \cdot b \cdot b$$
$$= 2^5 b^5 \quad \text{or} \quad 32b^5$$

Note that we "raise a product to a power" by raising each factor to that power.

Example 13. Simplify $(ax)^7$.

Solution. $(ax)^7 = a^7x^7$.

Example 14. Simplify $(-3y)^5$.

Solution

$$(-3y)^5 = (-3)^5y^5$$
$$= -243y^5$$

Example 15. Simplify $(-3y)^4$.

Solution

$$(-3y)^4 = (-3)^4y^4$$
$$= 81y^4$$

> To raise a product to a power we raise each factor to that power.

This rule is called the **Third Law of Exponents** and can be written in general as

> **Third Law of Exponents**
>
> $$(ab)^n = a^nb^n$$
>
> where n is a whole number

Putting our laws of exponents together, we may simplify some expressions that look complicated but merely call for orderly steps. In Chapter 1 we learned to perform multiplications and divisions before additions and subtractions. Now we expand our rule for order of operations:

> In the absence of symbols of grouping, first deal with powers, then perform multiplications and divisions from left to right, then perform additions and subtractions from left to right.

Example 16. Simplify $(x^2y)^3$.

Solution

$$(x^2y)^3 = (x^2)^3(y^1)^3$$
$$= x^{2\cdot3}y^{1\cdot3} = x^6y^3$$

Example 17. Simplify $(5x^3y^4)^2$.

Solution

$$(5x^3y^4)^2 = 5^2(x^3)^2(y^4)^2$$
$$= 5^2x^{3\cdot2}y^{4\cdot2}$$
$$= 5^2x^6y^8$$
$$= 25x^6y^8$$

Example 18. Simplify $x(x^2z)^5$.

Solution

$$
\begin{aligned}
x(x^2z)^5 &= x \cdot (x^2)^5 z^5 \\
&= x \cdot x^{2 \cdot 5} \cdot z^5 \\
&= x \cdot x^{10} z^5 \\
&= x^{11} z^5
\end{aligned}
$$

Example 19. Simplify $x^2y(x^3y) + 5x(x^2y)^2$.

Solution

$$
\begin{aligned}
&x^2y(x^3y) + 5x(x^2y)^2 \\
&= x^2 \cdot x^3 \cdot y \cdot y + 5x(x^2)^2y^2 \\
&= x^5y^2 + 5x \cdot x^4y^2 \\
&= x^5y^2 + 5x^5y^2 \\
&= 6x^5y^2 \qquad \text{(notice that we combined like terms)}
\end{aligned}
$$

Example 20. Simplify $(-a^2y^2)^3 + (2a^3y^3)^2$.

Solution

$$
\begin{aligned}
&(-a^2y^2)^3 + (2a^3y^3)^2 \\
&= (-1a^2y^2)^3 + (2a^3y^3)^2 \\
&= (-1)^3(a^2)^3(y^2)^3 + 2^2(a^3)^2(y^3)^2 \\
&= -1a^6y^6 + 4a^6y^6 \\
&= 3a^6y^6
\end{aligned}
$$

Trial Run *Simplify.*

———— 1. $(a^3)^4$

———— 2. $(xy)^5$

———— 3. $(-2y)^3$

———— 4. $(-3a)^4$

———— 5. $(a^2b)^3$

———— 6. $(3x^2y^4)^2$

———— 7. $x(x^2y^3)^5$

———— 8. $(-a^2b^3)^5$

———— 9. $xy^2(x^3y) + 3x^2(x^2y^3)$

ANSWERS
1. a^{12}.　　2. x^5y^5.　　3. $-8y^3$.　　4. $81a^4$.　　5. a^6b^3.　　6. $9x^4y^8$.　　7. $x^{11}y^{15}$.
8. $-a^{10}b^{15}$.　　9. $4x^4y^3$.

Name _____ **Date** _____

EXERCISE SET 4.2

Simplify.

_____ 1. $x^2 \cdot x^3$

_____ 2. $a^5 \cdot a^2$

_____ 3. $y^3 \cdot y^3$

_____ 4. $x^4 \cdot x^4$

_____ 5. $x^3 \cdot y^2$

_____ 6. $x^3 \cdot y^4$

_____ 7. $a^0 \cdot a^7$

_____ 8. $a^9 \cdot a^0$

_____ 9. $y^3 \cdot y \cdot y^5$

_____ 10. $y^2 \cdot y \cdot y^7$

_____ 11. $(2x^2)(5x^3)$

_____ 12. $(3x^3)(7x^2)$

_____ 13. $(-5x^7)(4x)$

_____ 14. $(-9x^5)(3x)$

_____ 15. $(-2a^5)(-5a^7)$

_____ 16. $(-3a^4)(-5a^6)$

_____ 17. $(9y^3)(-y)$

_____ 18. $(7y^4)(-y)$

_____ 19. $(3m^2)(-2m)(m^5)$

_____ 20. $(4m^3)(-3m)(m^2)$

_____ 21. $(4x^2y^3)(-3x^5y^2)$

_____ 22. $(8x^3y^4)(-2x^2y^5)$

_____ 23. $3x(-x^2y)(-2x^4y^3)$

_____ 24. $-5x(-x^3y)(4x^4y^3)$

_____ 25. $(7xy^2)(-2xy)(-3x^2y)$

_____ 26. $(-2xy^2)(8x^2y^3)(-3xy)$

_____ 27. $(5x^2y^3z)(7xyz^4)$

_____ 28. $(9x^2y^3z^5)(3xyz^2)$

_____ 29. $(-3x^2y)(8x^3z^2)$

_____ 30. $(-4x^3y^2)(9x^5z^3)$

_____ 31. $(a^2)^3$

_____ 32. $(a^3)^2$

_____ 33. $(xy)^5$

_____ 34. $(xy)^7$

_____ 35. $(7x)^2$

_____ 36. $(8x)^2$

_____ 37. $(-3y^2)^2$

_____ 38. $(-5y^2)^2$

_____ 39. $(-4x)^3$

_____ 40. $(-3x)^3$

_____ 41. $(-5a^3)^2$

_____ 42. $(-6a^4)^2$

_____ 43. $(x^2y)^3$

_____ 44. $(xy^3)^4$

_____ 45. $(9x^2y^3)^2$

_____ 46. $(8x^2y^4)^2$

_____ 47. $3a(a^3b^2)^5$

_____ 48. $7a(a^2b^3)^4$

_____ 49. $(-xy^3)^4$

_____ 50. $(-x^2y)^5$

_____ 51. $5x(2x^3)^2$

_____ 52. $6x(3x^4)^2$

_____ 53. $(5y^4)^2(-2y^5)^3$

_____ 54. $(7y^2)^3(-y^4)^5$

_____ 55. $(-6a^2b)(2ab^3)^2$

_____ 56. $(2a^2b)(-3a^4b^2)^3$

_____ 57. $xy^2(3xy^4) + 2y^3(4x^2y^3)$

_____ 58. $x^2y^3(5x^2y^4) - 3y^5(x^4y^2)$

_____ 59. $-a^2(ab^3)^4 - b^6(2a^2b^2)^3$

_____ 60. $a^5(ab^2)^3 - 4a^2(a^3b^3)^2$

CHECKUP 4.2

Simplify.

_____ **1.** $x^6 \cdot x^2$

_____ **2.** $y^3 \cdot y^2 \cdot y$

_____ **3.** $(3a^2)(7a^5)$

_____ **4.** $(-6x^7)(2x)(3x^2)$

_____ **5.** $(-5x^2y^3)(-2xy^4)$

_____ **6.** $(2xy^2)(-x^3y)(-7xy)$

_____ **7.** $(4x^2y^3z^5)(-3xy^4z^2)$

_____ **8.** $(3x)^2$

_____ **9.** $(-2y)^3$

_____ **10.** $2a^2(3a^2b)^3$

4.3 Multiplying Monomials Times Polynomials

The distributive property, the rules for signs, and the laws of exponents are all the tools we need to simplify products such as

$$x(x^2 + 3x + 1)$$
$$-5a^2(a^3 - 4)$$
$$4xy(2x + y - 7)$$

The distributive property tells us we must multiply each term inside the parentheses by the monomial outside the parentheses, and the rules for signs and the First Law of Exponents tell us how to simplify each of those multiplications.

$$\begin{aligned} x(x^2 + 3x + 1) &= x \cdot x^2 + x \cdot 3x + x \cdot 1 \\ &= x^1 \cdot x^2 + 3 \cdot x \cdot x + x \cdot 1 \\ &= x^3 + 3x^2 + x \end{aligned}$$

$$\begin{aligned} -5a^2(a^3 - 4) &= -5a^2 \cdot a^3 - 5a^2(-4) \\ &= -5a^5 + 20a^2 \end{aligned}$$

$$\begin{aligned} 4xy(2x + y - 7) &= 4xy \cdot 2x + 4xy \cdot y + 4xy(-7) \\ &= 4 \cdot 2 \cdot x \cdot x \cdot y + 4xy \cdot y + 4(-7)xy \\ &= 8x^2y + 4xy^2 - 28xy \end{aligned}$$

Remember that we learned to do such products without writing down each step, using arrows to keep our work straight.

Example 1. Multiply $2a(4a^3 - 5a - 3)$.

Solution. $2a(4a^3 - 5a - 3)$.

$$= 2a(4a^3 - 5a - 3)$$
$$= 8a^4 - 10a^2 - 6a$$

Example 2. Multiply $x^3y(x^2y^2 - 4xy + 3xy^5)$.

Solution. $x^3y(x^2y^2 - 4xy + 3xy^5)$.

$$= x^3y(x^2y^2 - 4xy + 3xy^5)$$
$$= x^5y^3 - 4x^4y^2 + 3x^4y^6$$

Example 3. Simplify $x^2(2x - y) + x(xy - 3x^2)$.

Solution

$$\begin{aligned} &x^2(2x - y) + x(xy - 3x^2) \\ &= 2x^3 - x^2y + x^2y - 3x^3 \\ &= 2x^3 - 3x^3 - x^2y + x^2y \\ &= -x^3 \end{aligned}$$

Example 4. Simplify $2a(x - 3xy) - 5x(a - 4ay)$.

Solution

$$\begin{aligned} &2a(x - 3xy) - 5x(a - 4ay) \\ &= 2ax - 6axy - 5ax + 20axy \\ &= 2ax - 5ax - 6axy + 20axy \\ &= -3ax + 14axy \end{aligned}$$

Trial Run *Simplify.*

_____ **1.** $x(x^2 - 5x + 2)$

_____ **2.** $-2a^2(a - 6)$

_____ **3.** $3x(4x^2 - 2x + 7)$

_____ **4.** $a^2b(a^2b^2 - 3ab + 4ab^4)$

_____ **5.** $x^2(3x + y) - x(xy - 2x^2)$

_____ **6.** $2a(x - 2xy) - 3x(a + 3ay)$

ANSWERS
1. $x^3 - 5x^2 + 2x$. **2.** $-2a^3 + 12a^2$. **3.** $12x^3 - 6x^2 + 21x$. **4.** $a^4b^3 - 3a^3b^2 + 4a^3b^5$.
5. $5x^3$. **6.** $-ax - 13axy$.

EXERCISE SET 4.3

Multiply.

_____ 1. $5(2x + 3)$

_____ 2. $4(x + 5)$

_____ 3. $-7(2x^2 - 4)$

_____ 4. $-3(4x^2 - 7)$

_____ 5. $3x(2x - 4)$

_____ 6. $5x(3x - 2)$

_____ 7. $y^2(2y - 5)$

_____ 8. $y^2(3y - 8)$

_____ 9. $-a^2(a^2 - 2)$

_____ 10. $-a^2(2a^2 - 3)$

_____ 11. $4x^2(9x^2 - 4)$

_____ 12. $5x^2(8x^2 - 1)$

_____ 13. $-3a^3(7a^2 - 3a)$

_____ 14. $-4a^3(6a^2 - 5a)$

_____ 15. $2xy(5x + 2y)$

_____ 16. $-5xy(3x + 5y)$

_____ 17. $-x^2y^2(11x - 4y)$

_____ 18. $-2x^2y^2(x - 7y)$

_____ 19. $2a^2b^2(7a^2 - b^2)$

_____ 20. $6a^2b^2(a^2 - 2b^2)$

_____ 21. $2m(m^2 - 5m + 6)$

_____ 22. $9m(m^2 - m - 6)$

_____ 23. $5x^2y^2(3x^2y^2 - xy - 4)$

_____ 24. $4x^2y^2(5x^2y^2 - xy - 6)$

_____ 25. $-a^2b(3 - 2ab + 4a^2b^2)$

_____ 26. $-a^2b(5 - 3ab + 2a^2b^2)$

_____ 27. $4mn(8m^3 - 2m^2n + mn^2 - 3n^3)$

_____ 28. $5mn(6m^3 - m^2n + 2mn^2 - 7n^3)$

_____ 29. $x^2(2xy - 6) + 2x^2(xy + 3)$

_____ 30. $x(3xy - x^2) - x^2(y - x)$

_____ 31. $x^2(x^2 + 5) - 3(x^2 - 8)$

_____ 32. $x^2(x^2 - 4) + 2(x^2 - 5)$

_____ 33. $3m(3m + n) - n(3m + n)$

_____ 34. $7m(2m - n) - 2n(4m + n)$

_____ 35. $3ax(2 - 3y) - 2ax(1 - 2y)$

_____ 36. $5ax(3 - y) - 7ax(4 - 3y)$

_____ 37. $5xy(3x - 4y) - 2xy(4x - 5y)$

_____ 38. $8xy(2x - y) - xy(x - 2y)$

_____ 39. $-2x^2y^3(3x + 5xy^4) + x^3y(y^2 - 3y^6)$

_____ 40. $-x^2y^3(8x - 3xy^4) + 2x^3y(4y^2 - 5y^6)$

CHECKUP 4.3

Multiply.

_____ **1.** $5x(2x - 4)$

_____ **2.** $y^2(3y - 2)$

_____ **3.** $-2a^2(a^2 - 7a)$

_____ **4.** $3xy(5x - 2y)$

_____ **5.** $4m(4m^2 - 7m + 2)$

_____ **6.** $6x^2y^2(2x^2y^2 - 3xy + 4)$

_____ **7.** $-2mn(4m^3 - m^2n + 2mn^2 - n^3)$

_____ **8.** $2x^2(3xy - 2) - x^2(7xy - 4)$

_____ **9.** $5x(3x + y) - y(4x - y)$

_____ **10.** $-3ax(1 - 2y) + 2ax(5 - 3y)$

4.4 Multiplying Polynomials

Now that we know how to multiply with monomials, we must learn how to multiply other polynomials. We shall use what we know about the distributive property and the First Law of Exponents.

4.4-1 Multiplying Binomials

Recall how we used the distributive property to multiply monomials times binomials.

$$2x(x + 3) = 2x^2 + 6x$$
$$5(x + 3) = 5x + 15$$

Suppose that we wish to find the product

$$(2x + 5)(x + 3)$$

According to the distributive property, each term in the first binomial $2x + 5$ must be multiplied times the second binomial, $x + 3$. Then the resulting terms must be combined if possible.

$$(2x + 5)(x + 3) = 2x(x + 3) + 5(x + 3)$$
$$= 2x^2 + 6x + 5x + 15$$
$$= 2x^2 + 11x + 15$$

> To multiply binomials, we multiply the first term in the first binomial times the second binomial; then we multiply the second term in the first binomial times the second binomial. Finally, we combine like terms.

Example 1. Multiply $(x - 2)(3x + 1)$.

Solution

$$(x - 2)(3x + 1)$$
$$= x(3x + 1) - 2(3x + 1)$$
$$= 3x^2 + x - 6x - 2$$
$$= 3x^2 - 5x - 2$$

Example 2. Multiply $(5 - x)(4 + 7x)$.

Solution

$$(5 - x)(4 + 7x)$$
$$= 5(4 + 7x) - x(4 + 7x)$$
$$= 20 + 35x - 4x - 7x^2$$
$$= 20 + 31x - 7x^2$$

Example 3. Multiply $(-3a - 5)(7a - 1)$.

Solution

$$(-3a - 5)(7a - 1)$$
$$= -3a(7a - 1) - 5(7a - 1)$$
$$= -21a^2 + 3a - 35a + 5$$
$$= -21a^2 - 32a + 5$$

Example 4. Multiply $(x - y)(x + 2y)$.

Solution

$$(x - y)(x + 2y)$$
$$= x(x + 2y) - y(x + 2y)$$
$$= x^2 + 2xy - yx - 2y^2$$
$$= x^2 + 2xy - xy - 2y^2$$
$$= x^2 + xy - 2y^2$$

Example 5. Find $(x + 5)^2$.

Solution. Remember that $(x + 5)^2$ means $(x + 5)(x + 5)$.

$$\begin{aligned}
(x + 5)^2 &= (x + 5)(x + 5) \\
&= x(x + 5) + 5(x + 5) \\
&= x^2 + 5x + 5x + 25 \\
&= x^2 + 10x + 25
\end{aligned}$$

As you can see from these examples, you need only know the distributive property and rules for multiplying monomials in order to be able to multiply binomials.

Notice also from these examples that if we multiply two binomials, each containing a variable term with an exponent of 1 and a constant term, our answer seems to contain a variable term with an exponent of 2, a variable term with an exponent of 1, and a constant term. This is not always true but it is something to note.

Example 6. Multiply $(a + b)(c + d)$.

Solution

$$\begin{aligned}
&(a + b)(c + d) \\
&= a(c + d) + b(c + d) \\
&= ac + ad + bc + bd \qquad \text{(no like terms)}
\end{aligned}$$

Example 7. Multiply $(x + 3)(x - 3)$.

Solution

$$\begin{aligned}
&(x + 3)(x - 3) \\
&= x(x - 3) + 3(x - 3) \\
&= x^2 - 3x + 3x - 9 \\
&= x^2 - 9
\end{aligned}$$

Example 8. Multiply $(a - 5)(a + 5)$.

Solution

$$\begin{aligned}
&(a - 5)(a + 5) \\
&= a(a + 5) - 5(a + 5) \\
&= a^2 + 5a - 5a - 25 \\
&= a^2 - 25
\end{aligned}$$

Example 9. Multiply $(7 + y)(7 - y)$.

Solution

$$\begin{aligned}
&(7 + y)(7 - y) \\
&= 7(7 - y) + y(7 - y) \\
&= 49 - 7y + 7y - y^2 \\
&= 49 - y^2
\end{aligned}$$

Examples 9 through 11 illustrate a very interesting "special product." In each case we multiplied the *sum* of two terms times the *difference* of the same two terms and our answer was just the **difference of the squares** of the original two terms.

Example 10. Multiply $(x - 6)(x + 6)$.

Solution

$$\underbrace{(x - 6)}_{\text{difference}}\underbrace{(x + 6)}_{\text{sum}}$$

$$\begin{aligned}
&= x(x + 6) - 6(x + 6) \\
&= x^2 + 6x - 6x - 36 \\
&= x^2 - 36 \qquad \text{(difference of the squares)}
\end{aligned}$$

Again, this fact should be noted, but you should continue to use the distributive property to multiply such binomials until you become more familiar with the procedure. Let's practice a few more products of binomials.

Example 11. Multiply $(2x^2 - 3)(x - 5)$.

Solution

$$(2x^2 - 3)(x - 5)$$
$$= 2x^2(x - 5) - 3(x - 5)$$
$$= 2x^3 - 10x^2 - 3x + 15 \qquad \text{(no like terms)}$$

Example 12. Multiply $(x^3 - 4)(7x^3 - 10)$.

Solution

$$(x^3 - 4)(7x^3 - 10)$$
$$= x^3(7x^3 - 10) - 4(7x^3 - 10)$$
$$= 7x^6 - 10x^3 - 28x^3 + 40$$
$$= 7x^6 - 38x^3 + 40$$

Trial Run   *Find the product.*

————— **1.** $(2x + 3)(x + 4)$

————— **2.** $(x - 5)(3x + 2)$

————— **3.** $(3 - x)(5 + 2x)$

————— **4.** $(-2a - 1)(3a + 2)$

————— **5.** $(2x - y)(x - 3y)$

————— **6.** $(y - 5)(y + 5)$

ANSWERS
1. $2x^2 + 11x + 12$.   **2.** $3x^2 - 13x - 10$.   **3.** $15 + x - 2x^2$.   **4.** $-6a^2 - 7a - 2$.
5. $2x^2 - 7xy + 3y^2$.   **6.** $y^2 - 25$.

4.4-2   Multiplying Binomials by FOIL

We shall now discover a shortcut that we may use in performing multiplication with two binomials. In Section 4.4-1 we used the distributive property to find products such as

$$(x + 7)(2x - 3) = x(2x - 3) + 7(2x - 3)$$
$$= 2x^2 - 3x + 14x - 21$$
$$= 2x^2 + 11x - 21$$

The shortcut offered here actually involves the same process with less writing. Notice what happens if we multiply our binomials using the following steps:

> **1.** Multiply the First terms in the binomials.
> **2.** Multiply the Outer terms in the binomials.
> **3.** Multiply the Inner terms in the binomials.
> **4.** Multiply the Last terms in the binomials.
> **5.** Combine the results of those multiplications.

We shall abbreviate First by F, Outer by O, Inner by I, and Last by L.

$$\overset{\text{F} \quad\quad \text{L}}{(x + 7)(2x - 3)} = \overset{\text{F} \quad \text{O} \quad \text{I} \quad\; \text{L}}{2x^2 - 3x + 14x - 21}$$

$$= 2x^2 + 11x - 21$$

This is called the **FOIL method.** Notice that the terms in our product on the right side of the sign are exactly the same terms that we found when we did the problem the longer way. We are doing exactly the same thing, skipping one step of writing.

Example 13. Multiply $(5x - 1)(3x + 3)$ using FOIL.

Solution

$$(5x - 1)(3x + 3)$$

$$= \overset{\text{F} \quad\quad \text{L}}{(5x - 1)(3x + 3)} = \overset{\text{F} \quad\; \text{O} \quad\; \text{I} \quad\; \text{L}}{15x^2 + 15x - 3x - 3}$$

$$= 15x^2 + 12x - 3$$

Example 14. Multiply $(2x + 3)(2x - 3)$ using FOIL.

Solution

$$(2x + 3)(2x - 3)$$

$$= \overset{\text{F} \quad\quad \text{L}}{(2x + 3)(2x - 3)} = \overset{\text{F} \quad\;\; \text{O} \quad\; \text{I} \quad\; \text{L}}{4x^2 - 6x + 6x - 9}$$

$$= 4x^2 - 9$$

Example 15. Multiply $(a + 2b)(3c - d)$ using FOIL.

Solution

$$(a + 2b)(3c - d)$$

$$= \overset{\text{F} \quad\quad \text{L}}{(a + 2b)(3c - d)} = \overset{\text{F} \quad\;\; \text{O} \quad\; \text{I} \quad\;\; \text{L}}{3ac - ad + 6bc - 2bd}$$

$$= 3ac - ad + 6bc - 2bd \qquad \text{(no like terms)}$$

Remember that the FOIL method is merely a shortcut for writing down the terms of the product of *two binomials*. The process still involves the distributive property.

Trial Run *Find the product.*

_____ **1.** $(3x - 7)(x + 2)$

_____ **2.** $(x^2 - 9)(x^2 + 7)$

_____ **3.** $(7y - x)(7y + x)$

_____ **4.** $(a - b)(2c + d)$

_____ **5.** $(3x + 2)^2$

_____ **6.** $(2x^3 - 7)(x^3 - 4)$

4.4-3 Multiplying Polynomials

Do you think that you have the tools necessary for finding products such as

$$(x + 2)(x^2 + 5x + 3)?$$

If you guess that this calls for the distributive property again, you are absolutely correct. In this problem, you must multiply x times $x^2 + 5x + 3$ and then multiply 2 times $x^2 + 5x + 3$. Let's perform this multiplication.

$$\begin{aligned}
(x + 2)(x^2 + 5x + 3) &= x(x^2 + 5x + 3) + 2(x^2 + 5x + 3) \\
&= x^3 + 5x^2 + 3x + 2x^2 + 10x + 6 \\
&= x^3 + 5x^2 + 2x^2 + 3x + 10x + 6 \\
&= x^3 + 7x^2 + 13x + 6
\end{aligned}$$

> To multiply polynomials, we use the distributive property and then combine like terms.

Example 16. Multiply $(x^2 - 3)(x^2 + 2x - 4)$.

Solution

$$\begin{aligned}
&(x^2 - 3)(x^2 + 2x - 4) \\
&= x^2(x^2 + 2x - 4) - 3(x^2 + 2x - 4) \\
&= x^4 + 2x^3 - 4x^2 - 3x^2 - 6x + 12 \\
&= x^4 + 2x^3 - 7x^2 - 6x + 12
\end{aligned}$$

Example 17. Multiply $(3x^2 - 5x + 1)(2x^2 + x - 7)$.

Solution

$$\begin{aligned}
&(3x^2 - 5x + 1)(2x^2 + x - 7) \\
&= 3x^2(2x^2 + x - 7) - 5x(2x^2 + x - 7) + 1(2x^2 + x - 7) \\
&= 6x^4 + 3x^3 - 21x^2 - 10x^3 - 5x^2 + 35x + 2x^2 + x - 7 \\
&= 6x^4 + 3x^3 - 10x^3 - 21x^2 - 5x^2 + 2x^2 + 35x + x - 7 \\
&= 6x^4 - 7x^3 - 24x^2 + 36x - 7
\end{aligned}$$

Suppose that we wish to find the product of three polynomials such as

$$2x(x - 3)(5x + 1)$$

or

$$(x - 2)(x - 1)(x + 4)$$

The associative property tells us that the order in which we perform the multiplications does not matter. Let's find the easiest way.

To multiply

$$2x(x - 3)(5x + 1)$$

it is easier to multiply the binomials together first and then multiply that product by the monomial.

$$2x(x - 3)(5x + 1) = 2x(5x^2 + x - 15x - 3)$$
$$= 2x(5x^2 - 14x - 3)$$
$$= 10x^3 - 28x^2 - 6x$$

To multiply

$$(x - 2)(x - 1)(x + 4)$$

let's multiply the last two binomials and then multiply that product by the first binomial.

$$(x - 2)(x - 1)(x + 4) = (x - 2)(x^2 + 4x - x - 4)$$
$$= (x - 2)(x^2 + 3x - 4)$$
$$= x(x^2 + 3x - 4) - 2(x^2 + 3x - 4)$$
$$= x^3 + 3x^2 - 4x - 2x^2 - 6x + 8$$
$$= x^3 + x^2 - 10x + 8$$

Example 18. Multiply $-3y^2(6 - y)(5 + y)$.

Solution

$$-3y^2(6 - y)(5 + y) = -3y^2(30 + 6y - 5y - y^2)$$
$$= -3y^2(30 + y - y^2)$$
$$= -90y^2 - 3y^3 + 3y^4$$

Example 19. Multiply $(x + 1)(x - 7)^2$.

Solution

$$(x + 1)(x - 7)^2 = (x + 1)(x - 7)(x - 7)$$
$$= (x + 1)(x^2 - 7x - 7x + 49)$$
$$= (x + 1)(x^2 - 14x + 49)$$
$$= x(x^2 - 14x + 49) + 1(x^2 - 14x + 49)$$
$$= x^3 - 14x^2 + 49x + x^2 - 14x + 49$$
$$= x^3 - 13x^2 + 35x + 49$$

Example 20. Find $(a + 2)^3$.

Solution

$$(a + 2)^3 = (a + 2)(a + 2)(a + 2)$$
$$= (a + 2)(a^2 + 2a + 2a + 4)$$
$$= (a + 2)(a^2 + 4a + 4)$$
$$= a(a^2 + 4a + 4) + 2(a^2 + 4a + 4)$$
$$= a^3 + 4a^2 + 4a + 2a^2 + 8a + 8$$
$$= a^3 + 6a^2 + 12a + 8$$

Unfortunately, there are no shortcuts to use here and lots of places to make errors. You should take your time and be careful when finding products of polynomials.

Trial Run   *Find the product.*

———— **1.** $3x(x - 2)(4x - 1)$

———— **2.** $(x - 2)(x - 4)(x + 7)$

———— **3.** $-2a^2(5 - a)(6 + a)$

———— **4.** $(x - 2)(x + 3)^2$

———— **5.** $(y - 3)^3$

ANSWERS
1. $12x^3 - 27x^2 + 6x$.   **2.** $x^3 + x^2 - 34x + 56$.   **3.** $-60a^2 + 2a^3 + 2a^4$.
4. $x^3 + 4x^2 - 3x - 18$.   **5.** $y^3 - 9y^2 + 27y - 27$.

EXERCISE SET 4.4

Find the product.

_____ **1.** $(x - 2)(x + 3)$ _____ **2.** $(x + 4)(x - 1)$

_____ **3.** $(4y - 1)(3y - 5)$ _____ **4.** $(5y - 2)(3y - 1)$

_____ **5.** $(2a + 7)(-3a + 5)$ _____ **6.** $(3a + 1)(-2a + 7)$

_____ **7.** $(6x - 5)(2x + 3)$ _____ **8.** $(5x - 3)(3x + 4)$

_____ **9.** $(9 - 2y)(2 - y)$ _____ **10.** $(8 - 3y)(3 - y)$

_____ **11.** $(2a - b)(a - 5b)$ _____ **12.** $(3a - b)(a - 4b)$

_____ **13.** $(x - 3y)(2x + y)$ _____ **14.** $(x - 2y)(3x + y)$

_____ **15.** $(-x + y)(2x - 3y)$ _____ **16.** $(-x + y)(3x - 5y)$

_____ **17.** $(a + 3b)^2$ _____ **18.** $(a + 5b)^2$

_____ **19.** $(2 + 5x)(2 - 5x)$ _____ **20.** $(3 - 2x)(3 + 2x)$

_____ **21.** $(4x - y)^2$ _____ **22.** $(3x - 2y)^2$

_____ **23.** $(a - 2b)(c + d)$ _____ **24.** $(x - y)(z + 2w)$

_____ **25.** $(a^2 - 4)(a - 3)$ _____ **26.** $(x^2 - 5)(x - 2)$

_____ **27.** $(6a^2 - 1)(a^2 - 5)$ _____ **28.** $(5a^2 - 2)(a^2 - 1)$

_____ **29.** $(5a^2 - 2)^2$ _____ **30.** $(4a^2 - 1)^2$

_____ **31.** $(a^3 - 3)(a^3 + 1)$ _____ **32.** $(x^3 - 4)(x^3 + 1)$

_____ **33.** $(xy - 10)(xy + 5)$ _____ **34.** $(xy - 7)(2xy + 1)$

_____ **35.** $(ab - 2c)(ab + 2c)$ _____ **36.** $(xy - 3z)(xy + 3z)$

_____ **37.** $(3yz - 2)^2$ _____ **38.** $(4ab - 3)^2$

_____ **39.** $(6a^2 + b)(6a^2 - b)$ _____ **40.** $(5x^2 - y)(5x^2 + y)$

_____ **41.** $(x^2 + 2y^2)^2$ _____ **42.** $(3x^2 - y^2)^2$

_____ **43.** $(x - 3)(x^2 - 2x + 1)$ _____ **44.** $(x - 5)(x^2 - 6x + 9)$

_____ **45.** $(b - 5)(b^2 + 5b + 25)$ _____ **46.** $(x - 2)(x^2 + 2x + 4)$

_____ **47.** $(x^2 - 4)(x^2 + 3x + 2)$ _____ **48.** $(x^2 - 1)(x^2 - x - 2)$

_____ **49.** $(1 - x + 2x^2)(2 + 3x)$ _____ **50.** $(1 - 3x + x^2)(2 + x)$

_____ **51.** $3x(x - 2)(x - 5)$ _____ **52.** $5x(x - 3)(x + 1)$

_____ **53.** $-y^2(8 - y)(7 + y)$ _____ **54.** $-y^2(5 - y)(4 + y)$

_____ **55.** $(a - b)(a + 2b)(a - 5b)$ _____ **56.** $(a - 2b)(a + b)(a - 3b)$

_____ **57.** $(x + 2)(x - 3)^2$ _____ **58.** $(x - 4)(x - 2)^2$

_____ **59.** $(4 - y)^2(5 + y)$ _____ **60.** $(3 - y)^2(4 + y)$

_____ **61.** $(a + 2)^3$ _____ **62.** $(a + 5)^3$

_____ **63.** $(3x - y)^3$ _____ **64.** $(2x - z)^3$

CHECKUP 4.4

Find the product.

_____ **1.** $(x - 2)(x + 5)$

_____ **2.** $(2y - 3)(y - 6)$

_____ **3.** $(5 - 2y)^2$

_____ **4.** $(2a - 5b)(3a + b)$

_____ **5.** $(x^2 - 5)(x^2 + 5)$

_____ **6.** $(-2x - 3y)(x - y)$

_____ **7.** $(a - 6)(a^2 + 7)$

_____ **8.** $(xy - 3)(xy + 2)$

_____ **9.** $(x + 2)(x^2 - 2x + 4)$

_____ **10.** $-2x(3 - x)(5 + x)$

4.5 Switching from Words to Variables

Let's continue our practice of switching word expressions to variable expressions, making use of some of the skills learned in this chapter. As before, we shall always begin by deciding what the variable represents. Then we translate the word expressions in the problem into variable expressions.

Example 1. If one side of a square is s, write an expression for the perimeter. Then write an expression for the area.

Solution. We may use a picture to help here.

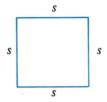

We must recall that the perimeter is just the sum of the lengths of all the sides, so our perimeter P is

$$s + s + s + s = 4s$$
$$P = 4s$$

To find the area of a square, we must multiply the length of one side times the length of another side, so our area A is

$$s \cdot s = s^2$$
$$A = s^2$$

Example 2. If the length of a rectangle is 5 feet longer than the width, write an expression for the perimeter. Then write an expression for the area.

Solution. Letting w represent the width, how could we represent the length?

$$w = \text{width}$$
$$w + 5 = \text{length}$$

Our perimeter becomes

$$w + w + (w + 5) + (w + 5) = 4w + 10$$
$$P = 4w + 10$$

The area of a rectangle is found by multiplying length times width, so here the area is

$$(w + 5)(w) = w^2 + 5w$$
$$A = w^2 + 5w$$

Example 3. If the length of a rectangle is twice the width, write an expression for the area.

Solution

$$\text{let } w = \text{width}$$
$$\text{then } 2w = \text{length}$$

and our expression for the area becomes

$$(2w)(w) = 2w^2$$
$$A = 2w^2$$

Consecutive integers are integers that are next to each other in the list of all integers. For instance,

$$22, 23, 24 \qquad \text{are consecutive integers}$$
$$-8, -7, -6 \qquad \text{are consecutive integers}$$
$$0, 1, 2 \qquad \text{are consecutive integers}$$

To get from one integer to the next consecutive integer do you see that we must add 1? So if a first integer is x, the next consecutive integer must be $x + 1$. What would be the next consecutive integer?

$$(x + 1) + 1 = x + 2$$

The next consecutive integer would be $x + 3$, and so on.

Example 4. Write an expression for the sum of two consecutive integers.

Solution

$$\text{let } x = \text{first integer}$$
$$\text{then } x + 1 = \text{next consecutive integer}$$

The sum of these two consecutive integers is

$$x + (x + 1) = 2x + 1$$

Example 5. Write an expression for the product of two consecutive integers.

Solution

$$\text{let } x = \text{first integer}$$
$$\text{then } x + 1 = \text{next consecutive integer}$$

The product of these integers is found by multiplying them together.

$$x(x + 1) = x^2 + x$$

A triangle in which one side is perpendicular to another side is called a **right triangle**. The perpendicular sides form a **right angle** (90°).

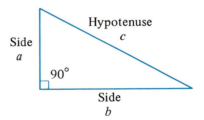

The side opposite the right angle (side c) is always called the **hypotenuse** of the right triangle. The other sides are often labeled a and b. A Greek mathematician named Pythagoras discovered that if the lengths of the sides of a right triangle are squared and then added together, their sum is equal to the square of the length of the hypotenuse.

> **Pythagorean Theorem**
> $$a^2 + b^2 = c^2$$

For instance, suppose that we find the hypotenuse of a right triangle whose sides are 3 feet and 4 feet long. Here

$a = 3$
$b = 4$

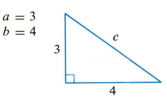

So

$$3^2 + 4^2 = c^2$$
$$9 + 16 = c^2$$
$$25 = c^2$$

What is c? What number squared is 25? Either 5 or -5, since $(5)^2 = 25$ and $(-5)^2 = 25$. Since c represents length, which is always positive, we choose

$$c = 5 \text{ feet}$$

Example 6. If one side of a right triangle is 3 times as long as the other side, write an expression for the square of the hypotenuse.

Solution

$$\text{let } a = \text{first side}$$
$$\text{then } 3a = \text{second side}$$
$$a^2 = \text{square of first side}$$
$$(3a)^2 = \text{square of second side}$$

So the square of the hypotenuse must be

$$a^2 + (3a)^2$$
$$= a^2 + 3^2a^2$$
$$= a^2 + 9a^2$$
$$= 10a^2$$

Example 7. If one side of a right triangle is 5 inches longer than the other side, write an algebraic expression for the square of the hypotenuse.

Solution

$$\text{let } a = \text{first side}$$
$$a + 5 = \text{second side}$$
$$a^2 = \text{square of first side}$$
$$(a + 5)^2 = \text{square of second side}$$

So the square of the hypotenuse is

$$a^2 + (a + 5)^2$$
$$= a^2 + (a + 5)(a + 5)$$
$$= a^2 + (a^2 + 5a + 5a + 25)$$
$$= a^2 + a^2 + 10a + 25$$
$$= 2a^2 + 10a + 25$$

Trial Run

_____ 1. If the length of a rectangle is 4 more than the width, write an expression for the perimeter of the rectangle. Then write an expression for the area.

_____ 2. One integer is 3 more than twice another. Write an expression for their sum. Then write an expression for their product.

_____ **3.** Write an expression for the product of two consecutive integers.

_____ **4.** If one side of a right triangle is 5 times as long as the other, write an expression for the square of the length of the hypotenuse.

_____ **5.** One number is two more than another. Write an expression for the sum of their squares.

_____ **6.** The length of a rectangle is 3 times the width. Write an expression for the area.

ANSWERS
1. $4x + 8$, $x^2 + 4x$.   **2.** $3x + 3$, $2x^2 + 3x$.   **3.** $x^2 + x$.   **4.** $26x^2$.   **5.** $2x^2 + 4x + 4$.
6. $3x^2$.

EXERCISE SET 4.5

_____ **1.** One integer is 12 more than another. Write an expression for their product.

_____ **2.** One integer is 3 less than 4 times another. Write an expression for their product.

_____ **3.** If the width of a rectangle is 7 less than the length, write an expression for the area.

_____ **4.** If the length of a rectangle is 10 more than the width, write an expression for the area.

_____ **5.** If the length of one side of a right triangle is 5 more than twice the length of the other side, write an expression for the square of the length of the hypotenuse.

_____ **6.** If the length of one side of a right triangle is 4 times the length of the other side, write an expression for the square of the length of the hypotenuse.

_____ **7.** Write an expression for the product of two consecutive integers.

_____ **8.** Write an expression for the sum of the squares of two consecutive integers.

_____ **9.** One number is 9 less than another. Write an expression for the square of the larger added to 5 times the smaller.

_____ **10.** One number is 2 more than another. Write an expression for the square of the smaller added to 7 times the larger.

_____ **11.** Write an expression for twice the square of an integer added to 4 times the integer.

_____ **12.** Write an expression for 3 times an integer subtracted from 5 times the square of the integer.

_____ **13.** Write an expression for the product of two consecutive integers increased by 3 times the smaller.

_____ **14.** Write an expression for the product of two consecutive integers decreased by twice the larger.

_____ **15.** Write an expression for the area of a triangle if the length of the base is 5 more than the height. $\left(A = \frac{1}{2}bh. \right)$

_____ **16.** Write an expression for the area of a triangle if the height of a triangle is 3 less than the length of the base. $\left(A = \frac{1}{2}bh. \right)$

_____ **17.** Wick plans to plow a rectangular field that is 15 rods longer than it is wide. Write an expression for the area of the field.

_____ **18.** Kaaryn is building a room on her house that is 5 feet longer than it is wide. Write an expression for the area of the room.

_____ **19.** If one number is 5 times the other, write an expression for the product of the two numbers increased by twice their sum.

_____ **20.** If one number is 2 less than twice another, write an expression for the product of the two numbers decreased by 3 times their sum.

CHECKUP 4.5

_____ 1. One number is 3 less than another. Write an expression for their product.

_____ 2. If the length of a rectangle is 3 less than 7 times the width, write an expression for the area.

_____ 3. Write an expression for the sum of the squares of two consecutive integers decreased by twice the smaller.

_____ 4. If one side of a right triangle is 7 more than the other, write an expression for the square of the hypotenuse.

_____ 5. If the base of a triangle is 4 times the height, write an expression for the area.

Summary

In this chapter we learned to name a polynomial according to the number of terms it contains.

Monomials contain *one* term.
$\qquad$ *examples:* x, $-5x$, $10x^2$, $-xy$
Binomials contain *two* terms.
$\qquad$ *examples:* $x + 2$, $6 - 5x$, $x^2 + y^2$
Trinomials contain *three* terms.
$\qquad$ *examples:* $x^2 + x - 2$, $a + b + c$

We used some very important definitions to help us work with exponents.

$$a^n = \underbrace{a \cdot a \cdot \cdots \cdot a}_{n \text{ factors}}$$

$$a^0 = 1 \qquad \text{when } a \neq 0$$

Three laws of exponents were used to find products of monomials.

First Law of Exponents: $a^m \cdot a^n = a^{m+n}$
Second Law of Exponents: $(a^m)^n = a^{m \cdot n}$
Third Law of Exponents: $(a \cdot b)^n = a^n \cdot b^n$

The distributive property was very important in multiplying polynomials. We learned a shortcut for multiplying two binomials together (FOIL) but decided that there was no such shortcut for other polynomial products.

Then we practiced switching from word expressions to variable expressions involving polynomials.

Name _____ Date _____

REVIEW EXERCISES

SECTION 4.1

1. Name the following expressions.
   _____ (a) $3a + 5b$ _____ (b) $7x - 2y + 5x$ _____ (c) $-3m^2$

2. Use exponents to rewrite each expression.
   _____ (a) $-2 \cdot x \cdot x \cdot y \cdot y \cdot y$ _____ (b) $(3a)(3a)(3a)$
   _____ (c) $5 \cdot m \cdot m - 3 \cdot n \cdot n \cdot n$

3. Write the following without exponents.
   _____ (a) $2x^4$ _____ (b) $(-5a)^3$ _____ (c) $-3m^3n^2$

4. Evaluate.
   _____ (a) $(-3)^3$ _____ (b) $-2a^2$, when $a = 1$
   _____ (c) $(5m^2n)^0$ when $m = 3$ and $n = -1$

Simplify.

_____ 5. $-5(2a + b) + 3(a + b)$

_____ 6. $2(x^2 - x - 9) - (5x^2 + 2x - 4)$

SECTION 4.2

_____ 7. $x^7 \cdot x^3$ _____ 8. $(2a^3)(-3a)(a^4)$

_____ 9. $(9xy^6)(-2x^4y^2)$ _____ 10. $(m^2)^3$

_____ 11. $(-3y)^4$ _____ 12. $(x^3y^2z)^2$

_____ 13. $3a(a^4b^2)^3$ _____ 14. $-x(3xy^2)^3 + 3x^2y^4(2xy)^2$

SECTION 4.3

_____ 15. $8(5x - 3)$ _____ 16. $7y(3y + 2)$

_____ 17. $3m^2(m^2 - 4mn - 5n^2)$ _____ 18. $5xy(7x - 11y)$

_____ 19. $-a^2b(6 - 3ab + 4a^2b^2)$ _____ 20. $4xy(3x - y) - 5xy(2x - y)$

Find the products.

SECTION 4.4

_____ 21. $(x + 9)(x - 8)$ _____ 22. $(3y - 2)(4y + 1)$

_____ 23. $(2a - 3b)(3a - b)$ _____ 24. $(x - 5y)^2$

_____ 25. $(m^2 - 3)(m^2 + 5)$ _____ 26. $(2xy - 1)^2$

_____ 27. $(y - 1)(y^2 - 3y - 4)$ _____ 28. $2x(x - 3)(x + 3)$

_____ 29. $(a - 2)(a - 3)(a + 1)$ _____ 30. $(2 - y)^2(3 + y)$

SECTION 4.5

_____ 31. One number is 8 more than twice the second. Write an expression for 5 times the smaller number subtracted from the square of the larger number.

_____ 32. Joyce is adding a family room to her house. If the width of the room is to be 10 feet less than the length, write an expression for the area of the room.

5 Factoring Polynomials

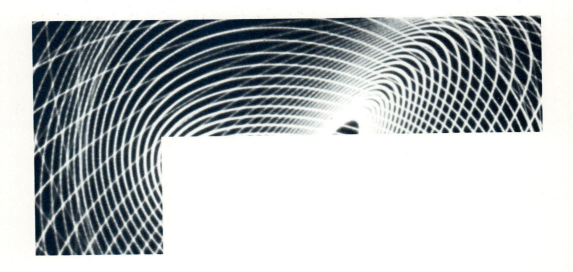

In Chapter 4 we learned how to multiply polynomials. In this chapter we learn how to go backwards and reverse what we did in Chapter 4. Recall that **factors** are quantities being multiplied together to give a product. Our purpose in this chapter is to learn procedures for figuring out what factors were multiplied together to give a certain polynomial.

In this chapter we learn how to

1. Find the common monomial factor for a polynomial.
2. Factor a difference of two squares.
3. Factor a general trinomial.
4. Factor a four-term polynomial by grouping.

5.1 Looking for Common Factors

Factoring is a process by which we try to rewrite a polynomial as a *product*. It is very much a trial-and-error, educated-guess procedure, but you will become very good at it with practice.

In order to figure out what quantities would multiply together to give a certain product, you must be very familiar with the multiplication of polynomials.

5.1-1 Looking for Common Numerical Factors

Recall the kinds of products we found when we multiplied quantities like

$$3(a + b)$$
$$5(x - 3)$$
$$-7(2x - y)$$
$$8(x^2 + 3x - 5)$$

Here we were multiplying constants times polynomials and the answers were found using the distributive property.

$$3(a + b) = 3a + 3b$$
$$5(x - 3) = 5x - 15$$
$$-7(2x - y) = -14x + 7y$$
$$8(x^2 + 3x - 5) = 8x^2 + 24x - 40$$

In words, we say that

$3a + 3b$ is the product of the factors 3 and $(a + b)$
$5x - 15$ is the product of the factors 5 and $(x - 3)$
$-14x + 7y$ is the product of the factors -7 and $(2x - y)$
$8x^2 + 24x - 40$ is the product of the factors 8 and $(x^2 + 3x - 5)$

Can you rewrite the following expressions as products of factors?

$$2x + 2y$$
$$5x - 10y$$
$$6x + 3$$
$$9z^2 + 27z - 18$$

In the first expression

$$2x + 2y$$

we notice that each term is 2 times some other factor. Would you agree that we could rewrite

$$2x + 2y = 2(x + y)?$$

In factoring, you can *always* decide whether your "guess" is correct. All you must do is multiply your factors together and see if your answer is the same as the original polynomial. Since we know by the distributive property that

$$2(x + y) = 2x + 2y$$

we know we have factored $2x + 2y$ correctly.

To factor $5x - 10y$, notice that 5 is a factor of both terms. We write

$$5x - 10y = 5(x - 2y)$$

Checking again using the distributive property on the right-hand side, we know that this is correct.

To factor $6x + 3$, we see that 3 is a factor of both terms and we write

$$6x + 3 = 3(2x + 1)$$

Checking again, we know that this is correct.

What about $9z^2 + 27z - 18$? We notice that 3 is a factor of each term, but 9 is also a factor of each term. Which do we choose? We must agree to choose the *largest possible* common factor. In this case we choose 9 and write

$$9z^2 + 27z - 18 = 9(z^2 + 3z - 2)$$

The process discussed here is called "factoring out the largest common numerical factor," and it is accomplished using the following steps.

To factor out the largest common numerical factor:

---

1. Look for the largest numerical factor common to all terms in the original polynomial.
2. Use the distributive property to rewrite the original polynomial as a product of that numerical factor and a new polynomial factor.
3. Check your factoring by multiplying your factors to see if the product is the original polynomial.

Example 1. Factor $100x^2 - 100y^2$.

  Solution. $100x^2 - 100y^2 = 100(x^2 - y^2)$.

Example 2. Factor $14x^2 + 7x - 70$.

  Solution. $14x^2 + 7x - 70 = 7(2x^2 + x - 10)$.

Example 3. Factor $-5x^2 + 10$.

  Solution. $-5x^2 + 10 = -5(x^2 - 2)$.

Notice from Example 3 that if the first numerical coefficient in the original polynomial is negative, we usually agree to factor out a negative common factor.

Example 4. Factor $-2x^2 + 28$.

  Solution. $-2x^2 + 28 = -2(x^2 - 14)$.

Example 5. Factor $19x - 19$.

  Solution. $19x - 19 = 19(x - 1)$.

Example 6. Factor $-7x + 7$.

  Solution. $-7x + 7 = -7(x - 1)$.

If you always remember to check your factors by the distributive property, you will not make errors in signs or forget to include a 1 where necessary in the new polynomial. Errors like these

$$-7x + 7 = -7(x + 1) \qquad wrong$$
$$-7x + 7 = -7(x - 1) \qquad right$$
$$2x^2 + 2 = 2(x^2) \qquad wrong$$
$$2x^2 + 2 = 2(x^2 + 1) \qquad right$$
$$3x + 6y = 3(x + 6y) \qquad wrong$$
$$3x + 6y = 3(x + 2y) \qquad right$$

are very common, but they can be avoided if you always multiply your factors to see if their product is the original polynomial.

5.1-2 Looking for Common Monomial Factors

The process used to factor polynomials such as

$$3x^2 + 6x$$
$$10xy - 2x$$
$$98a^2 - 28a$$
$$x^5 + 2x^3 + x^2$$

is similar to the process used in factoring out common numerical factors. Once again we are looking for the factors common to every term in the polynomial. Those common factors may contain a numerical part and/or a variable part.

In the product $3x^2 + 6x$, we notice that both terms contain a factor of 3 *and* a factor of x. Would you agree that we may write

$$3x^2 + 6x = 3x(x + 2)?$$

Does the product of the factors on the right equal the polynomial on the left? Certainly; so we have factored correctly.

Consider $10xy - 2x$. Both terms contain a factor of 2 *and* a factor of x, so we write

$$10xy - 2x = 2x(5y - 1)$$

Checking again by multiplying our factors, we find that this is correct.

The factors of $98a^2 - 28a$ may not be immediately obvious. We notice that 2 and a are factors contained in both terms, so we write

$$2a(49a - 14)$$

But then we look closely at $49a - 14$ and realize that 7 is a factor of both terms. It is not too late to factor out that 7.

$$98a^2 - 28a = 2a(49a - 14)$$
$$= 2a \cdot 7(7a - 2)$$
$$= 14a(7a - 2)$$

Indeed, $14a$ could have been factored out at the start, but you were not wrong if you did not spot it right away. It is never too late to remove a common factor.

In the polynomial $x^5 + 2x^3 + x^2$ we see that each term contains a power of x. The highest power of x that can be factored out is actually the lowest power contained in any one of the terms. In this case the lowest power is x^2, so

$$x^5 + 2x^3 + x^2 = x^2(\underline{\hspace{1cm}} + \underline{\hspace{1cm}} + \underline{\hspace{1cm}})$$

Now we must "fill in the blanks," using the distributive property and the First Law of Exponents.

We know that x^2 times the quantity in the first blank must equal x^5:

$$x^2 \cdot \underline{\hspace{1cm}} = x^5$$

so the first blank should be filled by x^3.

We know that x^2 times the quantity in the second blank must equal $2x^3$:

$$x^2 \cdot \underline{\hspace{1cm}} = 2x^3$$

so the second blank should be filled by $2x$.

We know that x^2 times the quantity in the third blank must equal x^2:

$$x^2 \cdot \underline{\hspace{1cm}} = x^2$$

so the third blank should be filled by 1.

$$x^5 + 2x^3 + x^2 = x^2(x^3 + 2x + 1)$$

Checking by the distributive property, we have

$$x^2(x^3 + 2x + 1) = x^2 \cdot x^3 + x^2 \cdot 2x + x^2 \cdot 1 = x^5 + 2x^3 + x^2$$

and our factors are correct.

Example 7. Factor $3x^2 - 15x$.

Solution. $3x^2 - 15x = 3x(x - 5)$.

Example 8. Factor $-x^3 + 3x^2 - 4x$.

Solution. $-x^3 + 3x^2 - 4x = -x(x^2 - 3x + 4)$.

Example 9. Factor $8x^7 - 2x^6 + 10x^5 - 2x^3$.

Solution. $8x^7 - 2x^6 + 10x^5 - 2x^3 = 2x^3(4x^4 - x^3 + 5x^2 - 1)$.

If all the terms of a polynomial contain powers of two (or more) variables, we must factor out some power of each of those variables. We inspect the terms, factor out the lowest power of each variable, and then proceed to "fill in the blanks" using the distributive property and the First Law of Exponents.

To factor $x^3y^4 + x^2y^3 + x^2y$, we see that the lowest power of x is x^2 and the lowest power of y is y^1. We write

$$x^3y^4 + x^2y^3 + x^2y = x^2y(\underline{\hspace{1cm}} + \underline{\hspace{1cm}} + \underline{\hspace{1cm}})$$

For the first blank, $x^2y \cdot \underline{\hspace{1cm}} = x^3y^4$: we need xy^3
For the second blank, $x^2y \cdot \underline{\hspace{1cm}} = x^2y^3$: we need y^2
For the third blank, $x^2y \cdot \underline{\hspace{1cm}} = x^2y$: we need 1

So

$$x^3y^4 + x^2y^3 + x^2y = x^2y(xy^3 + y^2 + 1)$$

Again we check our factors by multiplying:

$$x^2y(xy^3 + y^2 + 1) = x^2y \cdot xy^3 + x^2y \cdot y^2 + x^2y \cdot 1 = x^3y^4 + x^2y^3 + x^2y$$

Example 10. Factor $3a^2x^2 - 5a^2x$.

Solution. $3a^2x^2 - 5a^2x = a^2x(3x - 5)$.

Example 11. Factor $x^7y^8 - 2x^5y^6$.

Solution. $x^7y^8 - 2x^5y^6 = x^5y^6(x^2y^2 - 2)$.

Example 12. Factor $5x^3y^2z^2 - 10x^2y^2z^2 + 25x^2y^3z$.

Solution. $5x^3y^2z^2 - 10x^2y^2z^2 + 25x^2y^3z = 5x^2y^2z(xz - 2z + 5y)$.

Trial Run   *Factor.*

_____ **1.** $15x^2 - 10x - 50$

_____ **2.** $-3x^2 + 9$

_____ **3.** $13y - 13$

_____ **4.** $-5x - 5$

_____ **5.** $4x^2 - 16x$

_____ **6.** $-x^3 + 2x^2 - 10x$

_____ **7.** $x^5y^4 + 5x^4y^3$

_____ **8.** $2x^2y^2z^3 - 6x^2y^3z^2 + 4x^3y^2z^2$

ANSWERS
1. $5(3x^2 - 2x - 10)$.   **2.** $-3(x^2 - 3)$.   **3.** $13(y - 1)$.   **4.** $-5(x + 1)$.   **5.** $4x(x - 4)$.
6. $-x(x^2 - 2x + 10)$.   **7.** $x^4y^3(xy + 5)$.   **8.** $2x^2y^2z^2(z - 3y + 2x)$.

The process that we have just learned is called "factoring out a common monomial factor." Although we shall discover other types of factoring, the common monomial factor is always the very first thing we look for when factoring polynomials.

EXERCISE SET 5.1

Factor out the common monomial factor.

_____ 1. $7x + 14$

_____ 2. $5x - 15$

_____ 3. $21x - 28$

_____ 4. $9x + 21$

_____ 5. $42 - 6y$

_____ 6. $45 - 9y$

_____ 7. $6a^2 - 36$

_____ 8. $8a^2 - 40$

_____ 9. $-12x + 18y$

_____ 10. $-15x + 20y$

_____ 11. $30x^2 + 15y^2$

_____ 12. $60x^2 - 12y^2$

_____ 13. $11a^3 + 44b^3$

_____ 14. $13a^3 - 39b^3$

_____ 15. $7x^2 - 21x - 7$

_____ 16. $9x^2 + 27x + 9$

_____ 17. $10y^2 + 15y - 35$

_____ 18. $12y^2 - 18y - 33$

_____ 19. $28 - 7a + 35a^2$

_____ 20. $33 - 22a + 55a^2$

_____ 21. $50x^2 - 10xy + 20y^2$

_____ 22. $36x^2 - 24xy + 48y^2$

_____ 23. $2x^3 - 6x^2 + 4x + 8$

_____ 24. $8x^3 - 16x^2 + 24x + 8$

_____ 25. $x^2 + x$

_____ 26. $2x^2 + x$

_____ 27. $-y^3 + y^2 - 2y$

_____ 28. $-y^3 - y^2 + 4y$

_____ 29. $2b^3 - 6b^2 + 4b$

_____ 30. $5b^3 - 10b^2 + 25b$

_____ 31. $7b^3 + 2b^5 - 3b^7$

_____ 32. $9b^4 + 4b^6 - 2b^8$

_____ 33. $3x^3 - 6x^2y + 12xy^2$

_____ 34. $5x^3 + 15x^2y + 55xy^2$

_____ 35. $-a^4b + 6a^3b^2 - 5a^2b^3$

_____ 36. $-a^5b - 4a^4b^2 + 5a^3b^3$

_____ 37. $4x^3y^3 - 20x^2y^2 + 2xy$

_____ 38. $9x^4y^4 + 18x^3y^3 - 3x^2y^2$

_____ 39. $x^3y^2z - 2x^4y^3z + x^5y^4z$

_____ 40. $x^6y^4z^2 + 5x^4y^2z^2 - x^2z^2$

_____ 41. $63x^2 + 72x$

_____ 42. $36x^3 + 45x^2$

_____ 43. $100a^3 - 60b^3$

_____ 44. $75a^3 - 100b^3$

_____ 45. $16b^3 + 64b^4$

_____ 46. $12b^5 + 60b^7$

_____ 47. $6a^3 - 42a^2 - 78a$

_____ 48. $9a^3 - 72a^2 + 81a$

_____ 49. $5x^3y - 80xy^3$

_____ 50. $7x^4y^2 - 56x^3y^3$

_____ 51. $-7b^2 - 42b^3 + 35b^4$

_____ 52. $-12b^4 + 36b^5 - 48b^6$

_____ 53. $9x^3y - 12x^2y^2 + 20xy^3$

_____ 54. $27x^3y + 45x^2y^2 - 63xy^3$

_____ 55. $72a^4 - 48a^3 + 8a^2$

_____ 56. $72a^5 - 54a^4 - 6a^3$

_____ 57. $12x^3y - 8x^2y^2 - 4xy^3$

_____ 58. $18x^3y - 15x^2y^2 - 3xy^3$

_____ 59. $9x^4yz + 18x^3y^2z - 27x^2y^3z - 45xy^4z$

_____ 60. $35xy^4z - 28xy^3z - 14xy^2z + 7xyz$

CHECKUP 5.1

Factor.

_____ **1.** $3x + 15$

_____ **2.** $35 - 14y$

_____ **3.** $-x^2 + x$

_____ **4.** $9x^2 - 18x + 27$

_____ **5.** $15x^2 - 10xy + 25y^2$

_____ **6.** $8a^3 - 16a^2 - 40a$

_____ **7.** $10x^3y - 26x^2y^2 + 14xy^3$

_____ **8.** $9a^3b^3 - 27a^2b^2 + 3ab$

_____ **9.** $-8m^3 - 24m$

_____ **10.** $x^4yz - x^3y^2z + x^2y^3z$

5.2 Factoring the Difference of Two Squares

In Chapter 4 we noted that whenever we multiplied the sum of two terms times the difference of those same two terms, the product was the difference of the squares of the two terms. For instance,

$$(x + 3)(x - 3) = x^2 - 3x + 3x - 9 = x^2 - 9$$
$$(y - 5)(y + 5) = y^2 + 5y - 5y - 25 = y^2 - 25$$
$$(a + 1)(a - 1) = a^2 - a + a - 1 = a^2 - 1$$
$$(x - yz)(x + yz) = x^2 + xyz - xyz - y^2z^2 = x^2 - y^2z^2$$

This pattern allows us to find products of sums and differences of the same two terms without actually performing the term-by-term multiplication. For instance, to multiply a sum and difference such as

$$(x + 7)(x - 7)$$

we note that the first term is x and the second term is 7. The square of the first term is x^2 and the square of the second term is 49. So

$$(x + 7)(x - 7) = x^2 - 49$$

Example 1. Find $(x - 10)(x + 10)$.

Solution. $(x - 10)(x + 10) = x^2 - 100$.

Example 2. Find $(ab + c)(ab - c)$.

Solution. $(ab + c)(ab - c) = (ab)^2 - c^2$
$$= a^2b^2 - c^2$$

Example 3. Find $(x^2 - 1)(x^2 + 1)$.

Solution. $(x^2 - 1)(x^2 + 1) = (x^2)^2 - 1$
$$= x^4 - 1$$

In general, we notice that for any expressions A and B,

$$(A + B)(A - B) = A^2 - B^2$$

In this chapter we are concerned with *reversing* the operation of multiplication. Remember that factoring is a process by which we try to rewrite a polynomial as a product of factors.

In particular, we would now like to look at polynomials that are differences of two squares and rewrite those polynomials as products of sums and differences of two terms.

Let's try to factor

$$x^2 - 16$$

Is this a difference of two squares? Is x^2 a square? Of course. Is 16 a square? Yes, 16 is 4^2. You should agree that

$$x^2 - 16 = (x + 4)(x - 4)$$

Once again, we can check our factors by multiplication.

$$(x + 4)(x - 4) = x^2 - 4x + 4x - 16 = x^2 - 16$$

Let's factor

$$25a^2 - 121$$

Is this a difference of two squares? Is $25a^2$ a square? Yes, $25a^2$ is $(5a)^2$. Is 121 a square? Yes, 121 is 11^2. So we factor

$$25a^2 - 121 = (5a + 11)(5a - 11)$$

The difference of two squares is not difficult to spot. You must look for two terms with a minus sign in between. Then you must decide whether each term is a square of some expression. If so, one factor will be just the *sum* of those expressions and the other factor will be the *difference* of those expressions.

> **The Difference of Two Squares**
>
> $$A^2 - B^2 = (A + B)(A - B)$$

Example 4. Factor $25 - x^2$.

Solution. $25 - x^2 = (5 + x)(5 - x)$.

Example 5. Factor $x^2 - y^2$.

Solution. $x^2 - y^2 = (x + y)(x - y)$.

Example 6. Factor $100a^2 - 81y^2$.

Solution. $100a^2$ is $(10a)^2$ and $81y^2$ is $(9y)^2$:

$$100a^2 - 81y^2 = (10a + 9y)(10a - 9y)$$

Example 7. Factor $5x^2 - 20$.

Solution. $5x^2 - 20$ is a difference of two terms, but those terms are *not* squares. We must not forget to look for a common monomial factor first. Both terms contain a factor of 5.

$$5x^2 - 20 = 5(x^2 - 4)$$

Now we see that $x^2 - 4$ *is* a difference of two squares, and we factor it.

$$5x^2 - 20 = 5(x^2 - 4)$$
$$= 5(x + 2)(x - 2)$$

Example 8. Factor $9a - ab^2$.

Solution

$$9a - ab^2 = a(9 - b^2)$$
$$= a(3 - b)(3 + b)$$

Example 9. Factor $3x^3 - 3x$.

Solution

$$3x^3 - 3x = 3x(x^2 - 1)$$
$$= 3x(x + 1)(x - 1)$$

Removing the common monomial factor turns a problem that looks complicated into a problem that looks very familiar. You should be on the lookout for common monomials whenever you are asked to factor a polynomial completely.

Do not think that you should skip the step in which you factor out the common monomial factor. Even mathematicians who have been factoring for years would not dream of trying to jump from the original problem to the completely factored form. It is too easy to make mistakes keeping so much in your head.

Example 10. Factor $99x^2y^2 - 44x^2z^2$ completely.

Solution

$$99x^2y^2 - 44x^2z^2 = 11x^2(9y^2 - 4z^2)$$
$$= 11x^2(3y + 2z)(3y - 2z)$$

Example 11. Factor $100a^4b^2 - 4a^2$ completely.

   Solution

$$100a^4b^2 - 4a^2 = 4a^2(25a^2b^2 - 1)$$
$$= 4a^2(5ab + 1)(5ab + 1)$$

Example 12. Factor $x^2 + 4y^2$ completely.

   Solution. $x^2 + 4y^2$ is a *sum* of two squares, rather than a difference of two squares. It cannot be factored. So

$$x^2 + 4y^2 = x^2 + 4y^2$$

It is worthwhile to note here that a *sum of two squares* cannot be factored in the context of this course. When a polynomial cannot be factored we sometimes say that the polynomial is **prime**.

To assist us in recognizing numbers that are squares, the following table of squares of the first 20 natural numbers is included here for reference.

n	n^2	n	n^2
1	1	11	121
2	4	12	144
3	9	13	169
4	16	14	196
5	25	15	225
6	36	16	256
7	49	17	289
8	64	18	324
9	81	19	361
10	100	20	400

Trial Run   *Factor completely.*

_____ **1.** $x^2 - 25$

_____ **2.** $9x^2 - 49$

_____ **3.** $32 - 2x^2$

_____ **4.** $x^2 + y^2$

_____ **5.** $12a^2 - 75b^2$

_____ **6.** $45x^3 - 5x$

ANSWERS
1. $(x - 5)(x + 5)$.   **2.** $(3x - 7)(3x + 7)$.   **3.** $2(4 - x)(4 + x)$.   **4.** Prime.
5. $3(2a - 5b)(2a + 5b)$.   **6.** $5x(3x - 1)(3x + 1)$.

So far we have learned to

1. Look first for a common monomial factor.
2. Look for a difference of two squares.

in seeking the completely factored form for a polynomial.

EXERCISE SET 5.2

Factor the following binomials completely.

_____ **1.** $x^2 - 25$ _____ **2.** $x^2 - 16$

_____ **3.** $9y^2 - 1$ _____ **4.** $144y^2 - 1$

_____ **5.** $49 - a^2$ _____ **6.** $36 - a^2$

_____ **7.** $4a^2 - 121$ _____ **8.** $9a^2 - 100$

_____ **9.** $25b^2 - 169$ _____ **10.** $49b^2 - 64$

_____ **11.** $x^2 - y^2$ _____ **12.** $a^2 - b^2$

_____ **13.** $a^2 - 36b^2$ _____ **14.** $a^2 - 81b^2$

_____ **15.** $4x^2 - 225y^2$ _____ **16.** $9x^2 - 25y^2$

_____ **17.** $m^2n^2 - 144$ _____ **18.** $x^2y^2 - 169$

_____ **19.** $9a^2b^2 - c^2$ _____ **20.** $16a^2b^2 - c^2$

_____ **21.** $25x^2y^2 - 16z^2$ _____ **22.** $36x^2y^2 - 169z^2$

_____ **23.** $x^4 - 81$ _____ **24.** $x^4 - 16$

_____ **25.** $25x^4 - 121$ _____ **26.** $9x^4 - 64$

_____ **27.** $225 - a^4b^4$ _____ **28.** $289 - a^4b^4$

_____ **29.** $3x^2 - 75$ _____ **30.** $2x^2 - 32$

_____ **31.** $11x^2 - 44y^2$ _____ **32.** $5x^2 - 45y^2$

_____ **33.** $16y - 9x^2y$ _____ **34.** $25y - 4x^2y$

_____ **35.** $6m^3 - 150m$ _____ **36.** $7m^3 - 112m$

_____ **37.** $49a^2b - 4b^3$ _____ **38.** $64a^2b - 9b^3$

_____ **39.** $2x^2y^2 - 72y^4$ _____ **40.** $3x^2y^2 - 48y^4$

_____ **41.** $3m^4 - 243$ _____ **42.** $5m^4 - 80$

_____ **43.** $2x^3y - 8xy^3$ _____ **44.** $3x^3y - 27xy^3$

_____ **45.** $27x^3y^3 - 12xy$ _____ **46.** $50x^3y^3 - 2xy$

_____ **47.** $a^6 - 256a^2$ _____ **48.** $a^6 - 81a^2$

_____ **49.** $289x^2y^2z^2 - z^2$ _____ **50.** $256x^2y^2z^2 - z^2$

Name _____ Date _____

CHECKUP 5.2

Factor completely.

_____ **1.** $x^2 - 64$

_____ **2.** $25 - y^2$

_____ **3.** $9a^2 - 100$

_____ **4.** $x^2 - 49y^2$

_____ **5.** $16x^4 - 81$

_____ **6.** $3x^2 - 75y^2$

_____ **7.** $7x^2y^2 - 28y^4$

_____ **8.** $5m^4 - 405$

_____ **9.** $4x^3y - 400xy^3$

_____ **10.** $3a - 243a^3$

5.3 Factoring Trinomials (Leading Coefficient of 1)

Recall from Chapter 4 the FOIL method, which we used to multiply binomials.

$$\begin{array}{cc} \text{F} \quad \text{L} & \text{F} \quad \text{O} \quad \text{I} \quad \text{L} \\ (x + 3)(x + 4) = x^2 + 4x + 3x + 12 \\ \text{I} \\ \text{O} \end{array}$$

$$= x^2 + 7x + 12$$

$$\begin{array}{cc} \text{F} \quad \text{L} & \text{F} \quad \text{O} \quad \text{I} \quad \text{L} \\ (x + 5)(x - 2) = x^2 - 2x + 5x - 10 \\ \text{I} \\ \text{O} \end{array}$$

$$= x^2 + 3x - 10$$

$$\begin{array}{cc} \text{F} \quad \text{L} & \text{F} \quad \text{O} \quad \text{I} \quad \text{L} \\ (a - 6)(a - 1) = a^2 - a - 6a + 6 \\ \text{I} \\ \text{O} \end{array}$$

$$= a^2 - 7a + 6$$

It seems that whenever we multiply two binomials each of the form "coefficient times variable, plus or minus a constant," our product usually turns out to be a trinomials of the form "coefficient times variable squared, plus or minus coefficient times variable, plus or minus a constant."

In this section we must learn how to decide what binomials were multiplied together to give trinomials of the type just described. We must learn how to factor trinomials such as

$$x^2 + 4x + 3$$
$$x^2 - 6x + 5$$
$$x^2 - 2x - 3$$
$$y^2 + 6y - 7$$

5.3-1 Factoring Trinomials with Positive Last Term

Notice in our FOIL examples that the first term in our trinomial (the F term) came from multiplying the first terms of the binomials. The last term in the trinomial (the L term) came from multiplying the second terms of the binomials. The middle term in the trinomial was the sum of the O and I terms.

Keeping that in mind, let's try to factor

$$x^2 + 4x + 3$$

We are looking for two binomials

$$(\quad)(\quad)$$

and since the first term of our trinomial is x^2, we can be quite sure of the first terms of our binomials.

$$(x \quad)(x \quad)$$

Since the last term of our trinomial is 3, we know that the second terms of our binomials must be 1 and 3 (since $1 \cdot 3 = 3$) or -1 and -3 (since $-1 \cdot -3 = 3$). That gives us the two possibilities:

$$(x + 1)(x + 3)$$
$$(x - 1)(x - 3)$$

Do both of these give the correct product?

$$(x + 1)(x + 3) = x^2 + 3x + x + 3$$
$$= x^2 + 4x + 3$$

$$(x - 1)(x - 3) = x^2 - 3x - x + 3$$
$$= x^2 - 4x + 3$$

The trinomial we wished to factor was

$$x^2 + 4x + 3$$

so the first pair of binomials is correct but the second pair is not. Thus, in factored form, we write

$$x^2 + 4x + 3 = (x + 3)(x + 1)$$

To factor the trinomial

$$x^2 - 6x + 5$$

we carry out the same process.

$$x^2 - 6x + 5 = (x \qquad)(x \qquad)$$

Now we seek numbers that multiply to give 5. The only possibilities are 1 and 5 *or* -1 and -5, so the possible factors are

$$(x + 1)(x + 5)$$
$$(x - 1)(x - 5)$$

Testing each pair, we have

$$(x + 1)(x + 5) = x^2 + 5x + x + 5 = x^2 + 6x + 5$$
$$(x - 1)(x - 5) = x^2 - 5x - x + 5 = x^2 - 6x + 5$$

So the second pair of factors is correct; the first pair is not correct. We write

$$x^2 - 6x + 5 = (x - 1)(x - 5)$$

Because of the *commutative property* for multiplication, we know that it does not matter in what order we write the factors.

$$x^2 - 6x + 5 = (x - 1)(x - 5)$$
$$x^2 - 6x + 5 = (x - 5)(x - 1)$$

Let's try to factor the trinomial

$$3 + 4x + x^2$$

Although the terms in this trinomial are in reverse order from our other examples, the principle is the same. We know that our possible factors must be

$$(3 + x)(1 + x)$$
$$(3 - x)(1 - x)$$

Testing each pair, we have

$$(3 + x)(1 + x) = 3 + 3x + 1x + x^2 = 3 + 4x + x^2$$
$$(3 - x)(1 - x) = 3 - 3x - 1x + x^2 = 3 - 4x + x^2$$

Since the first pair gives the correct product, we write

$$3 + 4x + x^2 = (3 + x)(1 + x)$$

We can always decide whether a pair of factors is correct by multiplying them together. If their product is the original trinomial, we have factored correctly. If not, we must try another pair.

Example 1. Factor $x^2 + 8x + 7$.

Solution. The possibilities are

$$(x + 1)(x + 7)$$
$$(x - 1)(x - 7)$$

and, after multiplying, we see that the correct pair is the first pair. So

$$x^2 + 8x + 7 = (x + 1)(x + 7)$$

Example 2. Factor $5 - 6x + x^2$.

Solution. The possibilities are

$$(5 - x)(1 - x)$$
$$(5 + x)(1 + x)$$

and the first pair is correct, so

$$5 - 6x + x^2 = (5 - x)(1 - x)$$

From these examples we should notice that in factoring trinomials:

> 1. If the numerical coefficients of the first and last terms of the trinomial are *positive* and the coefficient of the middle term is *positive,* the numerical coefficients of both terms of both binomial factors will be positive.
> 2. If the numerical coefficients of the last terms of the trinomial are *positive* but the coefficient of the middle term is *negative,* the numerical coefficient of the first term of both binomials will be positive but the numerical coefficient of the second term of both binomials will be negative.

Example 3. Factor $x^2 - 12x + 11$.

Solution. Since the first and last terms are positive, but the coefficient of the middle term $(-12x)$ is negative, the factors must be

$$x^2 - 12x + 11 = (x - 11)(x - 1)$$

Example 4. Factor $x^2 + 2x + 1$.

Solution. Since the first and last terms are positive and the middle term $(+2x)$ has a positive coefficient, the factors must be

$$x^2 + 2x + 1 = (x + 1)(x + 1)$$

We may also write

$$x^2 + 2x + 1 = (x + 1)^2$$

Checking your factors by multiplying is a *must.* Suppose that we consider a few more problems where a choice must be made from more possibilities.

To factor the trinomial

$$x^2 + 7x + 12$$

we notice that both terms of both binomials must be positive, so we start out with

$$x^2 + 7x + 12 = (x + \quad)(x + \quad)$$

There are several possibilities:

$$(x + 1)(x + 12)$$
$$(x + 6)(x + 2)$$
$$(x + 3)(x + 4)$$

Multiplying out each of these pairs of factors, we find that only the third pair gives us the correct middle term of $+7x$. We conclude that

$$x^2 + 7x + 12 = (x + 3)(x + 4)$$

After some practice, you will learn to leave out the possibilities that do not seem likely. You need not list all the possibilities, but be sure to check your factors before deciding that they are correct.

Example 5. Factor $x^2 - 9x + 20$.

Solution. $x^2 - 9x + 20 = (x - 5)(x - 4)$.

Example 6. Factor $x^2 + 2xy + 20y^2$. ↓error

Solution. $x^2 + 12xy + 20y^2 = (x + 10y)(x + 2y)$.

Example 7. Factor $8 + 6x + x^2$.

Solution. $8 + 6x + x^2 = (4 + x)(2 + x)$.

Example 8. Factor $15 - 8y + y^2$.

Solution. $15 - 8y + y^2 = (5 - y)(3 - y)$.

5.3-2 Factoring Trinomials with Negative Last Term

What happens when the last term of our trinomial is *negative?* Let's try to factor

$$x^2 - 2x - 3$$

Since the first term is x^2, we may write

$$x^2 - 2x - 3 = (x \qquad)(x \qquad)$$

and we must consider the possibilities for the second term in each binomial, knowing that their product must be the constant term of -3. The possibilities are -1 and 3 (since $-1 \cdot 3 = -3$) *or* 1 and -3 (since $1 \cdot -3 = -3$). Our possible pairs of factors, therefore, are

$$(x - 1)(x + 3)$$
$$(x + 1)(x - 3)$$

and we must multiply to decide which is the correct pair.

$$(x - 1)(x + 3) = x^2 + 3x - x - 3 = x^2 + 2x - 3$$
$$(x + 1)(x - 3) = x^2 - 3x + x - 3 = x^2 - 2x - 3$$

Since the product of the second pair is the original trinomial, we know that

$$x^2 - 2x - 3 = (x + 1)(x - 3)$$

Let's factor the trinomial

$$y^2 + 6y - 7$$

Once again we decide on the first terms of the binomials

$$y^2 + 6y - 7 = (y \qquad)(y \qquad)$$

and consider the possible second terms. They must be -1 and 7 (since $-1 \cdot 7 = -7$) *or* 1 and -7 (since $1 \cdot -7 = -7$). We look at the possible pairs of factors.

$$(y - 1)(y + 7)$$
$$(y + 1)(y - 7)$$

After multiplying both pairs, we see that the first pair gives the desired product and we conclude that

$$y^2 + 6y - 7 = (y - 1)(y + 7)$$

Example 9. Factor $x^2 + 10x - 11$.

    Solution. The possibilities are

$$(x - 1)(x + 11)$$
$$(x + 1)(x - 11)$$

After multiplying, we see that the first pair is correct.

$$x^2 + 10x - 11 = (x - 1)(x + 11)$$

Example 10. Factor $x^2 - 4x - 5$.

    Solution. The possibilities are

$$(x - 1)(x + 5)$$
$$(x + 1)(x - 5)$$

After multiplying, we see that the second pair is correct.

$$x^2 - 4x - 5 = (x + 1)(x - 5)$$

From these examples we notice that in factoring trinomials:

> If the numerical coefficient of the first term of the trinomial is *positive* but the numerical coefficient of the last term is *negative,* the first terms of both binomial factors will have positive coefficients, but one of the second terms of the binomial factors will have a positive coefficient and one of the second terms will have a negative coefficient.

Even if there are more possibilities for the second terms, this rule for the signs will still apply. For instance, suppose that we wish to factor

$$x^2 - 5x - 24$$

There are several possibilities for factors of -24, and the possible pairs of factors are

$$(x - 1)(x + 24) \quad \text{or} \quad (x + 1)(x - 24)$$
$$(x - 2)(x + 12) \quad \text{or} \quad (x + 2)(x - 12)$$
$$(x - 3)(x + 8) \quad \text{or} \quad (x + 3)(x - 8)$$
$$(x - 4)(x + 6) \quad \text{or} \quad (x + 4)(x - 6)$$

In checking these eight possible pairs, notice that the only difference between each product in the left column and the product beside it in the right column is the *sign* of the middle term. The *units* for the middle term are exactly the same. If the units for the middle term in the product

$$(x - 1)(x + 24)$$

are not correct, you may automatically reject

$$(x + 1)(x - 24)$$

also, and move on to the next possible pair.

    If the *units* for the middle term in the product

$$(x - 2)(x + 12)$$

are not correct, you may reject that pair and also the pair

$$(x + 2)(x - 12)$$

If you multiply

$$(x - 3)(x + 8)$$

and find that the *units* are correct but the *sign* is not correct, that means that you need the opposite signs on the second terms. So you choose

$$(x + 3)(x - 8)$$

and a quick multiplication verifies this choice.

$$x^2 - 5x - 24 = (x + 3)(x - 8)$$

You will become more skilled at factoring only through practice. After working many problems, you will learn to reject possibilities without writing them all. But remember that you must always check your factors before deciding that they are correct.

Example 11. Factor $x^2 - x - 12$.

    Solution. $x^2 - x - 12 = (x - 4)(x + 3)$.

Example 12. Factor $x^2 + 4x - 12$.

    Solution. $x^2 + 4x - 12 = (x + 6)(x - 2)$

Example 13. Factor $x^2 - 11xy - 12y^2$.

    Solution. $x^2 - 11xy - 12y^2 = (x - 12y)(x + y)$

Example 14. Factor $7 - 6x - x^2$.

    Solution. $7 - 6x - x^2 = (7 + x)(1 - x)$

Example 15. Factor $16 + 6y - y^2$.

    Solution. $16 + 6y - y^2 = (8 - y)(2 + y)$.

Trial Run   *Factor.*

———— **1.** $x^2 + 7x + 12$

———— **2.** $12 + 13a + a^2$

———— **3.** $y^2 - 9y + 18$

———— **4.** $x^2 - 6xy + 9y^2$

———— **5.** $x^2 + x - 6$

———— **6.** $10 - 3a - a^2$

———— **7.** $m^2 - 5m - 24$

———— **8.** $a^2 - 14ab - 15b^2$

ANSWERS

1. $(x + 3)(x + 4)$.   **2.** $(1 + a)(12 + a)$.   **3.** $(y - 6)(y - 3)$.   **4.** $(x - 3y)(x - 3y)$.
5. $(x + 3)(x - 2)$.   **6.** $(5 + a)(2 - a)$.   **7.** $(m - 8)(m + 3)$.   **8.** $(a + b)(a - 15b)$.

5.3-3 Remembering to Look for Common Monomial Factors

To factor trinomials such as

$$7x^2 + 14x + 7$$
$$6a - 5ax + ax^2$$
$$x^3 - 13x^2 - 30x$$
$$2x^3y + 4x^2y^2 - 16xy^3$$

we only need the skills that we have already developed. Each of these trinomials looks more complicated than any we have tackled thus far, but a methodical approach will allow us to factor each one.

Consider

$$7x^2 + 14x + 7$$

Remember that the first type of factoring we studied dealt with using the distributive property to remove common monomial factors. In this example, each term contains a factor of 7.

$$7x^2 + 14x + 7 = 7(x^2 + 2x + 1)$$

Now we must factor $x^2 + 2x + 1$ by the methods just learned. We find that

$$7x^2 + 14x + 7 = 7(x^2 + 2x + 1)$$
$$= 7(x + 1)(x + 1)$$

Let's factor the trinomial

$$6a - 5ax + ax^2$$

Since each term contains a factor of a, we write

$$6a - 5ax + ax^2 = a(6 - 5x + x^2)$$
$$= a(3 - x)(2 - x)$$

Now consider the trinomial

$$x^3 - 13x^2 - 30x$$

and notice that each term contains a factor of x.

$$x^3 - 13x^2 - 30x = x(x^2 - 13x - 30)$$
$$= x(x - 15)(x + 2)$$

Finally, in

$$2x^3y + 4x^2y^2 - 16xy^3$$

we see that each term contains the factors $2xy$.

$$2x^3y + 4x^2y^2 - 16xy^3 = 2xy(x^2 + 2xy - 8y^2)$$
$$= 2xy(x + 4y)(x - 2y)$$

Remember to always look first for a common factor whenever you are asked to factor a polynomial completely.

Example 16. Factor completely $5x^2 - 15x + 10$.

Solution

$$5x^2 - 15x + 10 = 5(x^2 - 3x + 2)$$
$$= 5(x - 1)(x - 2)$$

Example 17. Factor completely $x^2y + 7xy + 6y$.

Solution

$$x^2y + 7xy + 6y = y(x^2 + 7x + 6)$$
$$= y(x + 1)(x + 6)$$

Example 18. Factor completely $60 + 2x - 2x^2$.

Solution

$$60 + 2x - 2x^2 = 2(30 + x - x^2)$$
$$= 2(6 - x)(5 + x)$$

Example 19. Factor completely $2xy^2 - 4xy - 126x$.

Solution

$$2xy^2 - 4xy - 126x = 2x(y^2 - 2y - 63)$$
$$= 2x(y - 9)(y + 7)$$

Remember that our factors must always be checked by multiplying. Remember also that in multiplying three factors like these we agreed to multiply the binomials first and then multiply that product by the monomial. To check our last factors, we multiply

$$2x(y - 9)(y + 7)$$
$$= 2x(y^2 + 7y - 9y - 63)$$
$$= 2x(y^2 - 2y - 63)$$
$$= 2xy^2 - 4xy - 126x$$

and our factors are correct.

In case you are thinking that every trinomial of the type we have studied will always factor, look for a moment at the following example.

To factor $x^2 + x + 2$, the only possibility is

$$(x + 1)(x + 2)$$

But multiplying, we find that

$$(x + 1)(x + 2) = x^2 + 2x + 1x + 2 = x^2 + 3x + 2$$

We must conclude that $x^2 + x + 2$ does *not* factor. Again, in such a case we sometimes say that $x^2 + x + 2$ is *prime*.

Trial Run *Factor completely.*

_____ **1.** $6x^2 + 12x + 6$

_____ **2.** $2x^2 - 2x - 40$

_____ **3.** $12 - 12x + 3x^2$

_____ **4.** $ab^2 - 2ab - 35a$

_____ **5.** $-x^2y + 3xy + 28y$

_____ **6.** $3x^3y - 9x^2y + 30xy$

[handwritten: error should be minus sign]
[handwritten: this expression does not factor as it is]
[handwritten: $3xy(x^2 - 3x + 10)$]

ANSWERS
1. $6(x + 1)(x + 1)$.　　**2.** $2(x - 5)(x + 4)$.　　**3.** $3(2 - x)(2 - x)$.　　**4.** $a(b - 7)(b + 5)$.
5. $-y(x - 7)(x + 4)$.　　**6.** $3xy(x - 5)(x + 2)$.

[handwritten: multiplying back to check]
[handwritten: $3xy(x^2 + 2x - 5x - 10)$]
[handwritten: $3xy(x^2 - 3x - 10)$]
[handwritten: $3x^3y - 9x^2y - 30xy$]

EXERCISE SET 5.3

Factor completely.

_____ 1. $x^2 - 7x + 12$

_____ 3. $x^2 + 9x + 20$

_____ 5. $x^2 - 24x + 144$

_____ 7. $a^2 + 9a + 14$

_____ 9. $16 + 8x + x^2$

_____ 11. $y^2 - y - 30$

_____ 13. $x^2 - 2x - 35$

_____ 15. $m^2 + 4m - 32$

_____ 17. $x^2 + x - 72$

_____ 19. $b^2 - 8b - 33$

_____ 21. $x^2 - 4xy + 4y^2$

_____ 23. $a^2 + 13ab + 42b^2$

_____ 25. $m^2 + mn - 6n^2$

_____ 27. $x^2 - 10xy + 25y^2$

_____ 29. $x^2 + 22xy + 121y^2$

_____ 31. $39 - 16m + m^2$

_____ 33. $x^2y^2 + 2xy - 15$

_____ 35. $a^2b^2 + 5abc - 6c^2$

_____ 37. $x^4 - 5x^2 + 6$

_____ 39. $3x^2 - 3x - 18$

_____ 41. $5a^3 - 5a^2 - 60a$

_____ 43. $-2y^3 + 6y^2 + 20y$

_____ 45. $7x^2 - 21xy + 14y^2$

_____ 47. $-10ab^4 + 60ab^2 - 90a$

_____ 49. $4m^4 + 4m^2n^2 - 24n^4$

_____ 51. $x^3y - 18x^2y^2 + 80xy^3$

_____ 53. $2a^4b - 28a^3b^2 - 64a^2b^3$

_____ 55. $-y^4 - 17y^3 + 38y^2$

_____ 57. $a^3bc - 2a^2b^2c - 63ab^3c$

_____ 2. $x^2 - 7x + 10$

_____ 4. $x^2 + 9x + 18$

_____ 6. $x^2 - 18x + 81$

_____ 8. $a^2 + 12a + 27$

_____ 10. $49 + 14x + x^2$

_____ 12. $y^2 - 4y - 5$

_____ 14. $x^2 + 11x - 26$

_____ 16. $m^2 + 7m - 18$

_____ 18. $x^2 + x - 30$

_____ 20. $b^2 - 4b - 21$

_____ 22. $x^2 - 8xy + 16y^2$

_____ 24. $a^2 + 11ab + 30b^2$

_____ 26. $m^2 - 4mn - 45n^2$

_____ 28. $x^2 - 16xy + 64y^2$

_____ 30. $x^2 + 30xy + 225y^2$

_____ 32. $72 - 17m + m^2$

_____ 34. $x^2y^2 + 5xy - 24$

_____ 36. $a^2b^2 + 3abc - 10c^2$

_____ 38. $x^4 - 11x^2 + 24$

_____ 40. $5x^2 - 20x + 15$

_____ 42. $4a^3 + 8a^2 - 60a$

_____ 44. $-3y^3 - 24y^2 + 27y$

_____ 46. $8x^2 - 16xy - 24y^2$

_____ 48. $-3ab^4 - 12ab^2 - 12a$

_____ 50. $5m^4 - 24m^2n^2 - 30n^4$

_____ 52. $x^3y - 15x^2y^2 + 26xy^3$

_____ 54. $3a^3b^2 - 15a^2b^3 - 450ab^4$

_____ 56. $-y^4 + 18y^3 - 45y^2$

_____ 58. $a^3bc - a^2b^2c - 56ab^3c$

CHECKUP 5.3

Factor.

_____ 1. $x^2 - 15x + 56$

_____ 2. $x^2 + 5x + 6$

_____ 3. $a^2 + 5a - 24$

_____ 4. $x^2 - 4xy + 4y^2$

_____ 5. $12 + 8x + x^2$

_____ 6. $x^2y^2 + 6xyz - 7z^2$

_____ 7. $-3x^2 + 27x - 60$

_____ 8. $a^2x^2 - 2abx^2 - 8b^2x^2$

_____ 9. $-x^3y - 11x^2y^2 - 28xy^3$

_____ 10. $3x^2y - 6xy^2 - 9y^3$

5.4 Factoring More Polynomials

Let's look at a few more polynomials and the methods to use in factoring them.

5.4-1 Factoring Trinomials (Leading Coefficient Not 1)

In Section 5.3 we learned to factor trinomials into products of two binomials. In each trinomial we factored you may have noticed that once any possible common monomial factor had been removed, the coefficient of the variable-squared term was always 1 or −1. In such cases, we sometimes say that "the leading coefficient of the polynomial is 1."

Of course, you remember that there are many trinomials with leading coefficients that are *not* 1, and we shall learn to factor such trinomials in this section. The techniques that we use here are exactly the same as the techniques already learned. We continue to look for a common monomial factor first and then for two binomials that multiply together to give the trinomial being factored.

Let's try to factor the trinomial

$$3x^2 + 4x + 1$$

Notice that 3 is not a factor common to all three terms, so the first terms in our binomial factors must multiply to give $3x^2$, and the second terms in our binomials must multiply to give 1. Since all the signs in our trinomial are positive, we try

$$(3x + 1)(x + 1)$$

Let's check the product of this pair of binomials.

$$(3x + 1)(x + 1) = 3x^2 + 3x + x + 1$$
$$= 3x^2 + 4x + 1$$

and we conclude that

$$3x^2 + 4x + 1 = (3x + 1)(x + 1)$$

Let's factor the trinomial

$$3x^2 + 7x + 2$$

Again noting that the first terms in the binomials must multiply to give $3x^2$, we know that

$$3x^2 + 7x + 2 = (3x \quad)(x \quad)$$

We also know that the last terms must multiply to give 2, so the only possibilities are 2 and 1. But which one goes where? The possibilities are

$$(3x + 2)(x + 1)$$
$$(3x + 1)(x + 2)$$

and multiplying our pairs, we see that

$$(3x + 2)(x + 1) = 3x^2 + 3x + 2x + 2 = 3x^2 + 5x + 2$$
$$(3x + 1)(x + 2) = 3x^2 + 6x + x + 2 = 3x^2 + 7x + 2$$

The second pair gives the correct product, so we conclude that

$$3x^2 + 7x + 2 = (3x + 1)(x + 2)$$

Problems of this type require a little more care, and you may have to try several possibilities before finding the correct factors.

For instance, let's factor

$$6x^2 - x - 12$$

Notice that there are two ways to obtain the first term of $6x^2$.

$$(2x \quad)(3x \quad)$$
$$(6x \quad)(x \quad)$$

and there are several ways to obtain a third term of -12.

$$(\quad + 3)(\quad - 4)$$
$$(\quad - 3)(\quad + 4)$$
$$(\quad + 6)(\quad - 2)$$
$$(\quad - 6)(\quad + 2)$$
$$(\quad + 12)(\quad - 1)$$
$$(\quad - 12)(\quad + 1)$$

Writing all the possibilities is a major task. It is almost easier to jump in and try some, hoping to hit the correct pair before too long. We might reject the factors containing 12 and 1 because they will yield a middle term that is much too large.

We should also reject

$$(2x - 6)(3x + 2)$$
$$(2x + 6)(3x - 2)$$
$$(2x + 2)(3x - 6)$$
$$(2x - 2)(3x + 6)$$

Why? In each of these four pairs, one of the binomials contains a common factor of 2. If these were the correct binomials, a common factor of 2 could have been removed from the original trinomial:

$$6x^2 - x - 12$$

Since the terms of this trinomial do *not* contain a common factor of 2, we may reject those four pairs.

We are left with the pairs

$$(2x + 3)(3x - 4)$$
$$(2x - 3)(3x + 4)$$
$$(2x + 4)(3x - 3)$$
$$(2x - 4)(3x + 3)$$

but the last two pairs are rejected because the binomials contain common factors again. The only factors we must try are

$$(2x + 3)(3x - 4)$$
$$(2x - 3)(3x + 4)$$

Multiplying these pairs, we see that the second pair is correct. So we conclude that

$$6x^2 - x - 12 = (2x - 3)(3x + 4)$$

Once again we note that the commutative property allows us to write

$$6x^2 - x - 12 = (2x - 3)(3x + 4)$$

or

$$6x^2 - x - 12 = (3x + 4)(2x - 3)$$

The order in which we write the factors is not important, but the factors themselves are **unique**, which means that if a trinomial can be factored, there is *one and only one* correct pair of factors.

After some practice, you will become skillful at finding the correct pair of binomials which are the factors for a trinomial.

Example 1. Factor $5x^2 + 12x + 4$.

Solution. $5x^2 + 12x + 4 = (5x + 2)(x + 2)$.

Example 2. Factor $7x^2 - 23x + 6$.

Solution. $7x^2 - 23x + 6 = (7x - 2)(x - 3)$.

Example 3. Factor $7x^2 + 19x - 6$.

Solution. $7x^2 + 19x - 6 = (7x - 2)(x + 3)$.

Example 4. Factor $7x^2 - x - 6$.

Solution. $7x^2 - x - 6 = (7x + 6)(x - 1)$.

Example 5. Factor $15 + 19x - 10x^2$.

Solution. $15 + 19x - 10x^2 = (3 + 5x)(5 - 2x)$.

Example 6. Factor $6y^3 - 5y^2 - 21y$.

Solution. First remove the common monomial factor y. Then factor the remaining trinomial.

$$6y^3 - 5y^2 - 21y = y(6y^2 - 5y - 21)$$
$$= y(3y - 7)(2y + 3)$$

Example 7. Factor $3x^3y + 6x^2y^2 + 3xy^3$.

Solution. First remove the common factor $3xy$.

$$3x^3y + 6x^2y^2 + 3xy^3 = 3xy(x^2 + 2xy + y^2)$$
$$= 3xy(x + y)(x + y)$$

Trial Run   *Factor completely.*

_____ **1.** $2x^2 + 7x - 15$

_____ **2.** $21x^2 - 11x - 2$

_____ **3.** $2 - 11m + 12m^2$

_____ **4.** $2x^2 - 9xy + 10y^2$

_____ **5.** $4ax^3 - 12ax^2 + 9ax$

_____ **6.** $20x^3y + 100x^2y^2 + 125xy^3$

ANSWERS
1. $(2x - 3)(x + 5)$.   **2.** $(7x + 1)(3x - 2)$.   **3.** $(2 - 3m)(1 - 4m)$.   **4.** $(2x - 5y)(x - 2y)$.
5. $ax(2x - 3)(2x - 3)$.   **6.** $5xy(2x + 5y)(2x + 5y)$.

5.4-2   Factoring Trinomials of the Form $ax^4 + bx^2 + c$

Trinomials such as

$$x^4 - 5x^2 - 6$$
$$x^4 - 8x^2 - 9$$
$$x^4 - 5x^2 + 4$$

may look very different from the type we have studied so far, but they are actually very similar. Notice in each of these expressions that the square of the variable part of the middle term matches the variable part of the first term. What kind of factors would multiply to give such a trinomial? Let's look at a few products and see if we can discover a pattern that would help us factor such expressions.

Multiply: $(x^2 + 5)(x^2 + 2) = x^4 + 2x^2 + 5x^2 + 10$
$$= x^4 + 7x^2 + 10$$

Multiply: $(x^2 - 3)(x^2 + 4) = x^4 + 4x^2 - 3x^2 - 12$
$$= x^4 + x^2 - 12$$

Multiply: $(x^2 - 7)(x^2 - 1) = x^4 - x^2 - 7x^2 + 7$
$$= x^4 - 8x^2 + 7$$

We see from these products that there is a logical method to use in factoring the trinomials presented.

Let's try to factor $x^4 - 5x^2 - 6$. From our multiplications, we can be pretty sure that this trinomial will factor as the product of two binomials, each having a first term of x^2.

$$x^4 - 5x^2 - 6 = (x^2 \quad)(x^2 \quad)$$

Since the constant term is *negative* (-6), we know we need one positive and one negative number as our second term in each binomial.

$$x^4 - 5x^2 - 6 = (x^2 + \quad)(x^2 - \quad)$$

Using our usual methods, we eventually decide upon the correct numbers and we have

$$x^4 - 5x^2 - 6 = (x^2 + 1)(x^2 - 6)$$

which can be checked by multiplying the factors.

How about factoring $x^4 - 8x^2 - 9$? Again we decide that

$$x^4 - 8x^2 - 9 = (x^2 + \quad)(x^2 - \quad)$$

and after some thought, we conclude that

$$x^4 - 8x^2 - 9 = (x^2 + 1)(x^2 - 9)$$

Is this the completely factored form for our trinomial? Look carefully at the factors on the right. Does either one factor again? We know that the sum of two squares does not factor, so $x^2 + 1$ is prime, but what about $x^2 - 9$? This is a difference of two squares, so it must be factored as $(x + 3)(x - 3)$. Our original problem becomes

$$x^4 - 8x^2 - 9 = (x^2 + 1)(x^2 - 9)$$
$$= (x^2 + 1)(x + 3)(x - 3)$$

which is now in completely factored form.

Let's factor the last of our original trinomials.

$$x^4 - 5x^2 + 4 = (x^2 - \quad)(x^2 - \quad)$$
$$= (x^2 - 4)(x^2 - 1)$$

Once again we look carefully at our factors and discover that each is a difference of two squares.

$$x^4 - 5x^2 + 4 = (x^2 - 4)(x^2 - 1)$$
$$= (x + 2)(x - 2)(x + 1)(x - 1)$$

and our trinomial has been factored completely.

Example 8. Factor completely $x^4 + 8x^2 + 7$.

Solution. $x^4 + 8x^2 + 7 = (x^2 + 7)(x^2 + 1)$.

Example 9. Factor completely $1 - 2x^2 + x^4$.

Solution

$$1 - 2x^2 + x^4 = (1 - x^2)(1 - x^2)$$
$$= (1 + x)(1 - x)(1 + x)(1 - x)$$

Example 10. Factor completely $3x^4 + 3x^2 - 60$.

Solution

$$3x^4 + 3x^2 - 60 = 3(x^4 + x^2 - 20)$$
$$= 3(x^2 + 5)(x^2 - 4)$$
$$= 3(x^2 + 5)(x + 2)(x - 2)$$

Trial Run *Factor completely.*

_____ **1.** $x^4 + 8x^2 + 12$

_____ **2.** $2x^4 + 8x^2 + 6$

_____ **3.** $x^4 + 2x^2 - 3$

_____ **4.** $x^4 - 13x^2 + 36$

_____ **5.** $5x^4 - 25x^2 + 20$

ANSWERS
1. $(x^2 + 6)(x^2 + 2)$.   **2.** $2(x^2 + 1)(x^2 + 3)$.   **3.** $(x^2 + 3)(x + 1)(x - 1)$.
4. $(x + 3)(x - 3)(x + 2)(x - 2)$.   **5.** $5(x + 1)(x - 1)(x + 2)(x - 2)$.

5.4-3 ## Factoring by Grouping

In an expression such as

$$3(x + 1) + y(x + 1)$$

notice that there are just two terms to consider. The first term is $3(x + 1)$ and the second term is $y(x + 1)$. You should see that both terms contain the common factor $(x + 1)$. To completely factor an expression of this type, recall how the distributive property allows us to write

$$3\underline{A} + y\underline{A} = \underline{A}(3 + y)$$

We removed the common factor, A. In our original example, the common factor is not A; instead, the common factor is the binomial $(x + 1)$. The distributive property still allows us to write

$$3\underline{(x + 1)} + y\underline{(x + 1)} = \underline{(x + 1)}(3 + y)$$

If you have any doubts about whether the left-hand side is the same as the right-hand side, you should simplify both quantities and check the results.

$$3(x + 1) + y(x + 1) = 3x + 3 + yx + y$$
$$= 3x + xy + y + 3$$

and

$$(x + 1)(3 + y) = x(3 + y) + 1(3 + y)$$
$$= 3x + xy + 3 + y$$
$$= 3x + xy + y + 3$$

The results are the same, so our factors must be correct. Let's try factoring another expression of this type.

$$y^2(x - 3) - 7(x - 3)$$

Notice that both terms contain the common binomial factor $(x - 3)$. Removing that common factor, we can write

$$y^2(x - 3) - 7(x - 3) = (x - 3)(y^2 - 7)$$

To deal with an expression such as

$$a(x + y) + (x + y)$$

we should agree that we may rewrite this expression as

$$a(x + y) + 1(x + y)$$

Then the binomial factor $(x + y)$ is obviously common to both terms and we write

$$a(x + y) + 1(x + y) = (x + y)(a + 1)$$

Example 11. Factor $3(a + b) + x(a + b)$.

Solution. $3(a + b) + x(a + b) = (a + b)(3 + x)$.

Example 12. Factor $y(x^2 + 3) - 2(x^2 + 3)$.

Solution. $y(x^2 + 3) - 2(x^2 + 3) = (x^2 + 3)(y - 2)$.

Example 13. Factor $x(a - b) + (a - b)$.

Solution

$$\begin{aligned}
x(a - b) + (a - b) &= x(a - b) + 1(a - b) \\
&= (a - b)(x + 1)
\end{aligned}$$

Example 14. Factor $(x + 5) + 2a(x + 5)$.

Solution

$$\begin{aligned}
(x + 5) + 2a(x + 5) &= 1(x + 5) + 2a(x + 5) \\
&= (x + 5)(1 + 2a)
\end{aligned}$$

Example 15. Factor $3x(x^2 + 1) - (x^2 + 1)$.

Solution

$$\begin{aligned}
3x(x^2 + 1) - (x^2 + 1) &= 3x(x^2 + 1) - 1(x^2 + 1) \\
&= (x^2 + 1)(3x - 1)
\end{aligned}$$

Example 16. Factor $y(x^2 - 1) + 3(x^2 - 1)$.

Solution

$$\begin{aligned}
y(x^2 - 1) + 3(x^2 - 1) &= (x^2 - 1)(y + 3) \\
&= (x + 1)(x - 1)(y + 3)
\end{aligned}$$

Notice that the difference of two squares was factored in the last step and our polynomial is now in completely factored form.

A closer look at each of the expressions just factored reveals that if each were multiplied out, it would result in a polynomial of *four* terms. For instance,

$$3(a + b) + x(a + b)$$

becomes

$$3a + 3b + xa + xb$$

and

$$a(x + y) + (x + y)$$

becomes

$$ax + ay + x + y$$

Suppose that we were asked to factor such a four-term polynomial. Our last examples should give us a clue. We must try to group the terms into twosomes, each twosome containing a common factor. For instance, to start to factor

$$5x + 5y + bx + by$$

we notice that the first two terms contain the common factor 5 and the second two terms contain the common factor b. We can rewrite our *four* terms as

$$5x + 5y + bx + by = 5(x + y) + b(x + y)$$

But we still have *two* terms, so the polynomial is not yet factored into a product. Can you see that the two terms contain the common binomial factor $(x + y)$?

$$5x + 5y + bx + by = 5(x + y) + b(x + y)$$
$$= (x + y)(5 + b)$$

and we have rewritten our four-term polynomial as a product of factors.

To factor a polynomial such as

$$x^3 + 2x + 5x^2 + 10$$

we notice that the first two terms contain a common factor of x and the second two terms contain a common factor of 5. We first rewrite the polynomial as

$$x^3 + 2x + 5x^2 + 10 = x(x^2 + 2) + 5(x^2 + 2)$$

and then noticing the binomial factor $(x^2 + 2)$ common to both terms, we write

$$x^3 + 2x + 5x^2 + 10 = x(x^2 + 2) + 5(x^2 + 2)$$
$$= (x^2 + 2)(x + 5)$$

How about factoring a polynomial such as

$$4ax - 12ay + x - 3y?$$

The first two terms have a common factor of $4a$, but the second two terms have no common factor except 1. We may write

$$4ax - 12ay + x - 3y = 4a(x - 3y) + 1(x - 3y)$$
$$= (x - 3y)(4a + 1)$$

Let's try factoring one more polynomial of four terms.

$$6y^2 + 3y - 2ay - a$$

The first two terms contain a common factor of $3y$ and the second two terms contain a common factor of $a \ or \ -a$. Which should we remove? Take a look.

$$6y^2 + 3y - 2ay - a = 3y(2y + 1) + a(-2y - 1)$$

or

$$6y^2 + 3y - 2ay - a = 3y(2y + 1) - a(2y + 1)$$

Both versions are correct, but which one will allow us to continue factoring? The second version leaves us with a common binomial factor of $(2y + 1)$ in both terms, so we choose that form.

$$6y^2 + 3y - 2ay - a = 3y(2y + 1) - a(2y + 1)$$
$$= (2y + 1)(3y - a)$$

Example 17. Factor $7bx - 14by + 5x - 10y$.

Solution

$$7bx - 14by + 5x - 10y = 7b(x - 2y) + 5(x - 2y)$$
$$= (x - 2y)(7b + 5)$$

Example 18. Factor $-2x^2 + 2y^2 + 9x^2y - 9y^3$.

Solution

$$-2x^2 + 2y^2 + 9x^2y - 9y^3 = -2(x^2 - y^2) + 9y(x^2 - y^2)$$
$$= (x^2 - y^2)(-2 + 9y)$$
$$= (x + y)(x - y)(-2 + 9y)$$

Example 19. Factor $8rx - 4r - 2x + 1$.

Solution

$$8rx - 4r - 2x + 1 = 4r(2x - 1) - 1(2x - 1)$$
$$= (2x - 1)(4r - 1)$$

Trial Run *Factor out the common binomial factor.*

_____ **1.** $2x(a - 3) - 5(a - 3)$

_____ **2.** $y(a - b) + (a - b)$

_____ **3.** $(m^2 - 1) + n(m^2 - 1)$

Factor.

_____ **4.** $4x + 4y + bx + by$

_____ **5.** $x^3 - 7x^2 + 3x - 21$

_____ **6.** $3ax - 15ay + 2x - 10y$

_____ **7.** $x^3 - 3x^2 - x + 3$

_____ **8.** $8a^2 + 4a - 2ab - b$

ANSWERS
1. $(a - 3)(2x - 5)$. **2.** $(a - b)(y + 1)$. **3.** $(m + 1)(m - 1)(1 + n)$. **4.** $(x + y)(4 + b)$.
5. $(x - 7)(x^2 + 3)$. **6.** $(x - 5y)(3a + 2)$. **7.** $(x - 3)(x - 1)(x + 1)$.
8. $(2a + 1)(4a - b)$.

EXERCISE 5.4

Factor the following trinomials.

_____ **1.** $2x^2 + 7x + 3$ _____ **2.** $3x^2 + 16x + 5$

_____ **3.** $5x^2 + 7x + 2$ _____ **4.** $7x^2 + 22x + 3$

_____ **5.** $3a^2 - 17a + 10$ _____ **6.** $11a^2 - 35a + 6$

_____ **7.** $10x^2 - 19x + 6$ _____ **8.** $6x^2 - 25x + 25$

_____ **9.** $10y^2 - 7y - 12$ _____ **10.** $14y^2 + 17y - 6$

_____ **11.** $8m^2 + 14m - 15$ _____ **12.** $8m^2 + 18m - 35$

_____ **13.** $6x^2 - 5xy - 56y^2$ _____ **14.** $16x^2 + 6xy - 27y^2$

_____ **15.** $81x^2 - 36xy + 4y^2$ _____ **16.** $25x^2 - 30xy + 9y^2$

_____ **17.** $39a^2 - 38ab + 8b^2$ _____ **18.** $30a^2 - 89ab + 35b^2$

_____ **19.** $121 + 132a + 36a^2$ _____ **20.** $144 + 168a + 49a^2$

In Exercises 21–40, completely factor each trinomial. When the trinomial cannot be factored using integers, write "prime."

_____ **21.** $30c^2 - 145cd + 45d^2$ _____ **22.** $20c^2 - 68cd + 24d^2$

_____ **23.** $-16y^2 - 28y + 30$ _____ **24.** $-21y^2 - 27y + 30$

_____ **25.** $9x^4 - 33x^3 + 28x^2$ _____ **26.** $4a^4 - 32a^3 + 63a^2$

_____ **27.** $2x^2 - 7x - 3$ _____ **28.** $5x^2 - x - 8$

_____ **29.** $4a^2b^4 - 25a^2b^3 + 6a^2b^2$ _____ **30.** $4a^4b^2 - 23a^3b^2 - 6a^2b^2$

_____ **31.** $-18x^2 + 9x - 27$ _____ **32.** $-30x^2 + 6x - 40$

_____ **33.** $24x^3y + 26x^2y - 70xy$ _____ **34.** $105x^3y + 57x^2y - 72xy$

_____ **35.** $12m^2 + 13mn - 55n^2$ _____ **36.** $12m^2 - 23mn + 10n^2$

_____ **37.** $4x^6 + 12x^4y^2 + 9x^2y^4$ _____ **38.** $9x^4y^2 + 30x^2y^4 + 25y^6$

_____ **39.** $5k^2 + 12hk + 24h^2$ _____ **40.** $3k^2 + 10hk - 14h^2$

In Exercises 41–50, factor each polynomial by grouping.

_____ **41.** $5a + ac + 5b + bc$ _____ **42.** $7x + 7y + xz + yz$

_____ **43.** $x^3 - 5x^2 + 6x - 30$ _____ **44.** $x^3 - 9x^2 + 9x - 81$

_____ **45.** $14ax + 21ay - 10x - 15y$ _____ **46.** $36ax + 45ay - 28x - 35y$

_____ **47.** $x^3 - 2x^2 - 25x + 50$ _____ **48.** $x^3 - 3x^2 - 4x + 12$

_____ **49.** $70a^2 - 40ab + 21a - 12b$ _____ **50.** $30a^2 - 75ab + 4a - 10b$

Factor.

_____ **51.** $3x^6 - 30x^4y^2 + 75x^2y^4$ _____ **52.** $4x^4y^2 - 88x^2y^4 + 484y^6$

_____ **53.** $x^4 - 9x^2 + 20$ _____ **54.** $x^4 - 8x^2 - 9$

Name _____ Date _____

CHECKUP 5.4

Factor completely.

_____ **1.** $3x^2 + 16x + 5$

_____ **2.** $2a^2 - 13a + 21$

_____ **3.** $4a^2 - 4ab - 15b^2$

_____ **4.** $24a^2 + 3a - 21$

_____ **5.** $6x^2 - 5x + 2$

_____ **6.** $-8x^2 - 44x + 24$

_____ **7.** $18x^4 - 60x^3y + 50x^2y^2$

_____ **8.** $3x^4 + 2x^2y^2 - 8y^4$

_____ **9.** $5m + 5n + pm + pn$

_____ **10.** $3ax + 3ay - 2bx - 2by$

Summary

In this chapter we learned to rewrite a polynomial as a product of factors. Different types of factoring were used in different situations and we agreed to look for the following.

1. **Common monomial factor:** Look for a factor common to all terms of the polynomial and remove it using the distributive property.
2. **Difference of two squares:** Look for two square terms with a minus sign in between. Then

$$A^2 - B^2 = (A + B)(A - B)$$

3. **General trinomial:** Use trial and error to find two binomials that multiply to give the trinomial.
4. **Four-term polynomial:** Group the terms two by two, remove a common factor from each twosome, and look for a common binomial factor to remove using the distributive property.

We discovered that the order in which factors are written is not important, but the factors themselves are unique. A polynomial can be factored correctly in only one way. If a polynomial cannot be written as a product of factors, we say that it is *prime*.

Name _____ **Date** _____

REVIEW EXERCISES

Completely factor the following expressions. If the expressions cannot be factored using integers, write "prime."

SECTION 5.1

_____ **1.** $10x - 15$ _____ **2.** $15x^2 - 12x$

_____ **3.** $2y^3 - 5y^2$ _____ **4.** $36x^4 - 16y^2$

_____ **5.** $10x^2y + 4xy^2$ _____ **6.** $2m^3 - 10m^2 + 12m$

_____ **7.** $-3a^2b + 2a^3b^2 - 4a^4b^3$

SECTION 5.2

_____ **8.** $x^2 - 49$ _____ **9.** $25 - y^2$

_____ **10.** $a^2 - 64b^2$ _____ **11.** $x^2y^2 - 121$

_____ **12.** $3m^2 - 48$ _____ **13.** $64a^2b - 25b^3$

_____ **14.** $18x^4 - 8x^2y^2$ _____ **15.** $x^4 - 81y^4$

SECTION 5.3

_____ **16.** $x^2 - 8x + 15$ _____ **17.** $a^2 + 12ab + 27b^2$

_____ **18.** $x^2y^2 - 3xy - 10$ _____ **19.** $63 - 2m - m^2$

_____ **20.** $a^2b^2 - abc - 56c^2$ _____ **21.** $4a^2 - 20ab + 25b^2$

SECTION 5.4

_____ **22.** $6x^2 - 11x - 7$ _____ **23.** $-2x^2 + 4xy + 6y^2$

_____ **24.** $10m^2 + 15m - 220$ _____ **25.** $16x^2 - 40xy + 25y^2$

_____ **26.** $169 - 4a^2$ _____ **27.** $30x^3y - 4x^2y^2 - 2xy^3$

_____ **28.** $8x^2 - 11x + 3$ _____ **29.** $x^4 - 14x^2 + 45$

_____ **30.** $3y^4 + 63y^2 - 300$ _____ **31.** $4x + 4y + xz + yz$

_____ **32.** $x^3 - 4x^2 - x + 4$

6 Solving Quadratic Equations

In Chapter 3 we learned how to solve equations such as

$$3x + 1 = 16$$
$$5 - 2x = 25$$
$$2(y + 3) = y - 1$$

by finding the value of the variable that satisfies the equation. We found our solutions using addition, subtraction, multiplication, and division to isolate the variable. In each equation, the variable appeared with an exponent of 1. (There were no terms containing x^2 or y^2 or a^2.) Such equations are called **first-degree equations** (or **linear equations**).

In this chapter we shall be looking at equations that *do* contain terms such as x^2 or y^2 or a^2. A polynomial equation containing one variable in which the largest exponent on the variable is a 2 is called a **second-degree equation** or **quadratic equation**. Examples of such equations are

$$x^2 + 6x + 5 = 0$$
$$x^2 + x = 12$$
$$x^2 = 7x$$
$$3x^2 - 14x = 5$$

In this chapter we learn how to

1. Work with the zero product rule.
2. Solve quadratic equations by factoring.
3. Switch from word statements to variable statements that result in quadratic equations.

6.1 Solving Quadratic Equations

Before we learn how to solve quadratic equations, we must make a very important observation about *zero* as the result of multiplication.

6.1-1 The Zero Product Rule

If we are told that two quantities are multiplied together and the product is zero, what can we conclude? Can you see that one or the other or both of those quantities must equal zero? Earlier we learned that for any number A,

$$A \cdot 0 = 0$$
$$0 \cdot A = 0$$

Now we are suggesting that if A and B are any two numbers and we know that

$$A \cdot B = 0$$

then it must be true that

$$A = 0 \quad \text{or} \quad B = 0 \quad \text{or} \quad \text{both}$$

For instance, earlier we solved equations such as

$$3x = 0$$
$$-2y = 0$$

In the first equation we are saying that 3 times some number x is zero. Solving, we find

$$3x = 0$$
$$\frac{3x}{3} = \frac{0}{3}$$
$$x = 0$$

In the second equation we are saying that -2 times some number y is zero. Solving, we find

$$-2y = 0$$
$$\frac{-2y}{-2} = \frac{0}{-2}$$
$$y = 0$$

Such problems may help us agree that

> If the product of two factors is zero, one or the other or both factors *must* be zero.

This could be called the **zero product rule**, which may be stated in general as follows.

> **Zero Product Rule**
>
> If $A \cdot B = 0$, then
>
> $A = 0$ or $B = 0$ or   both

This rule is very important in learning to solve quadratic equations. We shall practice solving a few problems before moving on.

$$x(x - 2) = 0$$
$$(x + 3)(x - 5) = 0$$
$$(x - 7)(x - 7) = 0$$
$$(y + 9)(y - 9) = 0$$
$$(1 - x)(6 + x) = 0$$

In each equation we know that if the product is zero, one or the other or both of the factors must be zero.

For instance, if

$$x(x - 2) = 0$$

then we know that

$$x = 0 \quad \text{or} \quad x - 2 = 0$$

so that

$$x = 0 \quad \text{or} \quad x = 2$$

are solutions to the equation. Let's check both solutions.

CHECK:
$$x = 0$$
$$x(x - 2) = 0$$
$$0(0 - 2) \overset{?}{=} 0$$
$$0(-2) \overset{?}{=} 0$$
$$0 = 0$$

CHECK:
$$x = 2$$
$$x(x - 2) = 0$$
$$2(2 - 2) \overset{?}{=} 0$$
$$2(0) \overset{?}{=} 0$$
$$0 = 0$$

Since both solutions are correct, we have found two solutions for the problem: $x = 0$ or $x = 2$.

Looking next at

$$(x + 3)(x - 5) = 0$$

we observe two factors and we know that either one of them could be zero.

$$x + 3 = 0 \quad \text{or} \quad x - 5 = 0$$

Solving as before,

$$
\begin{array}{lll}
x + 3 = 0 & \text{or} & x - 5 = 0 \\
 x = -3 & \text{or} & x = 5
\end{array}
$$

and we check.

CHECK:
$$x = -3$$
$$(x + 3)(x - 5) = 0$$
$$(-3 + 3)(-3 - 5) \stackrel{?}{=} 0$$
$$(0)(-8) \stackrel{?}{=} 0$$
$$0 = 0$$

CHECK:
$$x = 5$$
$$(x + 3)(x - 5) = 0$$
$$(5 + 3)(5 - 5) \stackrel{?}{=} 0$$
$$(8)(0) \stackrel{?}{=} 0$$
$$0 = 0$$

Now we shall solve

$$(x - 7)(x - 7) = 0$$

Since the factors are $(x - 7)$ and $(x - 7)$, we know that

$$
\begin{array}{lll}
x - 7 = 0 & \text{or} & x - 7 = 0 \\
 x = 7 & \text{or} & x = 7
\end{array}
$$

Both solutions are the same, so we say that the solution is $x = 7$ and we check.

CHECK:
$$x = 7$$
$$(x - 7)(x - 7) = 0$$
$$(7 - 7)(7 - 7) \stackrel{?}{=} 0$$
$$0(0) \stackrel{?}{=} 0$$
$$0 = 0$$

It seems that we could use the following steps whenever we know that a product of factors is zero.

> **1.** Set each of the variable factors equal to zero.
> **2.** Solve the resulting equations.
> **3.** Check the solutions in the original problem.

Example 1. Solve $(y + 9)(y - 9) = 0$.

Solution. $(y + 9)(y - 9) = 0$, so

$$
\begin{array}{lll}
y + 9 = 0 & \text{or} & y - 9 = 0 \\
 y = -9 & \text{or} & y = 9
\end{array}
$$

CHECK: $$y = -9$$
$$(y + 9)(y - 9) = 0$$
$$(-9 + 9)(-9 - 9) \overset{?}{=} 0$$
$$0(-18) \overset{?}{=} 0$$
$$0 = 0$$

CHECK: $$y = 9$$
$$(y + 9)(y - 9) = 0$$
$$(9 + 9)(9 - 9) \overset{?}{=} 0$$
$$18(0) \overset{?}{=} 0$$
$$0 = 0$$

Example 2. Solve $(1 - x)(6 + x) = 0$.

Solution. $(1 - x)(6 + x) = 0$, so

$$1 - x = 0 \qquad \text{or} \qquad 6 + x = 0$$
$$-x = -1 \qquad\qquad\qquad x = -6$$
$$x = 1 \qquad \text{or} \qquad x = -6$$

CHECK: $$x = 1$$
$$(1 - x)(6 + x) = 0$$
$$(1 - 1)(6 + 1) \overset{?}{=} 0$$
$$0(7) \overset{?}{=} 0$$
$$0 = 0$$

CHECK: $$x = -6$$
$$(1 - x)(6 + x) = 0$$
$$[1 - (-6)][6 + (-6)] \overset{?}{=} 0$$
$$7(0) \overset{?}{=} 0$$
$$0 = 0$$

Trial Run *Solve and check.*

_____ **1.** $(x - 11)(x - 2) = 0$

_____ **2.** $(y - 3)(y + 5) = 0$

_____ **3.** $(a + 12)(a + 5) = 0$

_____ **4.** $(1 - x)(2 + x) = 0$

ANSWERS
1. $x = 11$ or $x = 2$.   **2.** $y = 3$ or $y = -5$.   **3.** $a = -12$ or $a = -5$.
4. $x = 1$ or $x = -2$.

6.1-2 Solving Quadratic Equations by Factoring

We have just learned how to solve an equation in which a product of factors is zero. We simply set each factor equal to zero, solve, and check. Now we shall see how the zero product rule can help us solve equations such as

$$x^2 + 3x + 2 = 0$$

$$x^2 - 4x - 77 = 0$$

$$a^2 - 3a = 0$$

$$y^2 - 25 = 0$$

Recall that these are called quadratic equations. The expression on the left-hand side is a sum of variable terms and constant terms, and each equation contains a squared variable. How do we solve equations such as these? Guessing the solutions seems like an approach that could take lots of valuable time. Perhaps we can make use of the *zero product rule*.

But wait a minute; these are not *products* that equal zero. Instead, they are *sums* of terms. Do we know a process for rewriting sums as products? Chapter 5 was devoted to just that process, the process of *factoring*.

Look at the first equation,

$$x^2 + 3x + 2 = 0$$

We may factor the left-hand side and rewrite our equation as

$$(x + 2)(x + 1) = 0$$

and we are ready to use the zero product rule.

$$(x + 2)(x + 1) = 0$$

so

$$
\begin{array}{lll}
x + 2 = 0 & \text{or} & x + 1 = 0 \\
x = -2 & \text{or} & x = -1
\end{array}
$$

Now we must check our solutions in the original equation.

$$x^2 + 3x + 2 = 0$$

CHECK:
$$x = -2$$
$$x^2 + 3x + 2 = 0$$
$$(-2)^2 + 3(-2) + 2 \stackrel{?}{=} 0$$
$$(-2)(-2) + 3(-2) + 2 \stackrel{?}{=} 0$$
$$4 - 6 + 2 \stackrel{?}{=} 0$$
$$6 - 6 \stackrel{?}{=} 0$$
$$0 = 0$$

CHECK:
$$x = -1$$
$$x^2 + 3x + 2 = 0$$
$$(-1)^2 + 3(-1) + 2 \stackrel{?}{=} 0$$
$$(-1)(-1) + 3(-1) + 2 \stackrel{?}{=} 0$$
$$1 - 3 + 2 \stackrel{?}{=} 0$$
$$3 - 3 \stackrel{?}{=} 0$$
$$0 = 0$$

Checking solutions to quadratic equations takes some care. You will not be required to check your answers to every problem, but we shall do so in the next two examples.

Let's solve the equation

$$x^2 - 4x - 77 = 0$$

Once again, we must write our polynomial in factored form, then use the zero product rule.

$$x^2 - 4x - 77 = 0$$
$$(x - 11)(x + 7) = 0$$

so

$$x - 11 = 0 \quad \text{or} \quad x + 7 = 0$$
$$x = 11 \quad \text{or} \quad x = -7$$

CHECK:
$$x = 11$$
$$x^2 - 4x - 77 = 0$$
$$(11)^2 - 4(11) - 77 \overset{?}{=} 0$$
$$(11)(11) - 4(11) - 77 \overset{?}{=} 0$$
$$121 - 44 - 77 \overset{?}{=} 0$$
$$121 - 121 \overset{?}{=} 0$$
$$0 = 0$$

CHECK:
$$x = -7$$
$$x^2 - 4x - 77 = 0$$
$$(-7)^2 - 4(-7) - 77 \overset{?}{=} 0$$
$$(-7)(-7) - 4(-7) - 77 \overset{?}{=} 0$$
$$49 + 28 - 77 \overset{?}{=} 0$$
$$77 - 77 \overset{?}{=} 0$$
$$0 = 0$$

So our solutions are $x = 11$ or $x = -7$.

Now consider the equation

$$a^2 - 3a = 0$$

Remember to look for a common monomial factor.

$$a^2 - 3a = 0$$
$$a(a - 3) = 0$$

Now we must set each factor equal to zero.

$$a = 0 \quad \text{or} \quad a - 3 = 0$$
$$a = 0 \quad \text{or} \quad a = 3$$

CHECK:
$$a = 0$$
$$a^2 - 3a = 0$$
$$(0)^2 - 3(0) \overset{?}{=} 0$$
$$(0)(0) - 3(0) \overset{?}{=} 0$$
$$0 - 0 \overset{?}{=} 0$$
$$0 = 0$$

CHECK:
$$a = 3$$
$$a^2 - 3a = 0$$
$$(3)^2 - 3(3) \overset{?}{=} 0$$
$$(3)(3) - 3(3) \overset{?}{=} 0$$
$$9 - 9 \overset{?}{=} 0$$
$$0 = 0$$

So our solutions are $a = 0$ or $a = 3$.

We have observed that solving quadratic equations with zero on one side requires the following steps.

> 1. Rewrite the polynomial in factored form.
> 2. Set each factor equal to zero.
> 3. Solve for the variable.
> 4. Check the solutions in the original equation.

Example 3. Solve $x^2 - 25 = 0$.

Solution. $x^2 - 25 = 0$, so

$$(x + 5)(x - 5) = 0$$
$$x + 5 = 0 \quad \text{or} \quad x - 5 = 0$$
$$x = -5 \quad \text{or} \quad x = 5$$

Example 4. Solve $0 = x^2 - 12x + 32$.

Solution. $0 = x^2 - 12x + 32$, so

$$0 = (x - 8)(x - 4)$$
$$x - 8 = 0 \quad \text{or} \quad x - 4 = 0$$
$$x = 8 \quad \text{or} \quad x = 4$$

Example 5. Solve $5x^2 \overset{?}{-} 10x = 0$.

Solution. $5x^2 \overset{?}{+} 10x = 0$, so

$$5x(x + 2) = 0$$
$$5x = 0 \quad \text{or} \quad x + 2 = 0$$
$$x = 0 \quad \text{or} \quad x = -2$$

Example 6. Solve $x^2 + 12x + 36 = 0$.

Solution. $x^2 + 12x + 36 = 0$, so

$$(x + 6)(x + 6) = 0$$
$$x + 6 = 0 \quad \text{or} \quad x + 6 = 0$$
$$x = -6 \quad \text{or} \quad x = -6$$

Here we say that the solution is $x = -6$.

Example 7. Solve $3y^2 - 12 = 0$ and check.

Solution

$$3y^2 - 12 = 0$$
$$3(y^2 - 4) = 0$$
$$3(y + 2)(y - 2) = 0$$

Since the constant factor 3 can never be zero, we know that

$$y + 2 = 0 \quad \text{or} \quad y - 2 = 0$$
$$y = -2 \quad \text{or} \quad y = 2$$

CHECK:
$$y = -2$$
$$3y^2 - 12 = 0$$
$$3(-2)^2 - 12 \overset{?}{=} 0$$
$$3(4) - 12 \overset{?}{=} 0$$
$$12 - 12 \overset{?}{=} 0$$
$$0 = 0$$

CHECK:

$$y = 2$$
$$3y^2 - 12 = 0$$
$$3(2)^2 - 12 \overset{?}{=} 0$$
$$3(4) - 12 \overset{?}{=} 0$$
$$12 - 12 \overset{?}{=} 0$$
$$0 = 0$$

To solve quadratic equations such as these, you must be able to factor and use the zero product rule.

Trial Run *Solve.*

_____ 1. $x^2 - 4x - 5 = 0$

_____ 2. $y^2 - 36 = 0$

_____ 3. $x^2 - 7x = 0$

_____ 4. $a^2 + 7a + 12 = 0$

_____ 5. $3x^2 - 15x + 18 = 0$

_____ 6. $27 + 6m - m^2 = 0$

ANSWERS
1. $x = 5$ or $x = -1$.   2. $y = 6$ or $y = -6$.   3. $x = 0$ or $x = 7$.
4. $a = -3$ or $a = -4$.   5. $x = 3$ or $x = 2$.   6. $m = 9$ or $m = -3$.

EXERCISE SET 6.1

Solve for the variable.

_____ 1. $(x - 3)(x - 2) = 0$ _____ 2. $(x - 7)(x - 5) = 0$

_____ 3. $(x + 2)(x - 6) = 0$ _____ 4. $(x + 5)(x - 11) = 0$

_____ 5. $(a + 13)(a + 5) = 0$ _____ 6. $(a + 1)(a + 9) = 0$

_____ 7. $x(x - 8) = 0$ _____ 8. $x(x + 10) = 0$

_____ 9. $(7 - x)(9 + x) = 0$ _____ 10. $(2 - x)(8 + x) = 0$

_____ 11. $2(y - 3)(y + 1) = 0$ _____ 12. $-3(y + 6)(y - 1) = 0$

_____ 13. $x^2 - 6x + 8 = 0$ _____ 14. $x^2 - 9x + 8 = 0$

_____ 15. $x^2 - 16 = 0$ _____ 16. $x^2 - 49 = 0$

_____ 17. $y^2 + 9y + 18 = 0$ _____ 18. $y^2 + 8y + 15 = 0$

_____ 19. $m^2 + 15m = 0$ _____ 20. $m^2 - 9m = 0$

_____ 21. $z^2 + 3z - 54 = 0$ _____ 22. $z^2 + z - 42 = 0$

_____ 23. $2x^2 - 10x = 0$ _____ 24. $3x^2 - 9x = 0$

_____ 25. $y^2 - y - 56 = 0$ _____ 26. $y^2 - y - 90 = 0$

_____ 27. $z^2 - 30z + 225 = 0$ _____ 28. $z^2 - 26z + 169 = 0$

_____ 29. $5y^2 - 500 = 0$ _____ 30. $3y^2 - 192 = 0$

_____ 31. $18n - 9n^2 = 0$ _____ 32. $20n - 5n^2 = 0$

_____ 33. $33 - 8y - y^2 = 0$ _____ 34. $80 - 2y - y^2 = 0$

_____ 35. $5x^2 + 10x + 5 = 0$ _____ 36. $3x^2 + 12x + 12 = 0$

_____ 37. $162 - 2a^2 = 0$ _____ 38. $75 - 3a^2 = 0$

_____ 39. $z^2 + 20z + 99 = 0$ _____ 40. $z^2 + 15z + 50 = 0$

CHECKUP 6.1

Solve.

_____ **1.** $2x(x - 13) = 0$

_____ **2.** $a^2 - 2a - 63 = 0$

_____ **3.** $y^2 - 144 = 0$

_____ **4.** $x^2 + 8x - 20 = 0$

_____ **5.** $36 - 5x - x^2 = 0$

_____ **6.** $7x^2 - 28x = 0$

_____ **7.** $3x^2 + 30x + 63 = 0$

_____ **8.** $x^2 - 16x + 64 = 0$

_____ **9.** $5x^2 - 5 = 0$

_____ **10.** $36 - 14x - 2x^2 = 0$

6.2 More Quadratic Equations

Remember that solving quadratic equations depends very much on the zero product rule. To solve quadratic equations, we must be sure that our polynomial equals zero. Then we may factor, set each factor equal to zero, and solve.

Suppose that we are asked to solve equations such as

$$x^2 - x = 20$$

$$y^2 = 16$$

$$x^2 = 7x$$

$$x^2 = 15x - 50$$

How can we put these equations into a form that allows us to use the zero product rule? Using addition and subtraction, we must get all the terms on one side of the equation and *zero* on the other side.

For instance, to put the first equation $x^2 - x = 20$ in the proper form, we must subtract 20 from both sides of the equation.

$$x^2 - x = 20$$
$$x^2 - x - 20 = 20 - 20$$
$$x^2 - x - 20 = 0$$

Now factoring, we have

$$(x - 5)(x + 4) = 0$$

so

$$x - 5 = 0 \quad \text{or} \quad x + 4 = 0$$
$$x = 5 \quad \text{or} \quad x = -4$$

To deal properly with the second equation we subtract 16 from both sides.

$$y^2 = 16$$
$$y^2 - 16 = 16 - 16$$
$$y^2 - 16 = 0$$

Factoring, we have

$$(y + 4)(y - 4) = 0$$

so

$$y + 4 = 0 \quad \text{or} \quad y - 4 = 0$$
$$y = -4 \quad \text{or} \quad y = 4$$

The third equation,

$$x^2 = 7x$$

requires that we subtract $7x$ from both sides.

$$x^2 = 7x$$
$$x^2 - 7x = 7x - 7x$$
$$x^2 - 7x = 0$$

Factoring, we find

$$x(x - 7) = 0$$

so

$$x = 0 \quad \text{or} \quad x - 7 = 0$$
$$x = 0 \quad \text{or} \quad x = 7$$

In the last equation,

$$x^2 = 15x - 50$$

we must subtract $15x$ and add 50 to both sides.

$$x^2 = 15x - 50$$
$$x^2 - 15x = 15x - 50 - 15x$$
$$x^2 - 15x = -50$$
$$x^2 - 15x + 50 = -50 + 50$$
$$x^2 - 15x + 50 = 0$$

Factoring, we have

$$(x - 10)(x - 5) = 0$$

so

$$x - 10 = 0 \quad \text{or} \quad x - 5 = 0$$
$$x = 10 \quad \text{or} \quad x = 5$$

To solve quadratic equations by factoring, we must use the following procedure.

---

1. Get zero on one side of the equation and get all other terms on the other side.
2. Factor the polynomial.
3. Set each factor equal to zero.
4. Solve for the variable and check.

Example 1. Solve $x^2 - 5x = -4$.

Solution

$$x^2 - 5x = -4$$
$$x^2 - 5x + 4 = -4 + 4$$
$$x^2 - 5x + 4 = 0$$
$$(x - 4)(x - 1) = 0$$
$$x - 4 = 0 \quad \text{or} \quad x - 1 = 0$$
$$x = 4 \quad \text{or} \quad x = 1$$

The check is left to you.

Example 2. Solve $x^2 - 10 = 3x$.

Solution

$$x^2 - 10 = 3x$$
$$x^2 - 10 - 3x = 3x - 3x$$
$$x^2 - 10 - 3x = 0$$

For factoring purposes, we now rearrange the terms.

$$x^2 - 3x - 10 = 0$$
$$(x - 5)(x + 2) = 0$$
$$x - 5 = 0 \quad \text{or} \quad x + 2 = 0$$
$$x = 5 \quad \text{or} \quad x = -2$$

The check is left to you.

Example 3. Solve $y^2 = -11y$.

Solution

$$y^2 = -11y$$
$$y^2 + 11y = -11y + 11y$$
$$y^2 + 11y = 0$$
$$y(y + 11) = 0$$
$$y = 0 \quad \text{or} \quad y + 11 = 0$$
$$y = 0 \quad \text{or} \quad y = -11$$

The check is left to you.

Example 4. Solve $24 - 5a = a^2$ and check.

Solution. $24 - 5a = a^2$. Let's write all terms on the right-hand side so that the leading coefficient will be positive.

$$24 - 5a + 5a = a^2 + 5a$$
$$24 = a^2 + 5a$$
$$24 - 24 = a^2 + 5a - 24$$
$$0 = a^2 + 5a - 24$$
$$0 = (a + 8)(a - 3)$$
$$a + 8 = 0 \quad \text{or} \quad a - 3 = 0$$
$$a = -8 \quad \text{or} \quad a = 3$$

CHECK:
$$a = -8$$
$$24 - 5a = a^2$$
$$24 - 5(-8) \overset{?}{=} (-8)^2$$
$$24 - 5(-8) \overset{?}{=} (-8)(-8)$$
$$24 + 40 \overset{?}{=} 64$$
$$64 = 64$$

CHECK:
$$a = 3$$
$$24 - 5a = a^2$$
$$24 - 5(3) \overset{?}{=} (3)^2$$
$$24 - 5(3) \overset{?}{=} (3)(3)$$
$$24 - 15 \overset{?}{=} 9$$
$$9 = 9$$

Trial Run *Solve.*

———— 1. $x^2 + 2x = 35$

———— 2. $y^2 = 64$

———— 3. $x^2 = 3x$

———— 4. $y^2 = 7y - 12$

———— 5. $13 = x^2 - 12x$

———— 6. $15 - 2a = a^2$

ANSWERS
1. $x = -7$ or $x = 5$.   2. $y = 8$ or $y = 8$.   3. $x = 0$ or $x = 3$.   4. $y = 3$ or $y = 4$.
5. $x = 13$ or $x = -1$.   6. $a = -5$ or $a = 3$.

The method we have just learned will also work to solve equations containing parentheses or like terms that must be combined. Remember, to solve a quadratic equation, our aim is to have zero on one side of the equation and all other terms on the other side.

To solve the equation

$$2x^2 + 3x = x^2 + 28$$

we must get all the terms on one side. Let's subtract x^2 from both sides of the equation.

$$2x^2 + 3x - x^2 = x^2 + 28 - x^2$$
$$2x^2 + 3x - x^2 = 28$$

Now combine like terms on the left.

$$x^2 + 3x = 28$$

To obtain zero on the right, we subtract 28 from both sides.

$$x^2 + 3x - 28 = 28 - 28$$
$$x^2 + 3x - 28 = 0$$

Now we may factor and solve.

$$(x + 7)(x - 4) = 0$$

$$x + 7 = 0 \qquad \text{or} \qquad x - 4 = 0$$
$$x = -7 \qquad \text{or} \qquad x = 4$$

To solve the equation

$$3(x^2 - 1) = 2x(x + 1)$$

we recall that parentheses must be dealt with first. By the distributive property, our equation becomes

$$3x^2 - 3 = 2x^2 + 2x$$

Now we subtract $2x^2$ and $2x$ from both sides.

$$3x^2 - 3 - 2x^2 - 2x = 2x^2 + 2x - 2x^2 - 2x$$
$$3x^2 - 3 - 2x^2 - 2x = 0$$

Combining like terms and rearranging the terms, we have

$$x^2 - 2x - 3 = 0$$

Now we factor and solve.

$$(x - 3)(x + 1) = 0$$
$$x - 3 = 0 \qquad \text{or} \qquad x + 1 = 0$$
$$x = 3 \qquad \text{or} \qquad x = -1$$

To our list of procedures for solving quadratic equations by factoring, we must add one more step.

---

1. Work with parentheses if necessary.
2. Get zero on one side of the equation and all other terms on the other side (combining like terms if possible).
3. Factor the polynomial.
4. Set each factor equal to zero.
5. Solve for the variable and check.

Example 5. Solve $4x^2 + x = 3(x^2 - 4x)$.

Solution

$$4x^2 + x = 3(x^2 - 4x)$$
$$4x^2 + x = 3x^2 - 12x$$

$$4x^2 + x - 3x^2 = 3x^2 - 12x - 3x^2$$
$$x^2 + x = -12x$$
$$x^2 + x + 12x = -12x + 12x$$
$$x^2 + 13x = 0$$
$$x(x + 13) = 0$$
$$x = 0 \quad \text{or} \quad x + 13 = 0$$
$$x = 0 \quad \text{or} \quad x = -13$$

Example 6. Solve $5x(x - 2) - 6 = 4x^2 - 2(5x + 1)$.

Solution

$$5x(x - 2) - 6 = 4x^2 - 2(5x + 1)$$
$$5x^2 - 10x - 6 = 4x^2 - 10x - 2$$
$$5x^2 - 10x - 6 - 4x^2 = 4x^2 - 10x - 2 - 4x^2$$
$$x^2 - 10x - 6 = -10x - 2$$
$$x^2 - 10x - 6 + 10x = -10x - 2 + 10x$$
$$x^2 - 6 = -2$$
$$x^2 - 6 + 2 = -2 + 2$$
$$x^2 - 4 = 0$$
$$(x + 2)(x - 2) = 0$$
$$x + 2 = 0 \quad \text{or} \quad x - 2 = 0$$
$$x = -2 \quad \text{or} \quad x = 2$$

Example 7. Solve $3(x^2 - 1) - 9 = x(x + 10)$.

Solution

$$3(x^2 - 1) - 9 = x(x + 10)$$
$$3x^2 - 3 - 9 = x^2 + 10x$$
$$3x^2 - 12 = x^2 + 10x$$
$$3x^2 - 12 - x^2 = x^2 + 10x - x^2$$
$$2x^2 - 12 = 10x$$
$$2x^2 - 12 - 10x = 10x - 10x$$
$$2x^2 - 10x - 12 = 0$$
$$2(x^2 - 5x - 6) = 0$$
$$2(x - 6)(x + 1) = 0$$
$$x - 6 = 0 \quad \text{or} \quad x + 1 = 0$$
$$x = 6 \quad \text{or} \quad x = -1$$

CHECK:
$$x = 6$$
$$3(x^2 - 1) - 9 = x(x + 10)$$
$$3[(6)^2 - 1] - 9 \overset{?}{=} 6(6 + 10)$$
$$3[36 - 1] - 9 \overset{?}{=} 6(16)$$
$$3[35] - 9 \overset{?}{=} 96$$
$$105 - 9 \overset{?}{=} 96$$
$$96 = 96$$

CHECK:
$$x = -1$$
$$3(x^2 - 1) - 9 = x(x + 10)$$
$$3[(-1)^2 - 1] - 9 \overset{?}{=} -1(-1 + 10)$$
$$3[1 - 1] - 9 \overset{?}{=} -1(9)$$
$$3(0) - 9 \overset{?}{=} -9$$
$$0 - 9 \overset{?}{=} -9$$
$$-9 = -9$$

Do not be tempted to skip steps in working these problems. Shortcuts may cause errors.

Solve.

_____ **1.** $4(x^2 - 2x) = 3x^2 - 15$

_____ **2.** $7x^2 - 6x = 3x(2x + 3)$

_____ **3.** $3y(y - 6) - 21 = 2y^2 - 6(3y + 2)$

_____ **4.** $3(x^2 - 2) + 16 = x(x + 12)$

_____ **5.** $5a(a - 1) + 36 = 3(2a^2 - 15) - 5a$

ANSWERS
1. $x = 5$ or $x = 3$.　**2.** $x = 0$ or $x = 15$.　**3.** $y = 3$ or $y = -3$.　**4.** $x = 1$ or $x = 5$.
5. $a = 9$ or $a = -9$.

EXERCISE SET 6.2

Solve.

_____ 1. $x^2 + 3x = 18$ _____ 2. $x^2 - 7x = 18$

_____ 3. $y^2 = 81$ _____ 4. $y^2 = 121$

_____ 5. $x^2 = 5x$ _____ 6. $x^2 = 11x$

_____ 7. $y^2 = 13y - 36$ _____ 8. $y^2 = 15y - 56$

_____ 9. $72 = x^2 - x$ _____ 10. $54 = x^2 - 3x$

_____ 11. $10 - 3a = a^2$ _____ 12. $33 - 8a = a^2$

_____ 13. $x^2 - 16x = -64$ _____ 14. $x^2 - 20x = -100$

_____ 15. $x^2 - 52 = 9x$ _____ 16. $x^2 - 36 = 9x$

_____ 17. $144 = a^2$ _____ 18. $225 = a^2$

_____ 19. $y^2 = -17y$ _____ 20. $y^2 = -15y$

_____ 21. $4x + 21 = x^2$ _____ 22. $12x + 28 = x^2$

_____ 23. $100 = y^2 - 15y$ _____ 24. $34 = y^2 + 15y$

_____ 25. $x^2 - 14x = -45$ _____ 26. $x^2 - 11x = -28$

_____ 27. $9x = x^2$ _____ 28. $16x = x^2$

_____ 29. $x^2 + 6 = -5x$ _____ 30. $x^2 + 24 = -11x$

_____ 31. $3y^2 = 12$ _____ 32. $4y^2 = 100$

_____ 33. $2x^2 - 12x = -18$ _____ 34. $3x^2 - 24x = -48$

_____ 35. $5y^2 = 5y$ _____ 36. $11y^2 = 22y$

_____ 37. $54 + 12x = 2x^2$ _____ 38. $45 - 40x = 5x^2$

_____ 39. $10y^2 = 360$ _____ 40. $9y^2 = 9$

_____ 41. $5(x^2 - 1) = 4(x^2 + 5)$ _____ 42. $3(x^2 - 2) = 2(x^2 + 15)$

_____ 43. $9a^2 - 31a = 4a(2a - 7)$ _____ 44. $5a(2a - 3) = a(9a - 5)$

_____ 45. $2x(x - 4) - 28 = x^2 - 2(3x - 10)$ _____ 46. $3x(2x - 1) + 8 = 5x^2 - 3(3x - 5)$

_____ 47. $4(x^2 - 2x) + 25 = x(3x + 2)$ _____ 48. $5(x^2 - 4x) + 9 = 2x(2x - 7)$

_____ 49. $5x(x - 1) + 8 = 3(x^2 + 2) - 9x$ _____ 50. $7x(x - 1) - 60 = 2(2x^2 + 15) - 4x$

_____ 51. $3y(y - 5) = 2y(y - 3)$ _____ 52. $4(y^2 - 1) - 2y = 3y(y + 4) - 4$

CHECKUP 6.2

Solve.

_____ **1.** $x^2 - 3x = 40$

_____ **2.** $x^2 = 7x$

_____ **3.** $63 = a^2 + 2a$

_____ **4.** $x^2 - 14x = -49$

_____ **5.** $9 = a^2$

_____ **6.** $3x + 70 = x^2$

_____ **7.** $3x^2 + 18 = 15x$

_____ **8.** $5(x^2 - 4) = 4(x^2 + 11)$

_____ **9.** $5y(2y - 6) = 2y(3y - 1)$

_____ **10.** $4y(y + 1) + 7 = 3(y^2 - 1) - 3y$

6.3 Switching from Word Statements to Quadratic Equations

In Chapter 4 we practiced switching from word expressions to variable expressions. We began by deciding what the variable represented. Then we translated the word expressions in the problem into variable expressions. Now we shall put those same skills to work in solving for unknown quantities in word statements.

Problem 1

Suppose that we wish to find the length of one side of a square checkerboard with area 64 square inches. First we must decide on the variable.

$$\text{let } s = \text{length of one side}$$

To find area, we must multiply the length of one side times the length of another side, so

$$\text{area} = s \cdot s$$
$$= s^2$$

But here we know that the area is 64 square inches, so we may write the equation

$$s^2 = 64$$

Since this is a quadratic equation, we put it in the proper form with zero on one side.

$$s^2 = 64$$
$$s^2 - 64 = 64 - 64$$
$$s^2 - 64 = 0$$

Now factor and solve.

$$(s + 8)(s - 8) = 0$$
$$s + 8 = 0 \quad \text{or} \quad s - 8 = 0$$
$$s = -8 \quad \text{or} \quad s = 8$$

Do both answers make sense? The length of a side cannot be a negative number, so our only solution is

$$s = 8 \text{ inches}$$

Problem 2

The Watsons wish to build a rectangular concrete patio which is 6 feet longer than it is wide, but they only have enough concrete mix to cover 280 square feet. What dimensions should their patio have? Suppose that we let

$$w = \text{width of patio}$$

Then

$$w + 6 = \text{length of patio}$$

Remember that the area of a rectangle is found by

$$\text{area} = \text{length} \times \text{width}$$

so here

$$\text{area} = (w + 6)(w)$$

But we know the area must be covered by 280 square feet of concrete, so

$$(w + 6)w = 280$$

Removing parentheses and getting zero on one side, we have

$$w^2 + 6w = 280$$
$$w^2 + 6w - 280 = 280 - 280$$
$$w^2 + 6w - 280 = 0$$

Now we factor and solve.

$$(w + 20)(w - 14) = 0$$
$$w + 20 = 0 \quad \text{or} \quad w - 14 = 0$$
$$w = -20 \quad \text{or} \quad w = 14$$

Once again, we cannot use a negative number for width, so our only solution is

$$w = 14 \text{ feet}$$

What is the length of the Watsons' patio?

$$\text{length} = w + 6$$
$$= 14 + 6$$
$$\text{length} = 20 \text{ feet}$$

So the Watsons' patio should have dimensions 20 by 14 feet. Does our solution check with the words of the problem? If the patio is 14 feet wide and 20 feet long, the length is certainly 6 feet more than the width. What about the area?

$$A = \text{length} \times \text{width}$$
$$A = 20(14)$$
$$= 280 \text{ square feet}$$

Our solution checks.

Let's try some more word problems.

Example 1. If the product of two consecutive integers is 90, find the integers.

Solution. Recall from Chapter 4 that consecutive integers follow each other immediately in the list of integers.

$$\text{let } x = \text{first integer}$$
$$x + 1 = \text{next consecutive integer}$$
$$x(x + 1) = \text{product of the consecutive integers}$$

Since the product is 90, we know that

$$x(x + 1) = 90$$
$$x^2 + x = 90$$
$$x^2 + x - 90 = 0$$
$$(x + 10)(x - 9) = 0$$
$$x + 10 = 0 \quad \text{or} \quad x - 9 = 0$$
$$x = -10 \quad \text{or} \quad x = 9$$

We have two perfectly acceptable solutions. When

$$x = -10$$
$$\text{first integer} = x = -10$$
$$\text{next integer} = x + 1 = -10 + 1 = -9$$

So our first pair of consecutive integers is -10 and -9.

When

$$x = 9$$
$$\text{first integer} = x = 9$$
$$\text{next integer} = x + 1 = 9 + 1 = 10$$

So our second pair of consecutive integers is 9 and 10.

In each case, the product of the consecutive integers is 90, so our solutions check with the words of the original problem.

Example 2. One side of a right triangle is 7 centimeters shorter than the other side. If the hypotenuse is 13 centimeters long, find the lengths of the sides.

Solution. Recall from Chapter 4 that if a and b are the lengths of the sides of a right triangle with hypotenuse of length c, then the Pythagorean Theorem says that

$$a^2 + b^2 = c^2$$

In our problem, suppose that we let

$$
\begin{aligned}
a &= \text{longer side} \\
a - 7 &= \text{shorter side} \\
13 &= \text{hypotenuse}
\end{aligned}
$$

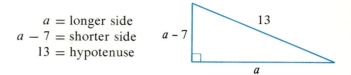

Then, by the Pythagorean Theorem,

$$
\begin{aligned}
a^2 + (a - 7)^2 &= 13^2 \\
a^2 + (a - 7)(a - 7) &= 13 \cdot 13 \\
a^2 + a^2 - 14a + 49 &= 169 \\
2a^2 - 14a + 49 &= 169 \\
2a^2 - 14a + 49 - 169 &= 169 - 169 \\
2a^2 - 14a - 120 &= 0
\end{aligned}
$$

Let's factor and solve.

$$
\begin{aligned}
2(a^2 - 7a - 60) &= 0 \\
2(a - 12)(a + 5) &= 0
\end{aligned}
$$

$$
\begin{array}{lcl}
a - 12 = 0 & \text{or} & a + 5 = 0 \\
a = 12 & \text{or} & a = -5
\end{array}
$$

Since a represents length, we reject the negative solution and conclude that $a = 12$ centimeters.

$$
\begin{aligned}
\text{longer side} &= a = 12 \text{ centimeters} \\
\text{shorter side} &= a - 7 = 12 - 7 = 5 \text{ centimeters} \\
\text{hypotenuse} &= 13 \text{ centimeters}
\end{aligned}
$$

Checking these lengths in the Pythagorean Theorem,

$$
\begin{aligned}
12^2 + 5^2 &\overset{?}{=} 13^2 \\
12 \cdot 12 + 5 \cdot 5 &\overset{?}{=} 13 \cdot 13 \\
144 + 25 &\overset{?}{=} 169 \\
169 &= 169
\end{aligned}
$$

Trial Run   *Solve.*

_____ 1. The length of the Abells' living room is the same as its width. If it takes 225 square feet of carpet to cover the room, find the length of a side of the room.

_____ 2. Find two consecutive integers whose product is 56.

_____ 3. One side of a right triangle is 2 centimeters longer than the other side. If the hypotenuse is 10 centimeters long, find the lengths of the two sides.

_____ 4. The width of a rectangle is 6 meters less than the width. The area of the rectangle is 40 square meters. Find the length and the width.

ANSWERS
1. 15 feet.   **2.** 7, 8 or -8, -7.   **3.** 6 centimeters, 8 centimeters.   **4.** 10 meters, 4 meters.

EXERCISE SET 6.3

Solve.

_____ **1.** One number is 5 more than another. The square of the smaller added to 3 times the larger is 43. Find the numbers.

_____ **2.** One number is 3 more than another. The square of the smaller added to 5 times the larger is 65. Find the numbers.

_____ **3.** The width of a rectangle is 1 less than the length. The area is 72 square feet. Find the length and the width.

_____ **4.** The width of a rectangle is 2 less than the length. The area is 63 square feet. Find the length and the width.

_____ **5.** Three times the square of an integer added to 6 times the integer is 105. Find the integer.

_____ **6.** If four times an integer is subtracted from twice the square of the integer, the result is 48. Find the integer.

_____ **7.** If the area of a square is 81 square centimeters, find the length of a side of the square.

_____ **8.** The area of a square is 36 square centimeters. Find the length of a side of the square.

_____ **9.** One side of a right triangle is 7 centimeters less than the other side and the hypotenuse is 17 centimeters. Find the lengths of the two sides.

_____ **10.** One side of a right triangle is 3 centimeters more than the other side and the hypotenuse is 15 centimeters. Find the lengths of the two sides.

_____ **11.** The length of a rectangle is twice its width. The area is 128 square feet. Find the dimensions of the rectangle.

_____ **12.** The length of a rectangle is 3 times its width. The area is 75 square feet. Find the dimensions of the rectangle.

_____ **13.** The Haskins plan to buy a rectangular lot that is 50 feet longer than it is wide. If the area of the lot is 30,000 square feet, find its dimensions.

_____ **14.** The Perrys plan to fence a dog pen that is to be 8 feet longer than it is wide. They have enough fence to enclose an area of 240 square feet. Find the dimensions of the pen.

_____ **15.** A tree was broken over by a storm in such a way that the part that was bending over was 26 feet. The part left standing was 14 feet less than the distance from the foot of the tree to the point where the broken part touched the ground. Find the height of the tree before the storm.

_____ **16.** A flower bed is in the shape of a right triangle. One side is 7 feet shorter than the other side and the hypotenuse is 17 feet. Find how many feet of picket fence will be needed to border the flower bed.

_____ **17.** The sum of the squares of two consecutive integers is 61. Find the integers.

_____ **18.** The sum of the squares of two consecutive integers is 113. Find the integers.

_____ **19.** The product of two consecutive even integers is 4 more than five times the smaller. Find the integers.

_____ **20.** The product of two consecutive odd integers is 7 more than their sum. Find the integers.

CHECKUP 6.3

Solve.

_____ 1. One number is 4 more than another. The square of the smaller added to twice the larger is 32. Find the numbers.

_____ 2. The width of a rectangle is 2 centimeters less than the length and the area is 35 square centimeters. Find the dimensions.

_____ 3. If the area of a square is 144 square centimeters, find the length of a side.

_____ 4. One side of a right triangle is 1 inch more than the other side. The hypotenuse is 5 inches. Find the lengths of the two sides.

_____ 5. The sum of the squares of two consecutive integers is 13. Find the integers.

Summary

In this chapter we learned to solve equations in which the largest exponent on the variable is 2. Such equations are called *quadratic* equations, and we developed the following procedure to solve quadratic equations.

1. Work with parentheses if necessary.
2. Get zero on one side of the equation and all other terms on the other side (combining like terms if possible).
3. Factor the polynomial.
4. Set each factor equal to zero.
5. Solve for the variable and check.

Success with this procedure depended on our ability to *factor* and work with the *zero product rule,* which states that if a product of factors is zero, one factor or the other (or both) must be zero.

We then practiced switching from word statements to variable statements that could be written as quadratic equations. We solved for the variable in the usual manner and checked to be sure that the solution made sense in the original word problem.

REVIEW EXERCISES

Solve for the variable.

SECTION 6.1

_____ **1.** $(x - 7)(x + 8) = 0$ _____ **2.** $a(a - 3) = 0$

_____ **3.** $(9 - y)(10 + y) = 0$ _____ **4.** $5(x - 8)(x - 2) = 0$

_____ **5.** $x^2 - 2x - 24 = 0$ _____ **6.** $x^2 - 169 = 0$

_____ **7.** $m^2 + 15m = 0$ _____ **8.** $162 - 2a^2 = 0$

_____ **9.** $30n - 10n^2 = 0$ _____ **10.** $z^2 + 17z + 52 = 0$

SECTION 6.2

_____ **11.** $y^2 + 11y = -24$ _____ **12.** $x^2 = 121$

_____ **13.** $y^2 = -23y$ _____ **14.** $40 = x^2 + 6x$

_____ **15.** $x^2 + 63 = 24x$ _____ **16.** $5x^2 = 405$

_____ **17.** $14z^2 - 9z = 5(2z^2 - z)$

_____ **18.** $5x(x - 2) - 18 = 4x^2 - 3(x - 4)$

_____ **19.** $4y(3y - 2) = 7y(y + 1)$

_____ **20.** $4(x^2 - 3) - 3x = 3x(x + 1) + 4$

SECTION 6.3

_____ **21.** The length of a rectangular bathroom is 4 feet more than the width. If 60 square feet of linoleum will cover the floor, what are the dimensions of the bathroom?

_____ **22.** If the area of a square is 225 square centimeters, find the length of a side of the square.

_____ **23.** One integer is 4 less than another and the product of the two is 117. Find the integers.

_____ **24.** Hannah wants to border her antique shawl with lace. The shawl is in the shape of a right triangle with one side 1 foot longer than the other side, and the hypotenuse is 5 feet. How many feet of lace should she buy?

_____ **25.** The product of two consecutive odd integers is 7 more than 4 times the larger. Find the integers.

7 Working with Rational Expressions

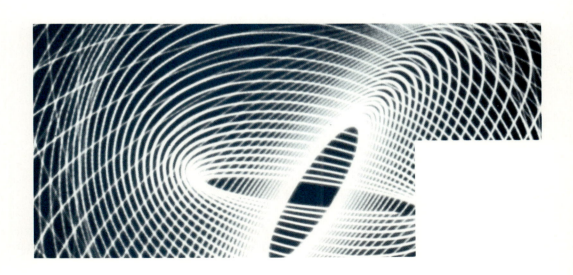

Up to this point, we have been working with numbers in the list of integers.

$$\ldots, -2, -1, 0, 1, 2, 3, 4, \ldots$$

We have learned the rules for adding, subtracting, multiplying, and dividing integers, and all our equations have had solutions that were integers.

It should come as no surprise to you that there are many useful numbers that are *not* integers, numbers such as

$$\frac{1}{2}, \quad -\frac{2}{3}, \quad 0.5, \quad \frac{17}{37}, \quad -0.23, \quad 7\frac{1}{2}$$

and so on. Although you have probably gained skill working with such numbers previously, we shall review those skills here and see how such numbers fit into the study of algebra.

The numbers listed above are called **rational numbers**, and in this chapter we shall learn how to

1. Recognize rational numbers.
2. Recognize rational algebraic expressions.
3. Reduce rational expressions.
4. Multiply and divide rational expressions.

7.1 Recognizing Rational Numbers

We have stated that numbers such as

$$\frac{1}{2}, \quad -\frac{2}{3}, \quad 0.5, \quad \frac{17}{37}, \quad -0.23, \quad 7\frac{1}{2}$$

and so on are called rational numbers, but how can we describe such numbers in general? Can you see that each can be written as a *fraction of two integers?* Let's check.

$$\frac{1}{2} = \frac{1}{2} \quad \begin{matrix} \leftarrow \text{integer} \\ \leftarrow \text{integer} \end{matrix}$$

$$-\frac{2}{3} = \frac{-2}{3} \quad \begin{matrix} \leftarrow \text{integer} \\ \leftarrow \text{integer} \end{matrix}$$

$$0.5 = \frac{5}{10} \quad \begin{matrix} \leftarrow \text{integer} \\ \leftarrow \text{integer} \end{matrix}$$

$$\frac{17}{37} = \frac{17}{37} \quad \begin{matrix} \leftarrow \text{integer} \\ \leftarrow \text{integer} \end{matrix}$$

$$-0.23 = \frac{-23}{100} \quad \begin{matrix} \leftarrow \text{integer} \\ \leftarrow \text{integer} \end{matrix}$$

$$7\frac{1}{2} = \frac{15}{2} \quad \begin{matrix} \leftarrow \text{integer} \\ \leftarrow \text{integer} \end{matrix}$$

A rational number is defined to be any number that can be written as a fraction with an integer in the numerator and an integer (not zero) in the denominator.

Definition of Rational Number. A number is rational if it can be written in the form

$$\frac{A}{B}$$

where A is an integer and B is an integer, but $B \neq 0$.

Notice how the integers themselves fit into this definition. Do you see that we could rewrite every integer as a fraction of integers?

$$2 = \frac{2}{1} \quad \text{or} \quad \frac{6}{3} \quad \text{or} \quad \frac{50}{25} \qquad \text{and so on}$$

$$-7 = \frac{-7}{1} \quad \text{or} \quad \frac{14}{-2} \quad \text{or} \quad \frac{-70}{10} \qquad \text{and so on}$$

$$0 = \frac{0}{5} \quad \text{or} \quad \frac{0}{16} \quad \text{or} \quad \frac{0}{-123} \qquad \text{and so on}$$

In other words, *integers are rational numbers.* To show that a number is rational, we must show that it can be written as a fraction of integers.

Example 1. Show that 0.2 is rational.

Solution. 0.2 is two-tenths:

$$0.2 = \frac{2}{10} \quad \begin{array}{l} \leftarrow \text{integer} \\ \leftarrow \text{integer} \end{array}$$

So 0.2 is a rational number.

Example 2. Show that -17 is rational.

Solution

$$-17 = \frac{-17}{1} \quad \begin{array}{l} \leftarrow \text{integer} \\ \leftarrow \text{integer} \end{array}$$

So -17 is a rational number.

Trial Run   *Show that the following numbers are rational.*

_____ **1.** 0.3

_____ **2.** $2\frac{1}{3}$

_____ **3.** -9

_____ **4.** 0.047

ANSWERS

1. $\frac{3}{10}$.   **2.** $\frac{7}{3}$.   **3.** $\frac{-9}{1}$.   **4.** $\frac{-47}{1000}$.

    It may seem to you that all numbers must be rational. Indeed, most numbers you work with every day *are* rational. The numbers π, $\sqrt{2}$, $\sqrt{3}$, which you may have seen before, are examples of numbers that are *not* rational. Numbers that are not rational are called **irrational numbers**, but we shall not discuss such numbers until Chapter 12.

7.1-1   Properties of Rational Numbers

You will be relieved to know that all the properties that you have learned for the integers will continue to work for the rational numbers. Even before we discuss how to add, subtract, multiply, and divide rational numbers, we shall agree to accept those properties. If A and B and C are rational numbers, the following properties hold.

Commutative Property for Addition

$$A + B = B + A$$

Associative Property for Addition

$$A + (B + C) = (A + B) + C$$

Commutative Property for Multiplication

$$A \cdot B = B \cdot A$$

Associative Property for Multiplication

$$A \cdot (B \cdot C) = (A \cdot B) \cdot C$$

Distributive Property

$$A \cdot (B + C) = A \cdot B + A \cdot C$$

Properties of Zero

$$A + 0 = A$$
$$A - 0 = A$$
$$A \cdot 0 = 0$$
$$\frac{0}{A} = 0 \qquad \text{if } A \neq 0$$
$$\frac{A}{0} \text{ is undefined}$$

Zero Product Rule

if $A \cdot B = 0$
then $A = 0$ or $B = 0$ or both

7.1-2 Solving Equations with Rational Solutions

We shall continue to use the methods that we have learned to solve equations. However, now that we are allowed to use all the rational numbers, we can solve some equations that do not have integer solutions. Let's try solving some first-degree equations. Remember: we must isolate the variable.

Example 3. Solve $2x + 3 = 4$ and check.

Solution

$$2x + 3 = 4$$
$$2x + 3 - 3 = 4 - 3$$
$$2x = 1$$
$$\frac{2x}{2} = \frac{1}{2}$$
$$x = \frac{1}{2}$$

CHECK:
$$x = \frac{1}{2}$$
$$2x + 3 = 4$$
$$2\left(\frac{1}{2}\right) + 3 \stackrel{?}{=} 4$$
$$1 + 3 \stackrel{?}{=} 4$$
$$4 = 4$$

Example 4. Solve $5y = 9$.

Solution

$$5y = 9$$
$$\frac{5y}{5} = \frac{9}{5}$$
$$y = \frac{9}{5}$$

Example 5. Solve $5x + 8 = 2x - 6$.

Solution

$$5x + 8 = 2x - 6$$
$$5x + 8 - 2x = 2x - 6 - 2x$$
$$3x + 8 = -6$$
$$3x + 8 - 8 = -6 - 8$$
$$3x = -14$$
$$\frac{3x}{3} = \frac{-14}{3}$$
$$x = \frac{-14}{3}$$

What about solving some *quadratic equations* when rational-number solutions are allowed? Remember that we shall make use of the zero product rule and our factoring skills.

Example 6. Solve $2x^2 + x - 1 = 0$.

Solution

$$2x^2 + x - 1 = 0$$
$$(2x - 1)(x + 1) = 0$$

$$2x - 1 = 0 \quad \text{or} \quad x + 1 = 0$$
$$2x = 1 \qquad\qquad x = -1$$
$$\frac{2x}{2} = \frac{1}{2}$$
$$x = \frac{1}{2} \quad \text{or} \quad x = -1$$

Example 7. Solve $4y^2 = 9$.

Solution

$$4y^2 = 9$$
$$4y^2 - 9 = 9 - 9$$
$$4y^2 - 9 = 0$$
$$(2y + 3)(2y - 3) = 0$$

$$2y + 3 = 0 \qquad \text{or} \qquad 2y - 3 = 0$$
$$2y = -3 \qquad\qquad 2y = 3$$
$$\frac{2y}{2} = \frac{-3}{2} \qquad\qquad \frac{2y}{2} = \frac{3}{2}$$
$$y = \frac{-3}{2} \quad \text{or} \quad y = \frac{3}{2}$$

Example 8. Solve $5x^2 = 13x - 6$.

Solution

$$5x^2 = 13x - 6$$
$$5x^2 - 13x = 13x - 6 - 13x$$
$$5x^2 - 13x = -6$$
$$5x^2 - 13x + 6 = -6 + 6$$
$$5x^2 - 13x + 6 = 0$$
$$(5x - 3)(x - 2) = 0$$

$$5x - 3 = 0 \quad \text{or} \quad x - 2 = 0$$
$$5x = 3 \qquad\qquad x = 2$$
$$\frac{5x}{5} = \frac{3}{5}$$
$$x = \frac{3}{5} \quad \text{or} \quad x = 2$$

Example 9. Solve $8z^2 = 7z$.

Solution

$$8z^2 = 7z$$
$$8z^2 - 7z = 7z - 7z$$
$$8z^2 - 7z = 0$$
$$z(8z - 7) = 0$$
$$z = 0 \quad \text{or} \quad 8z - 7 = 0$$
$$8z = 7$$
$$\frac{8z}{8} = \frac{7}{8}$$
$$z = 0 \quad \text{or} \quad z = \frac{7}{8}$$

Trial Run *Solve.*

_____ 1. $3x + 2 = 6$

_____ 2. $x + 7 = 3x - 2$

_____ 3. $3(z - 1) = 5(z + 2)$

_____ 4. $3x^2 + 20x - 7 = 0$

_____ 5. $25y^2 = 64$

_____ 6. $5z^2 = 3z$

_____ 7. $2(x^2 + 1) = 5x$

ANSWERS

1. $x = \frac{4}{3}$. 2. $x = \frac{9}{2}$. 3. $z = \frac{-13}{2}$. 4. $x = \frac{1}{3}$ or $x = -7$. 5. $y = \frac{8}{5}$ or $y = -\frac{8}{5}$.

6. $z = 0$ or $z = \frac{3}{5}$. 7. $x = \frac{1}{2}$ or $x = 2$.

7.1-3 Rational Algebraic Expressions

We have already spent a great deal of time working with polynomials. An algebraic expression that can be written as a fraction with a polynomial numerator and a polynomial denominator is called a **rational algebraic expression**. Some examples of rational algebraic expressions are

$$\frac{x^2 \leftarrow \text{polynomial}}{7x \leftarrow \text{polynomial}}$$

$$\frac{y + 3 \leftarrow \text{polynomial}}{y - 5 \leftarrow \text{polynomial}}$$

$$\frac{a^2 + 5 \leftarrow \text{polynomial}}{2a \leftarrow \text{polynomial}}$$

$$\frac{x^2 + 7x + 6 \leftarrow \text{polynomial}}{x^2 - 1 \leftarrow \text{polynomial}}$$

We should notice what happens when we evaluate each of these expressions using a particular value of the variable.

Example 10. Evaluate $\dfrac{x^2}{7x}$ when $x = 3$.

Solution. Substitute 3 for x in the expression.

$$\frac{x^2}{7x} = \frac{3^2}{7 \cdot 3}$$

$$= \frac{9}{21}$$

Example 11. Evaluate $\dfrac{y + 3}{y - 5}$ when $y = 0$.

Solution. Substitute 0 for y in the expression.

$$\frac{y + 3}{y - 5} = \frac{0 + 3}{0 - 5}$$

$$= \frac{3}{-5}$$

$$= \frac{-3}{5}$$

In Example 11, notice that we chose to write our final fraction with the negative sign in the numerator rather than in the denominator. In mathematics it is customary to "let the numerator carry the sign" for the fraction.

You should agree, for instance, that

$$\frac{-6}{3} = \frac{6}{-3} = -\frac{6}{3}$$

because each of these fractions has the value -2. Of the three ways to write this fraction, the first way

$$\frac{-6}{3}$$

is preferred for reasons that will become clear later in this chapter. In general, we note that if $b \neq 0$, then

$$\frac{-a}{b} = \frac{a}{-b} = -\frac{a}{b}$$

all represent the same fraction, but

$$\frac{-a}{b}$$

is the preferred form.

Example 12. Evaluate $\dfrac{a^2 + 5}{2a}$ when $a = -2$.

Solution. Substitute -2 for a in the expression.

$$\begin{aligned}
\frac{a^2 + 5}{2a} &= \frac{(-2)^2 + 5}{2(-2)} \\
&= \frac{4 + 5}{-4} \\
&= \frac{9}{-4} \\
&= \frac{-9}{4}
\end{aligned}$$

Example 13. Evaluate $\dfrac{y + 3}{y - 5}$ when $y = -3$.

Solution. Substitute -3 for y in the expression.

$$\begin{aligned}
\frac{y + 3}{y - 5} &= \frac{-3 + 3}{-3 - 5} \\
&= \frac{0}{-8} \\
&= 0
\end{aligned}$$

Example 14. Evaluate $\dfrac{y + 3}{y - 5}$ when $y = 5$.

Solution. Substitute 5 for y in the expression.

$$\begin{aligned}
\frac{y + 3}{y - 5} &= \frac{5 + 3}{5 - 5} \\
&= \frac{8}{0} \quad \text{is undefined}
\end{aligned}$$

Example 15. Evaluate $\dfrac{a^2 + 5}{2a}$ when $a = 0$.

Solution. Substitute 0 for a in the expression.

$$\begin{aligned}
\frac{a^2 + 5}{2a} &= \frac{0^2 + 5}{2 \cdot 0} \\
&= \frac{0 + 5}{0} \\
&= \frac{5}{0} \quad \text{is undefined}
\end{aligned}$$

Notice from these examples that whenever we obtained a numerical answer, that number was rational. Notice also that in some cases our answer was undefined, because *divi-*

sion by zero is always undefined. In working with rational algebraic expressions, we must avoid substituting for the variable any number that will make the denominator equal zero.

Example 16. What value of x will make $\dfrac{x}{x-3}$ undefined?

Solution. We look for a value of x that will make the denominator equal zero.

$$x - 3 = 0$$
$$x = 3$$

The value $x = 3$ will make $\dfrac{x}{x-3}$ undefined.

Example 17. What value of y will make $\dfrac{y+3}{y+2}$ undefined?

Solution. We look for a value of y that will make the denominator equal zero.

$$y + 2 = 0$$
$$y = -2$$

The value $y = -2$ will make $\dfrac{y+3}{y+2}$ undefined.

Example 18. What value of a will make $\dfrac{a^2-2}{5a}$ undefined?

Solution. We look for a value of a that will make the denominator equal zero.

$$5a = 0$$
$$\frac{5a}{5} = \frac{0}{5}$$
$$a = 0$$

The value $a = 0$ will make $\dfrac{a^2-2}{5a}$ undefined.

The process of finding the value of the variable that will make the denominator zero is called **restricting the variable**. Since a rational expression is undefined whenever the denominator is zero, we are saying that a rational expression will have meaning for any choice of a value for the variable provided that we restrict those choices. We must *not* choose a value for the variable that will make the denominator zero.

Example 19. Restrict the variable for $\dfrac{5x}{x-7}$.

Solution. Since the denominator is zero when

$$x - 7 = 0$$
$$x = 7$$

we restrict the choices for x by saying that $x \neq 7$.

Example 20. Restrict the variable for $\dfrac{x+3}{3x}$.

Solution. Since the denominator is zero when

$$3x = 0$$
$$\frac{3x}{3} = \frac{0}{3}$$
$$x = 0$$

we restrict the choices for x by saying that $x \neq 0$.

Trial Run *Evaluate.*

_____ 1. $\dfrac{x^2}{3x}$, $x = 2$

_____ 2. $\dfrac{y - 5}{y + 2}$, $y = 0$

_____ 3. $\dfrac{a - 4}{a^2 - 4}$, $a = 2$

Restrict the variable.

_____ 4. $\dfrac{x - 2}{x + 3}$

_____ 5. $\dfrac{y^2 + 1}{3y}$

_____ 6. $\dfrac{3a}{a - 9}$

ANSWERS

1. $\dfrac{4}{6}$.   2. $\dfrac{-5}{2}$.   3. Undefined.   4. $x \neq -3$.   5. $y \neq 0$.   6. $a \neq 9$.

Remember, the denominator of a rational algebraic expression cannot equal zero. It is perfectly all right for the numerator to equal zero, but never the denominator.

EXERCISE SET 7.1

Show that the following numbers are rational.

_____ 1. 4

_____ 2. 9

_____ 3. -0.9

_____ 4. -0.1

_____ 5. $5\frac{1}{3}$

_____ 6. $4\frac{2}{5}$

_____ 7. -13

_____ 8. -8

_____ 9. 0.13

_____ 10. 0.17

_____ 11. -0.031

_____ 12. -0.083

Solve for the variable.

_____ 13. $3x = x + 9$

_____ 14. $4x = x + 5$

_____ 15. $5y - 11 = 3$

_____ 16. $3y - 10 = 9$

_____ 17. $4 = 7a + 9$

_____ 18. $13 = 4a + 16$

_____ 19. $3z = 7z + 5$

_____ 20. $5z = 11z + 5$

_____ 21. $7x + 5 = 2x + 14$

_____ 22. $9y + 6 = 3y + 17$

_____ 23. $10y - 3 = 0$

_____ 24. $8y - 9 = 0$

_____ 25. $7(2y - 3) = 4(7 + 2y)$

_____ 26. $2(3y - 1) = 5(4 - y)$

_____ 27. $5(3a - 1) + 13 = -2(a - 4)$

_____ 28. $3(2a - 4) - 6 = 2(a - 9)$

_____ 29. $5x^2 + 13x - 6 = 0$

_____ 30. $6x^2 + 19x - 7 = 0$

_____ 31. $6z^2 + 7z = 20$

_____ 32. $8z^2 + 10z = 3$

_____ 33. $10x^2 = 22x - 4$

_____ 34. $12x^2 = 27x - 15$

_____ 35. $7a(3a - 1) = 6a - 2$

_____ 36. $5a(4a - 3) = 2a - 3$

_____ 37. $4(x^2 - 3x) = 3(x - 4) - 2$

_____ 38. $6(2x^2 - x) = 11(x - 1) + 5$

_____ **39.** $5(y^2 + 2) = 2(2y + 5)$ _____ **40.** $4(y^2 - 3) = 3(3y - 4)$

_____ **41.** $9z^2 = 4$ _____ **42.** $25z^2 = 144$

_____ **43.** $(4x + 1)(3x - 2) = 8 - 12x$ _____ **44.** $(5x + 1)(3x - 4) = 5x + 1$

Restrict the variable for each of the following.

_____ **45.** $\dfrac{6}{x^3}$ _____ **46.** $\dfrac{-4}{x^2}$

_____ **47.** $\dfrac{x^2}{x + 1}$ _____ **48.** $\dfrac{x^2}{x + 2}$

_____ **49.** $\dfrac{2(x + 2)}{3x}$ _____ **50.** $\dfrac{5(x - 1)}{4x}$

_____ **51.** $\dfrac{y - 1}{8 - y}$ _____ **52.** $\dfrac{y - 3}{5 - y}$

_____ **53.** $\dfrac{4x}{2x - 1}$ _____ **54.** $\dfrac{3x}{3x - 2}$

_____ **55.** $\dfrac{x - 7}{4}$ _____ **56.** $\dfrac{2x - 1}{5}$

CHECKUP 7.1

Show that the following are rational numbers.

_____ 1. -7

_____ 2. 0.31

Solve for the variable.

_____ 3. $2x = 5x + 1$

_____ 4. $20 = 6y + 9$

_____ 5. $8(2x - 3) = 6(x - 4)$

_____ 6. $20z^2 - 31z = 7$

_____ 7. $4x^2 = 25$

_____ 8. $8(y^2 - 2) = 4(3y - 4)$

Restrict the variable for each of the following.

_____ 9. $\dfrac{x - 3}{2x}$

_____ 10. $\dfrac{5y}{7 + y}$

7.2 Reducing Rational Expressions

Almost everyone will agree that

$$\frac{5}{10} = \frac{1}{2}$$

$$\frac{2}{6} = \frac{1}{3}$$

In other words, $\frac{5}{10}$ and $\frac{1}{2}$ represent the same rational number and $\frac{2}{6}$ and $\frac{1}{3}$ represent the same rational number. Before we can go on, we must agree as to what is meant when we say that two fractions are equal. Notice in the examples

$$\frac{5}{10} = \frac{1}{2} \quad \text{and} \quad \frac{2}{6} = \frac{1}{3}$$

that the product of the numerator on the left times the denominator on the right is *equal* to the product of the denominator on the left times the numerator on the right.

$$\frac{5}{10} = \frac{1}{2} \qquad 5(2) = 10(1) \qquad 10 = 10$$

$$\frac{2}{6} = \frac{1}{3} \qquad 2(3) = 6(1) \qquad 6 = 6$$

This fact will allow us to decide when two fractions are equal.

Equality of Fractions

$$\frac{A}{B} = \frac{C}{D}$$

if $A \cdot D = B \cdot C$

where $B \neq 0, \ D \neq 0$

Example 1. Are $\frac{2}{3}$ and $\frac{10}{15}$ equal fractions?

Solution

$$\frac{2}{3} \overset{?}{=} \frac{10}{15}$$

$$2(15) \overset{?}{=} 3(10)$$

$$30 = 30 \qquad \text{yes}$$

so

$$\frac{2}{3} = \frac{10}{15}$$

Example 2. Are $\frac{63}{39}$ and $\frac{21}{13}$ equal fractions?

Solution

$$\frac{63}{39} \overset{?}{=} \frac{21}{13}$$

$$63(13) \overset{?}{=} 39(21)$$

$$819 = 819 \qquad \text{yes}$$

so

$$\frac{63}{39} = \frac{21}{13}$$

Example 3. Are $\frac{72}{84}$ and $\frac{3}{4}$ equal fractions?

Solution

$$\frac{72}{84} \overset{?}{=} \frac{3}{4}$$
$$72(4) \overset{?}{=} 84(3)$$
$$288 \overset{?}{=} 252 \qquad \text{no}$$

so

$$\frac{72}{84} \neq \frac{3}{4}$$

7.2-1 Reducing Numerical Fractions

When a rational number is written as a fraction of constants, we sometimes refer to it as a **numerical fraction**. A numerical fraction contains no variables.

One way in which we can find a new fraction that is equivalent to a certain fraction is by a process called **reducing** the fraction. To reduce a fraction to an equivalent fraction, we are allowed to divide the numerator *and* denominator by any factor that they have in common.

For instance, consider the fraction

$$\frac{21}{33}$$

and look for a factor that the numerator and denominator have in common. The easiest way to do this is to factor the numerator and denominator.

$$\frac{21}{33} = \frac{3 \cdot 7}{3 \cdot 11}$$

The numerator and denominator contain a common factor of 3, so we divide the numerator and denominator by 3.

$$\frac{21}{33} = \frac{3 \cdot 7}{3 \cdot 11} = \frac{7}{11}$$

To indicate that we are dividing the numerator and denominator by 3, we may cross out those common factors of 3, like this:

$$\frac{21}{33} = \frac{\overset{1}{\cancel{3}} \cdot 7}{\underset{1}{\cancel{3}} \cdot 11} = \frac{7}{11}$$

A quick check will verify that

$$\frac{21}{33} = \frac{7}{11}$$
$$21(11) \overset{?}{=} 33(7)$$
$$233 = 233$$

Let's try to reduce the fraction

$$\frac{24}{84}$$

First we shall factor the numerator and denominator completely.

$$\frac{24}{84} = \frac{2 \cdot 2 \cdot 2 \cdot 3}{2 \cdot 2 \cdot 3 \cdot 7}$$

Looking for common factors, we see that the numerator and denominator contain $2 \cdot 2 \cdot 3$ in common. Dividing, we have

$$\frac{24}{84} = \frac{\cancel{2} \cdot \cancel{2} \cdot 2 \cdot \cancel{3}}{\cancel{2} \cdot \cancel{2} \cdot \cancel{3} \cdot 7} = \frac{2}{7}$$

> To reduce a fraction, we must
>
> **1.** Factor the numerator and denominator.
> **2.** Divide the numerator and denominator by any common factors.

Example 4. Reduce $\dfrac{25}{110}$.

Solution

$$\frac{25}{110} = \frac{5 \cdot 5}{2 \cdot 5 \cdot 11}$$

$$= \frac{\cancel{5} \cdot 5}{2 \cdot \cancel{5} \cdot 11}$$

$$= \frac{5}{2 \cdot 11}$$

$$= \frac{5}{22}$$

Example 5. Reduce $\dfrac{16}{39}$.

Solution

$$\frac{16}{39} = \frac{2 \cdot 2 \cdot 2 \cdot 2}{3 \cdot 13} \qquad \text{no common factors}$$

$$\frac{16}{39} \qquad \text{cannot be reduced}$$

Example 6. Reduce $\dfrac{-36}{180}$.

Solution

$$\frac{-36}{180} = \frac{-1 \cdot 2 \cdot 2 \cdot 3 \cdot 3}{2 \cdot 2 \cdot 3 \cdot 3 \cdot 5}$$

$$= \frac{-1 \cdot \cancel{2} \cdot \cancel{2} \cdot \cancel{3} \cdot \cancel{3}}{\cancel{2} \cdot \cancel{2} \cdot \cancel{3} \cdot \cancel{3} \cdot 5}$$

$$= \frac{-1}{5}$$

Decide if the pairs of fractions are equal.

_____ 1. $\dfrac{35}{40}, \dfrac{7}{8}$

_____ 2. $\dfrac{9}{16}, \dfrac{3}{4}$

_____ 3. $\dfrac{7}{9}, \dfrac{21}{36}$

Reduce.

_____ 4. $\dfrac{28}{44}$

_____ 5. $\dfrac{-70}{42}$

_____ 6. $\dfrac{51}{93}$

_____ 7. $\dfrac{25}{48}$

ANSWERS

1. Yes.   **2.** No.   **3.** No.   **4.** $\dfrac{7}{11}$.   **5.** $\dfrac{-5}{3}$.   **6.** $\dfrac{17}{31}$.   **7.** Cannot be reduced.

7.2-2 Reducing Rational Algebraic Expressions Containing Monomials

We turn our attention now to rational algebraic expressions containing a monomial in the numerator and a monomial in the denominator. The method we shall use to reduce such algebraic fractions is the same method used for numerical fractions. We must look for factors common to the numerator and denominator.

For instance, let's try to reduce

$$\frac{x^3}{x} \qquad \text{where } x \neq 0$$

In factored form, we may write

$$\frac{x^3}{x} = \frac{x \cdot x \cdot x}{x}$$

and dividing the numerator and denominator by the common factor x, we have

$$\frac{x^3}{x} = \frac{\overset{1}{\cancel{x}} \cdot x \cdot x}{\underset{1}{\cancel{x}}}$$

$$= \frac{x \cdot x}{1}$$

$$\frac{x^3}{x} = x^2 \qquad \text{where } x \neq 0$$

Let us try another:

$$\frac{a^7}{a^4} = \frac{\overset{1}{\cancel{a}} \cdot \overset{1}{\cancel{a}} \cdot \overset{1}{\cancel{a}} \cdot \overset{1}{\cancel{a}} \cdot a \cdot a \cdot a}{\underset{1}{\cancel{a}} \cdot \underset{1}{\cancel{a}} \cdot \underset{1}{\cancel{a}} \cdot \underset{1}{\cancel{a}}}$$

$$= \frac{a \cdot a \cdot a}{1}$$

$$\frac{a^7}{a^4} = a^3 \qquad \text{where } a \neq 0$$

And another:

$$\frac{y^5}{y^4} = \frac{\overset{1}{\cancel{y}} \cdot \overset{1}{\cancel{y}} \cdot \overset{1}{\cancel{y}} \cdot \overset{1}{\cancel{y}} \cdot y}{\underset{1}{\cancel{y}} \cdot \underset{1}{\cancel{y}} \cdot \underset{1}{\cancel{y}} \cdot \underset{1}{\cancel{y}}}$$

$$= \frac{y}{1}$$

$$\frac{y^5}{y^4} = y \qquad \text{where } y \neq 0$$

How about

$$\frac{x^2}{x^5} = \frac{\overset{1}{\cancel{x}} \cdot \overset{1}{\cancel{x}}}{\underset{1}{\cancel{x}} \cdot \underset{1}{\cancel{x}} \cdot x \cdot x \cdot x}$$

$$= \frac{1}{x \cdot x \cdot x}$$

$$\frac{x^2}{x^5} = \frac{1}{x^3} \qquad \text{where } x \neq 0$$

and

$$\frac{b^5}{b^6} = \frac{\overset{1}{\cancel{b}} \cdot \overset{1}{\cancel{b}} \cdot \overset{1}{\cancel{b}} \cdot \overset{1}{\cancel{b}} \cdot \overset{1}{\cancel{b}}}{\underset{1}{\cancel{b}} \cdot \underset{1}{\cancel{b}} \cdot \underset{1}{\cancel{b}} \cdot \underset{1}{\cancel{b}} \cdot \underset{1}{\cancel{b}} \cdot b}$$

$$\frac{b^5}{b^6} = \frac{1}{b} \qquad \text{where } b \neq 0$$

In each of these examples, we were reducing a fraction in which the numerator and denominator were powers of the same base. Let's look at our results and see if we can spot a pattern.

$$\frac{x^3}{x} = x^2$$

$$\frac{a^7}{a^4} = a^3$$

$$\frac{y^5}{y^4} = y$$

$$\frac{x^2}{x^5} = \frac{1}{x^3}$$

$$\frac{b^5}{b^6} = \frac{1}{b}$$

In each case, the exponent on the variable in the answer was the *difference* between the exponents in the original fraction. If the exponent in the numerator was larger than the exponent in the denominator, the variable ended up in the numerator. If the exponent in the denominator was larger than the exponent in the numerator, the variable ended up in the denominator.

So without writing out all the factors, we should see that if $x \neq 0$,

$$\frac{x^{20}}{x^9} = x^{11} \qquad (20 - 9 = 11)$$

$$\frac{x^{10}}{x} = x^9 \qquad (10 - 1 = 9)$$

$$\frac{x^{10}}{x^{15}} = \frac{1}{x^5} \qquad (15 - 10 = 5)$$

$$\frac{x^{37}}{x^{100}} = \frac{1}{x^{63}} \qquad (100 - 37 = 63)$$

This is actually the Fourth Law of Exponents, and it can be stated in general as follows:

Fourth Law of Exponents

$$\frac{x^m}{x^n} = x^{m-n} \qquad \text{if } m \text{ is larger or equal to } n, \text{ and } x \neq 0$$

$$\frac{x^m}{x^n} = \frac{1}{x^{n-m}} \qquad \text{if } n \text{ is larger than } m, \text{ and } x \neq 0$$

Example 7. Reduce $\dfrac{x^{17}}{x^{10}}$, where $x \neq 0$.

Solution

$$\frac{x^{17}}{x^{10}} = x^{17-10}$$
$$= x^7$$

Example 8. Reduce $\dfrac{y^7}{y}$, where $y \neq 0$.

Solution

$$\frac{y^7}{y} = \frac{y^7}{y^1}$$
$$= y^{7-1}$$
$$= y^6$$

Example 9. Reduce $\dfrac{x^{13}}{x^{13}}$, where $x \neq 0$.

Solution

$$\frac{x^{13}}{x^{13}} = x^{13-13}$$
$$= x^0$$
$$= 1$$

Remember that we defined $x^0 = 1$ earlier.

Example 10. Reduce $\dfrac{x^7}{x^{12}}$, where $x \neq 0$.

Solution

$$\frac{x^7}{x^{12}} = \frac{1}{x^{12-7}}$$

$$= \frac{1}{x^5}$$

Example 11. Reduce $\dfrac{x}{x^9}$, where $x \neq 0$.

Solution

$$\frac{x}{x^9} = \frac{x^1}{x^9}$$

$$= \frac{1}{x^{9-1}}$$

$$= \frac{1}{x^8}$$

Throughout the remainder of this chapter, we shall assume that variables do not represent numbers that would cause zero to occur as the denominator of any fraction. You will not be required to state restrictions on the variable unless specifically directed to do so.

If a fraction contains constant factors as well as variable factors in the numerator and denominator, we reduce the constant parts as before and then reduce the variable parts by the Fourth Law of Exponents.

Example 12. Reduce $\dfrac{10x^6}{2x^3}$.

Solution

$$\frac{10x^6}{2x^3} = \frac{\overset{1}{\cancel{2}} \cdot 5 \cdot x^6}{\underset{1}{\cancel{2}} \cdot x^3}$$

$$= 5x^{6-3}$$

$$= 5x^3$$

Example 13. Reduce $\dfrac{9a^8}{27a^3}$.

Solution

$$\frac{9a^8}{27a^3} = \frac{\overset{1}{\cancel{3}} \cdot \overset{1}{\cancel{3}} \cdot a^8}{\underset{1}{\cancel{3}} \cdot \underset{1}{\cancel{3}} \cdot 3 \cdot a^3} = \frac{a^{8-3}}{3} = \frac{a^5}{3}.$$

$$= \frac{a^{8-3}}{3}$$

$$= \frac{a^5}{3}$$

Example 14. Reduce $\dfrac{-42x^{10}}{7x}$.

Solution

$$\frac{-42x^{10}}{7x} = \frac{-1 \cdot 2 \cdot 3 \cdot \overset{1}{\cancel{7}} \cdot x^{10}}{\underset{1}{\cancel{7}} x^1}$$

$$= -6x^{10-1}$$

$$= -6x^9$$

Example 15. Reduce $\dfrac{3x^2}{12x^4}$.

Solution

$$\frac{3x^2}{12x^4} = \frac{\overset{1}{\cancel{3}} \cdot x^2}{\underset{1}{\cancel{3}} \cdot 2 \cdot 2 \cdot x^4}$$

$$= \frac{1}{4x^{4-2}}$$

$$= \frac{1}{4x^2}$$

If a fraction of monomials contains more than one variable, we continue to follow the same procedure. We reduce the constant parts first, then each of the variable parts in turn. For instance, we may reduce

$$\frac{36x^7y^5}{9x^4y^2} = \frac{4 \cdot \overset{1}{\cancel{9}} \cdot x^7 \cdot y^5}{\underset{1}{\cancel{9}} \cdot x^4y^2}$$

$$= 4x^{7-4}y^{5-2}$$

$$= 4x^3y^3$$

Example 16. Reduce $\dfrac{-2x^5y^2}{4x^4y^7}$.

Solution

$$\frac{-2x^5y^2}{4x^4y^7} = \frac{-1 \cdot \overset{1}{\cancel{2}} \cdot x^5 \cdot y^2}{2 \cdot \underset{1}{\cancel{2}} \cdot x^4 \cdot y^7}$$

$$= \frac{-1x^{5-4}}{2y^{7-2}}$$

$$= \frac{-x}{2y^5}$$

Example 17. Reduce $\dfrac{63a^2b^5}{77a^3b^6}$.

Solution

$$\frac{63a^2b^5}{77a^3b^6} = \frac{9 \cdot \overset{1}{\cancel{7}} \cdot a^2 \cdot b^5}{11 \cdot \underset{1}{\cancel{7}} \cdot a^3 \cdot b^6}$$

$$= \frac{9}{11a^{3-2}b^{6-5}}$$

$$= \frac{9}{11ab}$$

Example 18. Reduce $\dfrac{-x^7y^3z}{x^5y^8z}$.

Solution

$$\frac{-x^7y^5z}{x^5y^8z} = \frac{-1 \cdot x^{7-5}z^{1-1}}{y^{8-5}}$$

$$= \frac{-1x^2z^0}{y^3}$$

$$= \frac{-x^2}{y^3}$$

Trial Run *Reduce.*

———— 1. $\dfrac{-15x^7}{3x^2}$

———— 2. $\dfrac{2x^4}{10x^5}$

———— 3. $\dfrac{-3x^5y^3}{12x^4y^7}$

———— 4. $\dfrac{64a^2b^5}{72a^3b^5}$

———— 5. $\dfrac{-x^2y^3z}{x^9y^3z^5}$

———— 6. $\dfrac{24a^7b^2c^3}{8a^2b^2c}$

ANSWERS

1. $-5x^5$.　　2. $\dfrac{1}{5x}$.　　3. $\dfrac{-x}{4y^4}$.　　4. $\dfrac{8}{9a}$.　　5. $\dfrac{-1}{x^7z^4}$.　　6. $3a^5c^2$.

7.2-3 Reducing Rational Algebraic Expressions Containing Polynomials

Factoring provided the key to reducing fractions, and it shall continue to do so as we consider fractions that contain polynomials in the numerator and denominator. Remember, we reduce fractions by dividing the numerator and denominator by any *factors* they have in common.

Reducing a fraction such as

$$\frac{x^3 + 7x^2}{x^2}$$

may look complicated, but it will not be difficult if we remember to factor.

$$\frac{x^3 + 7x^2}{x^2} = \frac{x^2(x + 7)}{x^2}$$

Do the numerator and denominator have any factors in common? Certainly: x^2.

$$\frac{x^3 + 7x^2}{x^2} = \frac{\overset{1}{\cancel{x^2}}(x + 7)}{\underset{1}{\cancel{x^2}}}$$

$$= x + 7$$

Let's try some more examples, remembering to factor the numerator and denominator completely before looking for common factors.

Example 19. Reduce $\dfrac{2x + 2y}{2a - 4b}$.

Solution

$$\frac{2x + 2y}{2a - 4b} = \frac{\overset{1}{\cancel{2}}(x + y)}{\underset{1}{\cancel{2}}(a - 2b)}$$

$$= \frac{x + y}{a - 2b}$$

Example 20. Reduce $\dfrac{2x + 2y}{3x + 3y}$.

Solution

$$\frac{2x + 2y}{3x + 3y} = \frac{2\overset{1}{\cancel{(x + y)}}}{3\underset{1}{\cancel{(x + y)}}}$$

$$= \frac{2}{3}$$

Example 21. Reduce $\dfrac{z^2 - 9}{2z + 6}$.

Solution

$$\frac{z^2 - 9}{2z + 6} = \frac{\overset{1}{\cancel{(z + 3)}}(z - 3)}{2\underset{1}{\cancel{(z + 3)}}}$$

$$= \frac{z - 3}{2}$$

Example 22. Reduce $\dfrac{x^2 + xy}{5x + 5y}$.

Solution

$$\frac{x^2 + xy}{5x + 5y} = \frac{x\overset{1}{\cancel{(x + y)}}}{5\underset{1}{\cancel{(x + y)}}}$$

$$= \frac{x}{5}$$

Example 23. Reduce $\dfrac{x^2 + 3x + 2}{x^2 - 4}$.

Solution

$$\frac{x^2 + 3x + 2}{x^2 - 4} = \frac{\overset{1}{\cancel{(x + 2)}}(x + 1)}{\underset{1}{\cancel{(x + 2)}}(x - 2)}$$

$$= \frac{x + 1}{x - 2}$$

Example 24. Reduce $\dfrac{x^2 - 5x - 14}{x^2 - 14x + 49}$.

Solution

$$\frac{x^2 - 5x - 14}{x^2 - 14x + 49} = \frac{\overset{1}{\cancel{(x - 7)}}(x + 2)}{\underset{1}{\cancel{(x - 7)}}(x - 7)}$$

$$= \frac{x + 2}{x - 7}$$

Example 25. Reduce $\dfrac{3x^2 - 3x - 6}{x^3 - x}$.

Solution

$$\frac{3x^2 - 3x - 6}{x^3 - x} = \frac{3(x^2 - x - 2)}{x(x^2 - 1)}$$

$$= \frac{3(x - 2)\overset{1}{\cancel{(x + 1)}}}{x(x - 1)\underset{1}{\cancel{(x + 1)}}}$$

$$= \frac{3(x - 2)}{x(x - 1)}$$

There are no safe shortcuts for reducing algebraic fractions. We must always factor the numerator and denominator and then divide by common factors. To reduce a fraction such as

$$\frac{x - 5}{5 - x}$$

we should first rewrite the numerator and denominator with terms appearing in similar order.

$$\frac{x - 5}{5 - x} = \frac{x - 5}{-x + 5}$$

Then we may factor -1 from the expression in the denominator.

$$\frac{x - 5}{5 - x} = \frac{x - 5}{-x + 5}$$

$$= \frac{x - 5}{-1(x - 5)}$$

Now that the numerator and denominator contain the common factor $x - 5$, we may reduce the fraction in the usual way.

$$\frac{x - 5}{5 - x} = \frac{\overset{1}{\cancel{x - 5}}}{-1\underset{1}{\cancel{(x - 5)}}}$$

$$= \frac{1}{-1}$$

$$= -1$$

Let's try another such problem and reduce

$$\frac{2 - 5a}{5a - 2}$$

We shall use the same approach.

$$\frac{2 - 5a}{5a - 2} = \frac{-5a + 2}{5a - 2}$$

$$= \frac{-1(\cancel{5a - 2})^{1}}{\cancel{5a - 2}_{1}}$$

$$= -1$$

Indeed, it will always be true that

$$\boxed{\begin{array}{c} \dfrac{A - B}{B - A} = -1 \\[1em] \text{provided that } A \neq B \end{array}}$$

Whenever you see factors of $A - B$ and $B - A$ in the numerator and denominator, you will know that those factors will reduce to -1, provided that $A \neq B$.

Example 26. Reduce $\dfrac{x - y}{y - x}$.

Solution

$$\frac{x - y}{y - x} = -1$$

Example 27. Reduce $\dfrac{9 - 3x}{x - 3}$.

Solution

$$\frac{9 - 3x}{x - 3} = \frac{3\cancel{(3 - x)}^{-1}}{\cancel{x - 3}_{1}}$$

$$= 3(-1) = -3$$

Example 28. Reduce $\dfrac{36 - x^2}{x^2 - 3x - 18}$.

Solution

$$\frac{36 - x^2}{x^2 - 3x - 18} = \frac{(6 + x)(6 - x)}{(x - 6)(x + 3)}$$

$$= \frac{(6 + x)\cancel{(6 - x)}^{-1}}{\cancel{(x - 6)}_{1}(x + 3)}$$

$$= \frac{-1(6 + x)}{x + 3}$$

$$= \frac{-6 - x}{x + 3}$$

As usual, we have let the numerator carry the sign in each fraction.

Trial Run   *Reduce.*

_____ 1. $\dfrac{4x + 4y}{8}$

_____ 2. $\dfrac{3x + 3y}{3a - 3b}$

_____ 3. $\dfrac{4x + 4y}{8x + 8y}$

_____ 4. $\dfrac{a^2 - 16}{3a - 12}$

_____ 5. $\dfrac{x^2 + 3x}{2x + 6}$

_____ 6. $\dfrac{x^2 - 5x + 6}{x^2 - 4}$

_____ 7. $\dfrac{2 - 2x}{x - 1}$

ANSWERS

1. $\dfrac{x + y}{2}$.   2. $\dfrac{x + y}{a - b}$.   3. $\dfrac{1}{2}$.   4. $\dfrac{a + 4}{3}$.   5. $\dfrac{x}{2}$.   6. $\dfrac{x - 3}{x + 2}$.   7. -2.

Name _____ Date _____

EXERCISE SET 7.2

Decide if the pairs of fractions are equal.

_____ 1. $\dfrac{2}{5}, \dfrac{6}{15}$ _____ 2. $\dfrac{3}{7}, \dfrac{12}{28}$

_____ 3. $\dfrac{6}{7}, \dfrac{18}{42}$ _____ 4. $\dfrac{7}{8}, \dfrac{13}{14}$

_____ 5. $\dfrac{8}{10}, \dfrac{72}{90}$ _____ 6. $\dfrac{9}{15}, \dfrac{63}{105}$

_____ 7. $\dfrac{16}{25}, \dfrac{4}{5}$ _____ 8. $\dfrac{81}{100}, \dfrac{9}{10}$

_____ 9. $\dfrac{56}{24}, \dfrac{7}{3}$ _____ 10. $\dfrac{63}{28}, \dfrac{9}{4}$

Reduce the fractions. Leave the answers in factored form.

_____ 11. $\dfrac{3}{15}$ _____ 12. $\dfrac{6}{18}$

_____ 13. $\dfrac{-21}{-15}$ _____ 14. $\dfrac{-30}{-25}$

_____ 15. $\dfrac{9}{16}$ _____ 16. $\dfrac{25}{36}$

_____ 17. $\dfrac{-25}{35}$ _____ 18. $\dfrac{-49}{63}$

_____ 19. $\dfrac{18}{12}$ _____ 20. $\dfrac{24}{21}$

_____ 21. $\dfrac{-7}{49}$ _____ 22. $\dfrac{-9}{81}$

_____ 23. $\dfrac{8x^5}{4x^2}$ _____ 24. $\dfrac{9x^7}{3x^3}$

_____ 25. $\dfrac{-5a^6}{20a^5}$ _____ 26. $\dfrac{-4a^7}{12a^6}$

_____ 27. $\dfrac{6x^5y^2}{2x^3y^4}$ _____ 28. $\dfrac{21x^3y}{7x^2y^6}$

_____ 29. $\dfrac{-15ab^4}{20a^3b^7}$ _____ 30. $\dfrac{-18a^2b}{27a^8b^2}$

_____ 31. $\dfrac{25x^2yz}{20x^4yz^2}$ _____ 32. $\dfrac{28x^3yz^2}{35x^5y^4z^2}$

_____ 33. $\dfrac{2x + 2y}{6}$ _____ 34. $\dfrac{4x - 4y}{12}$

_____ 35. $\dfrac{3a + 15b}{3}$ _____ 36. $\dfrac{7a + 21b}{7}$

_____ 37. $\dfrac{18xy - 14x}{2x}$ _____ 38. $\dfrac{12xy - 6x}{3x}$

_____ 39. $\dfrac{8x}{64x + 4x^2}$ _____ 40. $\dfrac{15x}{27x + 3x^2}$

_____ 41. $\dfrac{3x}{9x^2 - 15xy}$ _____ 42. $\dfrac{5x}{25x^2 - 20xy}$

_____ 43. $\dfrac{36a^2 + 36b^2}{30}$ _____ 44. $\dfrac{21a^2 + 21b^2}{49}$

_____ 45. $\dfrac{-25x^2 - 25y^2}{-5}$ _____ 46. $\dfrac{-24x^2 - 24y^2}{-8}$

_____ 47. $\dfrac{49x^3}{7x^2 + 7x}$ _____ 48. $\dfrac{81x^4}{9x^2 + 9x}$

_____ 49. $\dfrac{-12a^2b - 15abc}{18ab}$ _____ 50. $\dfrac{-10a^2b - 18abc}{22ab}$

_____ 51. $\dfrac{8x^2 + 12x + 4}{40}$ _____ 52. $\dfrac{15x^2 + 12x + 9}{39}$

_____ 53. $\dfrac{-15x^2y^3z^3 + 25x^4y^2z^3 - 35x^2yz^3}{25x^2y^3z^3}$ _____ 54. $\dfrac{-21x^5y^2z^3 + 49x^7y^3z^3 - 35x^3yz^3}{49x^3yz^3}$

_____ 55. $\dfrac{4a + 4b}{4x + 4y}$ _____ 56. $\dfrac{3a + 3b}{3x + 3y}$

_____ 57. $\dfrac{3x - 3y}{6x + 6y}$ _____ 58. $\dfrac{4x - 4y}{12x + 12y}$

_____ 59. $\dfrac{9x + 9y}{18x + 18y}$ _____ 60. $\dfrac{5x + 5y}{15x + 15y}$

_____ 61. $\dfrac{-4a - 4b}{12a + 12b}$

_____ 62. $\dfrac{-3a - 3b}{15a + 15b}$

_____ 63. $\dfrac{x^2 - y^2}{3x + 3y}$

_____ 64. $\dfrac{x^2 - y^2}{8x - 8y}$

_____ 65. $\dfrac{x^2 - 49}{4x + 28}$

_____ 66. $\dfrac{x^2 - 25}{4x + 20}$

_____ 67. $\dfrac{15x - 15y}{5x^2 - 5y^2}$

_____ 68. $\dfrac{12x - 12y}{3x^2 - 3y^2}$

_____ 69. $\dfrac{-7a - 7b}{3a^2 - 3b^2}$

_____ 70. $\dfrac{-5a - 5b}{2a^2 - 2b^2}$

_____ 71. $\dfrac{x^2 - 4y^2}{3x + 6y}$

_____ 72. $\dfrac{x^2 - 9y^2}{2x + 6y}$

_____ 73. $\dfrac{x^2 + 9x + 20}{x^2 - 16}$

_____ 74. $\dfrac{x^2 + 7x + 12}{x^2 - 9}$

_____ 75. $\dfrac{x^2 + 6x + 8}{x^2 + 7x + 10}$

_____ 76. $\dfrac{x^2 + 8x + 15}{x^2 + 5x + 6}$

_____ 77. $\dfrac{25x^2 - 225}{5x^2 - 50x + 105}$

_____ 78. $\dfrac{16x^2 - 144}{4x^2 - 20x + 24}$

_____ 79. $\dfrac{3x^2 + 14xy - 5y^2}{3x^2 + 2xy - y^2}$

_____ 80. $\dfrac{6x^2 + xy - y^2}{2x^2 - xy - y^2}$

_____ 81. $\dfrac{4a^2 - 8ab - 12b^2}{2a^2 - 12ab + 18a^2}$

_____ 82. $\dfrac{5a^2 + 4ab - b^2}{3a^2 - 6ab + 3b^2}$

_____ 83. $\dfrac{x^2 - 25}{x^2 + 25}$

_____ 84. $\dfrac{4x^2 - 9}{4x^2 + 9}$

_____ 85. $\dfrac{2b - 1}{4b^2 - 4b + 1}$

_____ 86. $\dfrac{3b + 1}{9b^2 + 6b + 1}$

_____ 87. $\dfrac{x^4 - 16y^4}{x^2 + 2xy - 8y^2}$

_____ 88. $\dfrac{x^4 - 81y^4}{x^2 + 6xy - 27y^2}$

_____ 89. $\dfrac{a - b}{9b - 9a}$

_____ 90. $\dfrac{2a - b}{5b - 10a}$

_____ 91. $\dfrac{x^2 - 25}{5 - x}$

_____ 92. $\dfrac{x^2 - 49}{7 - x}$

CHECKUP 7.2

Reduce. Leave the answers in factored form.

_____ 1. $\dfrac{6}{10}$

_____ 2. $\dfrac{-49}{63}$

_____ 3. $\dfrac{a^6}{a^2}$

_____ 4. $\dfrac{-5x^4}{20x^5}$

_____ 5. $\dfrac{16xy - 20x}{4x}$

_____ 6. $\dfrac{6xy}{30x^2y - 24xy^2}$

_____ 7. $\dfrac{3a + 3b}{12a - 12b}$

_____ 8. $\dfrac{5y - 5x}{15x^2 - 15y^2}$

_____ 9. $\dfrac{m^2 - 4n^2}{2m - 4n}$

_____ 10. $\dfrac{x^2 + 6x + 8}{x^2 + 7x + 10}$

7.3 Multiplying and Dividing Rational Expressions

Now that we have learned how to reduce rational expressions, we must learn how to add, subtract, multiply, and divide fractions. The last two operations follow nicely from our study of reducing fractions, so we shall tackle them first.

7.3-1 Multiplying and Dividing Numerical Fractions

Perhaps you remember from previous work with fractions that to multiply fractions, we simply multiply the numerators and multiply the denominators. For instance,

$$\frac{1}{3} \cdot \frac{2}{5} = \frac{1 \cdot 2}{3 \cdot 5}$$

$$= \frac{2}{15}$$

$$\frac{-5}{8} \cdot \frac{7}{2} = \frac{-5 \cdot 7}{8 \cdot 2}$$

$$= \frac{-35}{16}$$

$$\frac{-4}{9} \cdot \frac{-11}{3} = \frac{(-4)(-11)}{9 \cdot 3}$$

$$= \frac{44}{27}$$

Notice that our answers are left in rational-number form. We do not change $\frac{-35}{16}$ or $\frac{44}{27}$ into mixed numbers $\left(-2\frac{3}{16} \text{ or } 1\frac{17}{27}\right)$. For work in algebra, fractional form is always preferred, even if the numerator is larger than the denominator.

Notice also that each of our answers is in simplest form; none of them could be reduced. Let's look at some more fractions to multiply.

$$\frac{3}{8} \cdot \frac{4}{9} = \frac{3 \cdot 4}{8 \cdot 9}$$

$$= \frac{12}{72}$$

But wait a minute; that answer should be reduced.

$$\frac{12}{72} = \frac{12 \cdot 1}{12 \cdot 6}$$

$$= \frac{1}{6}$$

Could we have spotted the common factors earlier—maybe even before we multiplied the numerators and denominators? Take a closer look.

$$\frac{3}{8} \cdot \frac{4}{9} = \frac{3 \cdot 4}{8 \cdot 9} = \frac{\overset{1}{\cancel{3}} \cdot \overset{1}{\cancel{4}}}{2 \cdot \underset{1}{\cancel{4}} \cdot \underset{1}{\cancel{3}} \cdot 3}$$

$$= \frac{1 \cdot 1}{2 \cdot 3}$$

$$= \frac{1}{6}$$

Let us multiply:

$$\frac{15}{36}\cdot\frac{44}{75}=\frac{15\cdot44}{36\cdot75}=\frac{\overset{1}{\cancel{3}}\cdot\overset{1}{\cancel{5}}\cdot\overset{1}{\cancel{4}}\cdot11}{\cancel{4}\cdot9\cdot\underset{1}{\cancel{3}}\cdot\underset{1}{\cancel{5}}\cdot5}$$

$$=\frac{1\cdot11}{9\cdot5}$$

$$=\frac{11}{45}$$

Since fractions are multiplied by multiplying numerators and multiplying denominators, it seems that we could factor the original numerators, factor the original denominators and look for any factors contained in one numerator and one denominator. Such common factors could then be divided out, as we learned in reducing fractions. For instance in our last problem,

$$\frac{15}{36}\cdot\frac{44}{75}=\frac{\overset{1}{\cancel{3}}\cdot\overset{1}{\cancel{5}}}{\cancel{4}\cdot9}\cdot\frac{\overset{1}{\cancel{4}}\cdot11}{\underset{1}{\cancel{3}}\cdot\underset{1}{\cancel{5}}\cdot5}$$

$$=\frac{1\cdot11}{9\cdot5}$$

$$=\frac{11}{45}$$

Since this answer matches our earlier answer, the method of reducing first, then multiplying, seems to be perfectly acceptable.

Example 1. Multiply $\dfrac{28}{27}\cdot\dfrac{18}{21}$.

Solution

$$\frac{28}{27}\cdot\frac{18}{21}=\frac{4\cdot\overset{1}{\cancel{7}}}{3\cdot\cancel{9}}\cdot\frac{2\cdot\overset{1}{\cancel{9}}}{3\cdot\cancel{7}}$$

$$=\frac{4\cdot2}{3\cdot3}$$

$$=\frac{8}{9}$$

Example 2. Multiply $\dfrac{-2}{3}\cdot\dfrac{15}{10}$.

Solution

$$\frac{-2}{3}\cdot\frac{15}{10}=\frac{-1\cdot\overset{1}{\cancel{2}}}{1\cdot\cancel{3}}\cdot\frac{\overset{1}{\cancel{3}}\cdot\overset{1}{\cancel{5}}}{\cancel{2}\cdot\cancel{5}}$$

$$=\frac{-1\cdot1}{1\cdot1}$$

$$=-1$$

Example 3. Multiply $\dfrac{-7}{8}\cdot16$.

Solution

$$\frac{-7}{8} \cdot 16 = \frac{-7}{8} \cdot \frac{16}{1}$$

$$= \frac{-1 \cdot 7}{1 \cdot \cancel{8}} \cdot \frac{\overset{1}{\cancel{8}} \cdot 2}{1}$$

$$= \frac{-1 \cdot 7 \cdot 2}{1}$$

$$= \frac{-14}{1}$$

$$= -14.$$

Example 4. Multiply $\dfrac{-4}{7} \cdot \dfrac{-14}{36}$.

Solution

$$\frac{-4}{7} \cdot \frac{-14}{36} = \frac{-1 \cdot \overset{1}{\cancel{4}}}{1 \cdot \cancel{7}} \cdot \frac{-2 \cdot \overset{1}{\cancel{7}}}{\cancel{4} \cdot 9}$$

$$= \frac{(-1)(-2)}{1 \cdot 9}$$

$$= \frac{2}{9}$$

Example 5. Multiply $\dfrac{3}{2} \cdot \dfrac{-5}{9} \cdot \dfrac{4}{15}$.

Solution

$$\frac{3}{2} \cdot \frac{-5}{9} \cdot \frac{4}{15} = \frac{\overset{1}{\cancel{3}}}{\cancel{2}} \cdot \frac{-1 \cdot \overset{1}{\cancel{5}}}{\cancel{3} \cdot 3} \cdot \frac{\overset{1}{\cancel{2}} \cdot 2}{3 \cdot \cancel{5}}$$

$$= \frac{1(-1)(2)}{3 \cdot 3}$$

$$= \frac{-2}{9}$$

Before we discuss the division of rational numbers, let's review some language that we shall be using. We must remember how to name the parts of a division problem such as $6 \div 2 = 3$.

The number being divided is called the **dividend**; the number doing the dividing is called the **divisor**; the number resulting from the division is called the **quotient**.

Dividing fractions is no more complicated than multiplying fractions. Recall from arithmetic that dividing a quantity by, say, 2, meant the same thing as finding $\dfrac{1}{2}$ of that quantity. In other words, to divide a number by 2, we can multiply the number by $\dfrac{1}{2}$.

Similarly, to divide by 3, we can multiply by $\frac{1}{3}$. Let us see if this idea really works.

$$6 \div 2 = 3 \quad\quad \text{and} \quad\quad 6 \cdot \frac{1}{2} = \frac{6}{1} \cdot \frac{1}{2}$$

$$= \frac{\overset{1}{\cancel{2}} \cdot 3}{1} \cdot \frac{1}{\underset{1}{\cancel{2}}}$$

$$= 3$$

$$15 \div 3 = 5 \quad\quad \text{and} \quad\quad 15 \cdot \frac{1}{3} = \frac{15}{1} \cdot \frac{1}{3}$$

$$= \frac{\overset{1}{\cancel{3}} \cdot 5}{1} \cdot \frac{1}{\underset{1}{\cancel{3}}}$$

$$= 5$$

For the rational number 2, the rational number $\frac{1}{2}$ is called its **reciprocal**. For the rational number 3, the rational number $\frac{1}{3}$ is called its reciprocal. In fact, every rational number (except zero) has its own reciprocal which can be found by inverting the number (or turning it upside down).

Find the reciprocal of 5. Since 5 can be written as $\frac{5}{1}$, its reciprocal must be $\frac{1}{5}$.

Find the reciprocal of $\frac{2}{3}$. The reciprocal of $\frac{2}{3}$ is found by inverting, so the reciprocal of $\frac{2}{3}$ is $\frac{3}{2}$.

Find the reciprocal of $\frac{-1}{9}$. By inverting, we find the reciprocal of $\frac{-1}{9}$ is $\frac{9}{-1}$ or $\frac{-9}{1}$. We usually agree to let the numerator carry the sign when a fraction is negative, so $\frac{-9}{1}$ is preferred.

Returning to our division problem, we now have the words to describe the procedure used in writing

$$6 \div 2 = \frac{6}{1} \div \frac{2}{1}$$

$$= \frac{6}{1} \cdot \frac{1}{2}$$

$$= \frac{3 \cdot 2}{1} \cdot \frac{1}{2}$$

$$= 3$$

Notice that we found the reciprocal of the divisor and multiplied.

To divide fractions, multiply the dividend by the reciprocal of the divisor, reducing where possible.

or

To divide fractions, leave the dividend as it is, invert the divisor, and multiply, reducing where possible.

Example 6. Divide $\dfrac{6}{17} \div \dfrac{2}{3}$.

Solution

$$\frac{6}{17} \div \frac{2}{3} = \frac{6}{17} \cdot \frac{3}{2}$$

$$= \frac{3 \cdot \overset{1}{\cancel{2}}}{17} \cdot \frac{3}{\underset{1}{\cancel{2}}}$$

$$= \frac{3 \cdot 3}{17 \cdot 1}$$

$$= \frac{9}{17}$$

Example 7. Divide $\dfrac{9}{5} \div 2$.

Solution

$$\frac{9}{5} \div 2 = \frac{9}{5} \div \frac{2}{1}$$

$$= \frac{9}{5} \cdot \frac{1}{2}$$

$$= \frac{9 \cdot 1}{5 \cdot 2}$$

$$= \frac{9}{10}$$

Example 8. Divide $11 \div \dfrac{5}{3}$.

Solution

$$11 \div \frac{5}{3} = \frac{11}{1} \div \frac{5}{3}$$

$$= \frac{11}{1} \cdot \frac{3}{5}$$

$$= \frac{11 \cdot 3}{1 \cdot 5}$$

$$= \frac{33}{5}$$

Example 9. Divide $\dfrac{1}{12} \div \dfrac{-3}{8}$.

Solution

$$\frac{1}{12} \div \frac{-3}{8} = \frac{1}{12} \cdot \frac{-8}{3}$$

$$= \frac{1}{3 \cdot \underset{1}{\cancel{4}}} \cdot \frac{-2 \cdot \overset{1}{\cancel{4}}}{3}$$

$$= \frac{1(-2)}{3 \cdot 3}$$

$$= \frac{-2}{9}$$

Symbols of inclusion continue to tell us what belongs with what, so we shall try a few problems involving parentheses.

Example 10. Find $\left(\dfrac{5}{6} \cdot \dfrac{24}{25}\right) \div 9$.

Solution

$$\left(\frac{5}{6} \cdot \frac{24}{25}\right) \div 9 = \left(\frac{\overset{1}{\cancel{5}}}{\cancel{6}} \cdot \frac{4 \cdot \overset{1}{\cancel{6}}}{\cancel{6} \cdot 5}\right) \div 9$$

$$= \left(\frac{1 \cdot 4}{1 \cdot 5}\right) \div 9$$

$$= \frac{4}{5} \div \frac{9}{1}$$

$$= \frac{4}{5} \cdot \frac{1}{9} = \frac{4 \cdot 1}{5 \cdot 9} = \frac{4}{45}$$

Example 11. Find $\dfrac{1}{8} \div \left(\dfrac{2}{3} \cdot \dfrac{-9}{14}\right)$.

Solution

$$\frac{1}{8} \div \left(\frac{2}{3} \cdot \frac{-9}{14}\right) = \frac{1}{8} \div \left(\frac{\overset{1}{\cancel{2}}}{\cancel{3}} \cdot \frac{-3 \cdot \overset{1}{\cancel{3}}}{7 \cdot \cancel{2}}\right)$$

$$= \frac{1}{8} \div \left(\frac{-3}{7}\right)$$

$$= \frac{1}{8} \cdot \frac{-7}{3}$$

$$= \frac{-7}{24}$$

Example 12. Find $\left(\dfrac{10}{11} \div \dfrac{5}{3}\right) \div \left(\dfrac{1}{2} \cdot 10\right)$.

Solution

$$\left(\frac{10}{11} \div \frac{5}{3}\right) \div \left(\frac{1}{2} \cdot 10\right) = \left(\frac{10}{11} \cdot \frac{3}{5}\right) \div \left(\frac{1}{2} \cdot \frac{10}{1}\right)$$

$$= \left(\frac{2 \cdot \overset{1}{\cancel{5}}}{11} \cdot \frac{3}{\cancel{5}}\right) \div \left(\frac{1}{\cancel{2}} \cdot \frac{\overset{1}{\cancel{2}} \cdot 5}{1}\right)$$

$$= \left(\frac{2 \cdot 3}{11}\right) \div \left(\frac{5}{1}\right)$$

$$= \frac{6}{11} \div \frac{5}{1}$$

$$= \frac{6}{11} \cdot \frac{1}{5}$$

$$= \frac{6}{55}$$

Trial Run   *Multiply.*

_____ **1.** $\dfrac{20}{21} \cdot \dfrac{14}{15}$

_____ **2.** $\dfrac{-4}{5} \cdot 15$

_____ **3.** $\dfrac{-3}{8} \cdot \dfrac{-16}{27}$

_____ **4.** $\dfrac{2}{3} \cdot \dfrac{-5}{14} \cdot \dfrac{33}{25}$

Divide.

_____ **5.** $\dfrac{8}{17} \div \dfrac{4}{5}$

_____ **6.** $\dfrac{4}{5} \div -3$

_____ **7.** $\dfrac{1}{15} \div \dfrac{-5}{9}$

_____ **8.** $\left(\dfrac{2}{3} \cdot \dfrac{15}{22} \right) \div \dfrac{5}{11}$

ANSWERS

1. $\dfrac{8}{9}$.   **2.** -12.   **3.** $\dfrac{2}{9}$.   **4.** $\dfrac{-11}{35}$.   **5.** $\dfrac{10}{17}$.   **6.** $\dfrac{-4}{15}$.   **7.** $\dfrac{-3}{25}$.   **8.** 1.

7.3-2 Multiplying and Dividing Monomial Fractions

Recall from our discussion of multiplication of numerical fractions that the product of fractions is found by multiplying their numerators and multiplying their denominators, and then reducing.

For instance, to multiply

$$\frac{6x^2}{y} \cdot \frac{5y^3}{2x}$$

we first multiply numerators and denominators.

$$\frac{6x^2}{y} \cdot \frac{5y^3}{2x} = \frac{6 \cdot 5 \cdot x^2 \cdot y^3}{2 \cdot x \cdot y}$$

Now our problem requires that we reduce a fraction of monomials as we did earlier. We reduce the numerical part, then the variable parts in turn using the fourth law of exponents.

$$\frac{6 \cdot 5 \cdot x^2 \cdot y^3}{2 \cdot x \cdot y} = \frac{\overset{1}{\cancel{2}} \cdot 3 \cdot 5 \cdot x^2 y^3}{\underset{1}{\cancel{2}} \cdot x \cdot y}$$

$$= 3 \cdot 5 \cdot x^{2-1} y^{3-1}$$

$$= 15xy^2$$

Example 13. Multiply $\dfrac{-6x^2y}{3z} \cdot \dfrac{15xz^2}{2y^3}$.

Solution

$$\dfrac{-6x^2y}{3z} \cdot \dfrac{15xz^2}{2y^4}$$

$$= \dfrac{-6 \cdot 15 \cdot x^2 \cdot x \cdot y \cdot z^2}{3 \cdot 2 \cdot y^4 \cdot z}$$

$$= \dfrac{-1 \cdot \overset{1}{\cancel{2}} \cdot \overset{1}{\cancel{3}} \cdot 15 \cdot x^3 y \cdot z^2}{\underset{1}{\cancel{3}} \cdot \underset{1}{\cancel{2}} \cdot y^4 \cdot z}$$

$$= \dfrac{-15x^3yz^2}{y^4z}$$

$$= \dfrac{-15 \cdot x^3 \cdot z^{2-1}}{y^{4-1}}$$

$$= \dfrac{-15x^3z}{y^3}$$

Example 14. Multiply $\dfrac{abc}{5} \cdot \dfrac{10}{a^3b^2}$.

Solution

$$\dfrac{abc}{5} \cdot \dfrac{10}{a^3b^2}$$

$$= \dfrac{10 \cdot a \cdot b \cdot c}{5 \cdot a^3b^2}$$

$$= \dfrac{2 \cdot \overset{1}{\cancel{5}} \cdot a \cdot b \cdot c}{\underset{1}{\cancel{5}} \cdot a^3b^2}$$

$$= \dfrac{2c}{a^{3-1}b^{2-1}}$$

$$= \dfrac{2c}{a^2b}$$

Example 15. Multiply $\dfrac{7a^2xy}{2bz} \cdot \dfrac{4b^3x^4}{9yz} \cdot \dfrac{6z^7}{14ax}$.

Solution

$$\dfrac{7a^2xy}{2bz} \cdot \dfrac{4b^3x^4}{9yz} \cdot \dfrac{6z^7}{14ax} = \dfrac{7 \cdot 4 \cdot 6 \cdot a^2 \cdot b^3 \cdot x \cdot x^4 \cdot y \cdot z^7}{2 \cdot 9 \cdot 14 \cdot a \cdot b \cdot x \cdot y \cdot z \cdot z}$$

$$= \dfrac{\overset{1}{\cancel{7}} \cdot \overset{1}{\cancel{2}} \cdot \overset{1}{\cancel{2}} \cdot 2 \cdot \overset{1}{\cancel{3}} \cdot a^2 \cdot b^3 \cdot x^5 \cdot y \cdot z^7}{\underset{1}{\cancel{2}} \cdot \underset{1}{\cancel{3}} \cdot 3 \cdot \underset{1}{\cancel{2}} \cdot \underset{1}{\cancel{7}} \cdot a \cdot b \cdot x \cdot y \cdot z^2}$$

$$= \dfrac{2a^2b^3x^5yz^7}{3abxyz^2}$$

$$= \dfrac{2a^{2-1}b^{3-1}x^{5-1}z^{7-2}}{3}$$

$$= \dfrac{2ab^2x^4z^5}{3}$$

Since we know that division of fractions requires multiplication by the reciprocal of the divisor, we may use our multiplication skills to divide monomial fractions. Just remember to invert the divisor and multiply.

Example 16. Divide $\dfrac{6a^2x^4}{5y} \div \dfrac{9x^3}{10y^4}$.

Solution

$$\dfrac{6a^2x^4}{5y} \div \dfrac{9x^3}{10y^4} = \dfrac{6a^2x^4}{5y} \cdot \dfrac{10y^4}{9x^3}$$

$$= \dfrac{6 \cdot 10 \cdot a^2 \cdot x^4 \cdot y^4}{5 \cdot 9 \cdot x^3 \cdot y}$$

$$= \dfrac{2 \cdot \overset{1}{\cancel{3}} \cdot 2 \cdot \overset{1}{\cancel{5}} \cdot a^2 \cdot x^4 \cdot y^4}{\underset{1}{\cancel{5}} \cdot \underset{1}{\cancel{3}} \cdot 3 \cdot x^3 \cdot y}$$

$$= \dfrac{2 \cdot 2 \cdot a^2 \cdot x^{4-3}y^{4-1}}{3}$$

$$= \dfrac{4a^2xy^3}{3}$$

Example 17. Divide $\dfrac{-7x^2y^4}{8z} \div \dfrac{5x^5y}{12z}$.

Solution

$$\dfrac{-7x^2y^4}{8z} \div \dfrac{5x^5y}{12z} = \dfrac{-7x^2y^4}{8z} \cdot \dfrac{12z}{5x^5y}$$

$$= \dfrac{-7 \cdot 12 \cdot x^2 \cdot y^4 \cdot z}{8 \cdot 5 \cdot x^5 \cdot y \cdot z}$$

$$= \dfrac{-7 \cdot 3 \cdot \overset{1}{\cancel{4}} \cdot x^2 \cdot y^4 \cdot z}{2 \cdot \underset{1}{\cancel{4}} \cdot 5 \cdot x^5 \cdot y \cdot z}$$

$$= \dfrac{-7 \cdot 3 \cdot y^{4-1}}{2 \cdot 5 \cdot x^{5-2}}$$

$$= \dfrac{-21y^3}{10x^3}$$

Example 18. Find $\left(\dfrac{11x^2y^5}{2z} \cdot \dfrac{10xy^4}{3z}\right) \div \dfrac{-x^2y^2}{z^2}$.

Solution

$$\left(\dfrac{11x^2y^5}{2z} \cdot \dfrac{10xy^4}{3z}\right) \div \dfrac{-x^2y^2}{z^2} = \dfrac{11 \cdot 10 \cdot x^2 \cdot x \cdot y^5 \cdot y^4}{2 \cdot 3 \cdot z \cdot z} \div \dfrac{-x^2y^2}{z^2}$$

$$= \dfrac{11 \cdot \overset{1}{\cancel{2}} \cdot 5 \cdot x^3 \cdot y^9}{\underset{1}{\cancel{2}} \cdot 3 \cdot z^2} \cdot \dfrac{-z^2}{x^2y^2}$$

$$= \dfrac{11 \cdot 5 \cdot x^3 \cdot y^9}{3 \cdot z^2} \cdot \dfrac{-1 \cdot z^2}{x^2 \cdot y^2}$$

$$= \frac{-1 \cdot 11 \cdot 5 \cdot x^3 \cdot y^9 \cdot z^2}{3 \cdot x^2 \cdot y^2 \cdot z^2}$$

$$= \frac{-1 \cdot 11 \cdot 5 \cdot x^{3-2} y^{9-2}}{3}$$

$$= \frac{-55xy^7}{3}$$

Trial Run *Simplify.*

_____ 1. $\dfrac{-8x^3y}{3z} \cdot \dfrac{9xz^3}{4y^4}$

_____ 2. $\dfrac{a^2bc^3}{7} \cdot \dfrac{14}{a^4c^5}$

_____ 3. $\dfrac{9x^2y}{5xz^3} \cdot \dfrac{25xz^2}{3xy} \cdot \dfrac{xy^2}{15}$

_____ 4. $\dfrac{4x}{y} \div \dfrac{3y}{-x}$

_____ 5. $\dfrac{12a^3x^5}{5y^2} \div \dfrac{16x^5}{10y^2}$

_____ 6. $\left(\dfrac{2x}{3} \div \dfrac{35x}{y} \right) \cdot \dfrac{15}{2xy}$

ANSWERS

1. $\dfrac{-6x^4z^2}{y^3}$.   2. $\dfrac{2b}{a^2c^2}$.   3. $\dfrac{x^2y^2}{z}$.   4. $\dfrac{-4x^2}{3y^2}$.   5. $\dfrac{3a^3}{2}$.   6. $\dfrac{1}{7x}$.

7.3-3 Multiplying and Dividing Polynomial Fractions

The methods we have learned for multiplying monomial fractions are all that we need to multiply fractions with polynomial numerators and denominators.

> 1. Factor all numerators and denominators.
> 2. Reduce, by dividing out any factor common to a numerator and a denominator.
> 3. Multiply remaining numerator factors together, multiply remaining denominators together.

Example 19. Multiply $\dfrac{x^2 - 4}{5x + 5} \cdot \dfrac{x + 1}{x + 2}$.

Solution

$$\frac{x^2 - 4}{5x + 5} \cdot \frac{x + 1}{x + 2} = \frac{\overset{1}{\cancel{(x + 2)}}(x - 2)}{5\cancel{(x + 1)}} \cdot \frac{\overset{1}{\cancel{x + 1}}}{\cancel{x + 2}}$$

$$= \frac{x - 2}{5}$$

Example 20. Multiply $\dfrac{x^2 + 7x + 10}{x^2 + 2x} \cdot \dfrac{x^3}{4x + 20}$.

Solution

$$\frac{x^2 + 7x + 10}{x^2 + 2x} \cdot \frac{x^3}{4x + 20} = \frac{(x + 5)(x + 2)}{x(x + 2)} \cdot \frac{x^3}{4(x + 5)}$$

$$= \frac{x^3}{4x}$$

$$= \frac{x^{3-1}}{4}$$

$$= \frac{x^2}{4}$$

Example 21. Multiply $\dfrac{5x + 5y}{3x - 6a} \cdot \dfrac{6x - 12a}{10x + 10y}$.

Solution

$$\frac{5x + 5y}{3x - 6a} \cdot \frac{6x - 12a}{10x + 10y}$$

$$= \frac{5(x + y)}{3(x - 2a)} \cdot \frac{6(x - 2a)}{10(x + y)}$$

$$= \frac{5 \cdot 6}{3 \cdot 10}$$

$$= \frac{\overset{1}{5} \cdot \overset{1}{2} \cdot \overset{1}{3}}{\underset{1}{3} \cdot \underset{1}{2} \cdot \underset{1}{5}}$$

$$= 1$$

Example 22. Multiply $\dfrac{x^2 - 2x - 3}{x^2 - 1} \cdot \dfrac{2x^2 + 5x - 3}{x^2 - 9}$.

Solution

$$\frac{x^2 - 2x - 3}{x^2 - 1} \cdot \frac{2x^2 + 5x - 3}{x^2 - 9} = \frac{(x - 3)(x + 1)}{(x + 1)(x - 1)} \cdot \frac{(2x - 1)(x + 3)}{(x - 3)(x + 3)}$$

$$= \frac{2x - 1}{x - 1}$$

Once again, we declare that if we can multiply polynomial fractions, we can surely divide them. We simply must remember to invert the divisor and multiply.

Example 23. Divide $\dfrac{x^2 - 25}{5} \div (x - 5)$.

Solution

$$\frac{x^2 - 25}{5} \div (x - 5) = \frac{x^2 - 25}{5} \div \frac{x - 5}{1} = \frac{(x + 5)(x - 5)}{5} \cdot \frac{1}{x - 5}$$

$$= \frac{(x + 5)(x - 5)}{5} \cdot \frac{1}{x - 5}$$

$$= \frac{x + 5}{5}$$

Example 24. Divide $\dfrac{x^2 + 12x + 36}{5x^3 + 30x^2} \div \dfrac{7x + 42}{10x^4}$.

Solution

$$\frac{x^2 + 12x + 36}{5x^3 + 30x^2} \cdot \frac{7x + 42}{10x^4} = \frac{x^2 + 12x + 36}{5x^3 + 30x^2} \cdot \frac{10x^4}{7x + 42}$$

$$= \frac{\overset{1}{\cancel{(x + 6)}}\,\overset{1}{\cancel{(x + 6)}}}{5x^2\cancel{(x + 6)}} \cdot \frac{10x^4}{7\cancel{(x + 6)}}$$

$$= \frac{10x^4}{5 \cdot 7 \cdot x^2}$$

$$= \frac{2 \cdot \overset{1}{\cancel{5}} \cdot x^4}{\cancel{5} \cdot 7 \cdot x^2}$$

$$= \frac{2x^{4-2}}{7}$$

$$= \frac{2x^2}{7}$$

Example 25. Divide $(9x - 72) \div \dfrac{x - 8}{5}$.

Solution

$$(9x - 72) \div \frac{x - 8}{5} = \frac{9x - 72}{1} \cdot \frac{5}{x - 8}$$

$$= \frac{9\overset{1}{\cancel{(x - 8)}}}{1} \cdot \frac{5}{\cancel{x - 8}}$$

$$= 9 \cdot 5$$

$$= 45$$

Example 26. Find $\dfrac{3x^2 - 12}{5x + 10} \cdot \dfrac{x^2 + 9x + 14}{3x + 21} \div \dfrac{x^2 - 4x + 4}{3x + 1}$.

Solution

$$\frac{3x^2 - 12}{5x + 10} \cdot \frac{x^2 + 9x + 14}{3x + 21} \div \frac{x^2 - 4x + 4}{3x + 1}$$

$$= \frac{3(x^2 - 4)}{5(x + 2)} \cdot \frac{(x + 7)(x + 2)}{3(x + 7)} \div \frac{x^2 - 4x + 4}{3x + 1}$$

$$= \frac{\overset{1}{\cancel{3}}(x - 2)(x + 2)}{5\cancel{(x + 2)}} \cdot \frac{\overset{1}{\cancel{(x + 7)}}\,\overset{1}{\cancel{(x + 2)}}}{\cancel{3}\cancel{(x + 7)}} \div \frac{x^2 - 4x + 4}{3x + 1}$$

$$= \frac{(x - 2)(x + 2)}{5} \cdot \frac{3x + 1}{x^2 - 4x + 4}$$

$$= \frac{\overset{1}{\cancel{(x - 2)}}(x + 2)}{5} \cdot \frac{3x + 1}{\cancel{(x - 2)}(x - 2)}$$

$$= \frac{(x + 2)(3x + 1)}{5(x - 2)}$$

Trial Run *Simplify.*

_____ 1. $\dfrac{2x + 2}{x + 3} \cdot \dfrac{1}{x + 1}$

_____ 2. $\dfrac{x^2 + 5x + 6}{x^2 + 3x} \cdot \dfrac{x^3}{3x + 6}$

_____ 3. $\dfrac{3x + 3y}{4a - 8b} \cdot \dfrac{8a - 16b}{12x + 12y}$

_____ 4. $\dfrac{x^2 - 3x + 2}{x^2 - 1} \cdot \dfrac{2x^2 + x - 1}{x^2 - 4x - 4}$

_____ 5. $\dfrac{x^2 - 16}{3} \div \dfrac{x^2 - 8x + 16}{3x - 12}$

_____ 6. $\dfrac{x^2 - 2x - 35}{3x^2 + 15x} \div \dfrac{4x - 28}{9x^3}$

ANSWERS

1. $\dfrac{2}{x + 3}$.　　2. $\dfrac{x^2}{3}$.　　3. $\dfrac{1}{2}$.　　4. $\dfrac{2x - 1}{x - 2}$.　　5. $x + 4$.　　6. $\dfrac{3x^2}{4}$.

By now you should agree that factoring is an important tool in our work with fractions. Once the numerators and denominators are in factored form, we may reduce or multiply or divide with ease. It is important that you recognize the difference between factors and terms so that you will avoid making errors in reducing fractions.

Remember that *terms* are quantities that are being added together, but *factors* are quantities that are being multiplied together. We reduce fractions only by dividing numerators and denominators by common *factors*.

$\dfrac{x + 5}{5}$　　*cannot* be reduced

$\dfrac{5x}{5}$　　can be reduced (to x)

$\dfrac{x + 3}{x + 2}$　　*cannot* be reduced

$\dfrac{3x}{2x}$　　can be reduced $\left(\text{to } \dfrac{3}{2}\right)$

EXERCISE SET 7.3

Simplify.

_____ 1. $\dfrac{1}{3} \cdot \dfrac{2}{7}$

_____ 2. $\dfrac{2}{5} \cdot \dfrac{1}{3}$

_____ 3. $\dfrac{-1}{2} \cdot 6$

_____ 4. $\dfrac{-1}{5} \cdot 10$

_____ 5. $\dfrac{-2}{3} \cdot \dfrac{9}{10}$

_____ 6. $\dfrac{-3}{5} \cdot \dfrac{25}{33}$

_____ 7. $\dfrac{-2}{3} \cdot \dfrac{3}{5} \cdot \dfrac{-1}{6}$

_____ 8. $\dfrac{-4}{7} \cdot \dfrac{7}{8} \cdot \dfrac{-1}{3}$

_____ 9. $\dfrac{1}{2} \div \dfrac{1}{6}$

_____ 10. $\dfrac{1}{3} \div \dfrac{1}{12}$

_____ 11. $\dfrac{2}{3} \div \dfrac{5}{7}$

_____ 12. $\dfrac{2}{5} \div \dfrac{7}{9}$

_____ 13. $\dfrac{4}{5} \div \dfrac{12}{35}$

_____ 14. $\dfrac{3}{8} \div \dfrac{15}{16}$

_____ 15. $\dfrac{-3}{5} \div 9$

_____ 16. $\dfrac{-7}{8} \div 21$

_____ 17. $\dfrac{-20}{7} \div \dfrac{-16}{77}$

_____ 18. $\dfrac{-18}{5} \div \dfrac{-27}{55}$

_____ 19. $\dfrac{8}{25} \div \left(\dfrac{24}{35} \cdot \dfrac{14}{27}\right)$

_____ 20. $\dfrac{10}{66} \cdot \left(\dfrac{8}{9} \div \dfrac{25}{33}\right)$

_____ 21. $\dfrac{x}{y} \cdot \dfrac{1}{y}$

_____ 22. $\dfrac{-x}{y} \cdot \dfrac{1}{3}$

_____ 23. $\dfrac{3xy}{4z} \cdot \dfrac{16z}{12y}$

_____ 24. $\dfrac{5xy}{9z} \cdot \dfrac{27z}{15y}$

_____ 25. $\dfrac{3x}{2y} \cdot \dfrac{10y^2}{x^2} \cdot \dfrac{7}{x}$

_____ 26. $\dfrac{5x}{3y} \cdot \dfrac{21y^2}{x^2} \cdot \dfrac{2}{x}$

_____ 27. $\dfrac{1}{a} \div \dfrac{1}{ab}$

_____ 28. $\dfrac{1}{a} \div \dfrac{1}{a^2}$

_____ 29. $\dfrac{x}{3} \div -3$

_____ 30. $\dfrac{x}{5} \div -15$

_____ 31. $\dfrac{x^2y}{2z} \div \dfrac{xy}{8z^2}$

_____ 32. $\dfrac{x^3y}{3z^2} \div \dfrac{x^2y}{18z}$

_____ 33. $\dfrac{4ab^3}{c} \div \dfrac{4a^3}{5b^2c}$

_____ 34. $\dfrac{9a^3b^2}{c} \div \dfrac{9a}{7b^3c}$

_____ 35. $\left(\dfrac{2x}{3} \div \dfrac{4y}{5}\right) \div \dfrac{2xy}{45}$

_____ 36. $\dfrac{3x}{2} \div \left(\dfrac{9y}{7} \div \dfrac{3xy}{28}\right)$

_____ 37. $\left(\dfrac{34}{81a^3} \div \dfrac{17a^2}{18a^4b}\right) \cdot \dfrac{a}{b^3}$

_____ 38. $\left(\dfrac{38}{25a^4} \div \dfrac{19a}{10a^3b}\right) \cdot \dfrac{a^2}{b}$

_____ 39. $\dfrac{15x^2}{16y^3} \div \left(\dfrac{9y^2}{8x} \cdot \dfrac{y^6}{x^3}\right)$

_____ 40. $\dfrac{12x^3}{49y^3} \div \left(\dfrac{16y^2}{21x} \cdot \dfrac{y^7}{x^2}\right)$

_____ 41. $\dfrac{2x+2}{x+3} \cdot \dfrac{1}{x+1}$

_____ 42. $\dfrac{3x+3}{x+4} \cdot \dfrac{1}{x+1}$

_____ 43. $\dfrac{3x+6}{x^2} \cdot \dfrac{x}{8x+16}$

_____ 44. $\dfrac{5x+10}{x^3} \cdot \dfrac{x^4}{15x+30}$

_____ 45. $\dfrac{2a+6b}{a^2b} \cdot \dfrac{ab^2}{a+3b}$

_____ 46. $\dfrac{7a+14b}{a^3b^2} \cdot \dfrac{a^2b^3}{a+2b}$

_____ 47. $\dfrac{m^2+n^2}{6} \cdot \dfrac{42}{3m^2+3n^2}$

_____ 48. $\dfrac{m^2+n^2}{19} \cdot \dfrac{38}{5m^2+5n^2}$

_____ 49. $\dfrac{x^2-9y^2}{x^2+9y^2} \cdot \dfrac{x^2+9y^2}{x+3y}$

_____ 50. $\dfrac{x^2-25y^2}{x^2+25y^2} \cdot \dfrac{x^2+25y^2}{x+5y}$

_____ 51. $\dfrac{x^2-4}{3x-9} \cdot \dfrac{2x-6}{x^2+4x+4}$

_____ 52. $\dfrac{x^2-9}{2x-4} \cdot \dfrac{3x-6}{x^2-6x+9}$

_____ 53. $\dfrac{(x+3)^2}{64yz} \cdot \dfrac{40y^2z}{x^2-9}$

_____ 54. $\dfrac{(x+5)^2}{81yz^3} \cdot \dfrac{72y^5z}{x^2-25}$

_____ 55. $\dfrac{2x^2+3x+1}{2x^2+5x+3} \cdot \dfrac{2x^2+11x+12}{2x^2+13x+6}$

_____ 56. $\dfrac{20x^2-7x-3}{4x^2-7x-2} \cdot \dfrac{3x^2-5x-2}{15x^2-4x-3}$

_____ 57. $\dfrac{6x^2+xy-y^2}{2x^2-9xy-5y^2} \cdot \dfrac{x^2-25y^2}{15x-5y}$

_____ 58. $\dfrac{8x^2+18y-5y^2}{16x^2-8xy+y^2} \cdot \dfrac{4x^2+3xy-y^2}{12x+30y}$

_____ 59. $\dfrac{x^2-49}{y} \div \dfrac{x+7}{y^2}$

_____ 60. $\dfrac{x^2-64}{y^4} \div \dfrac{x+8}{y^3}$

_____ 61. $\dfrac{x^2 + 7x + 10}{63x} \div \dfrac{7x + 14}{9x}$

_____ 62. $\dfrac{x^2 - 12x + 32}{60x} \div \dfrac{5x - 20}{12x}$

_____ 63. $\dfrac{x + 2}{4x - 8} \div \dfrac{x^2 - 7x - 18}{2x - 4}$

_____ 64. $\dfrac{x + 7}{8x - 56} \div \dfrac{x^2 + 4x - 21}{4x - 28}$

_____ 65. $\dfrac{x - 2}{3} \div \dfrac{(x - 2)^2}{9}$

_____ 66. $\dfrac{x - 5}{4} \div \dfrac{(x - 5)^2}{16}$

_____ 67. $\dfrac{x^2 - 8x + 15}{x^2 - 25} \div \dfrac{x^2 - 9}{x + 5}$

_____ 68. $\dfrac{x^2 - 7x - 18}{x^2 - 81} \div \dfrac{x^2 + 4x + 4}{x + 9}$

_____ 69. $\dfrac{a^2 - 2ab + b^2}{10a^5b^5} \div \dfrac{a - b}{50a^3b^3}$

_____ 70. $\dfrac{a^2 - 6ab + 9b^2}{14a^4b^7} \div \dfrac{a - 3b}{70a^3b^2}$

_____ 71. $\dfrac{xy - y^2}{x^2 - 2xy + y^2} \div \dfrac{5xy - 3y^2}{5x^2 - 8xy + 3y^2}$

_____ 72. $\dfrac{x^2 - 7xy}{x^2 - 14xy + 49y^2} \div \dfrac{3x^2 + 2xy}{3x^2 - 19xy - 14y^2}$

_____ 73. $\left(\dfrac{x^2 - 25}{10x} \cdot \dfrac{50}{3x + 15}\right) \div \dfrac{2x - 10}{3x}$

_____ 74. $\left(\dfrac{x^2 - 9}{3x} \cdot \dfrac{45}{4x - 12}\right) \div \dfrac{5x + 15}{7x}$

_____ 75. $\dfrac{4x^2 - 36}{x^2 + 2x - 15} \cdot \left(\dfrac{2x^2 + 11x + 5}{16x^2 + 16x - 96} \div \dfrac{6x^2 - 7x + 2}{3x^2 - 4x - 4}\right)$

_____ 76. $\dfrac{9x^2 - 1}{3x^2 - 26x - 9} \cdot \left(\dfrac{x^2 - 13x + 36}{2x^2 - 21x + 55} \div \dfrac{3x^2 - 13x + 4}{x^2 - 10x + 25}\right)$

CHECKUP 7.3

Simplify.

_____ 1. $\dfrac{1}{3} \cdot \dfrac{2}{7}$

_____ 2. $\dfrac{2}{3} \div \dfrac{10}{9}$

_____ 3. $-12 \div \dfrac{4}{5}$

_____ 4. $\dfrac{3x}{2y} \cdot \dfrac{10y^2}{x^2} \cdot \dfrac{x}{15}$

_____ 5. $\dfrac{4x}{y} \div \dfrac{-2x^2}{y^3}$

_____ 6. $\dfrac{2x + 8y}{xy^2} \cdot \dfrac{x^2y}{x^2 + 4xy}$

_____ 7. $\dfrac{x^2 - 9}{3x - 12} \cdot \dfrac{2x - 8}{x^2 + 6x + 9}$

_____ 8. $\dfrac{3}{x - y} \cdot \dfrac{4x - 4y}{12}$

_____ 9. $\dfrac{x^2 - 64}{y^2} \div \dfrac{x + 8}{y}$

_____ 10. $\dfrac{6x + 18}{54x} \div \dfrac{x^2 + 9x + 18}{3x}$

Summary

In this chapter we have learned to recognize rational expressions.

A rational number is a number that can be written as a fraction with an integer in the numerator and a nonzero integer in the denominator.

A rational algebraic expression is a fraction with a polynomial numerator and a polynomial denominator.

To reduce algebraic expressions, we learned to factor numerators and denominators, and then divide by common factors. In dealing with fractions of monomials, we used the Fourth Law of Exponents, which states that

$$\frac{x^m}{x^n} = x^{m-n} \quad \text{if } m \text{ is larger than or equal to } n \quad (x \neq 0)$$

$$\frac{x^m}{x^n} = \frac{1}{x^{n-m}} \quad \text{if } n \text{ is larger than } m \quad (x \neq 0)$$

We then discovered that multiplying and dividing fractions depended on our ability to factor and reduce.

To multiply fractions we learned to

1. Factor numerators and denominators.
2. Reduce where possible.
3. Multiply remaining numerator factors and remaining denominator factors to arrive at the product.

To divide fractions we learned to

1. Invert the divisor and change the operation to multiplication.
2. Proceed as with multiplication.

Name _____ Date _____

REVIEW EXERCISES

Section 7.1

1. Show that the following numbers are rational.

_____ **(a)** $3\frac{1}{8}$ _____ **(b)** 0.021 _____ **(c)** -7

2. Restrict the variable for each of the following.

_____ **(a)** $\dfrac{3}{x^2}$ _____ **(b)** $\dfrac{x^2}{x+4}$ _____ **(c)** $\dfrac{5x}{2x-1}$

Solve for the variable.

_____ **3.** $3x = x + 5$ _____ **4.** $4(y-7) = 9(y-5)$

_____ **5.** $4(2a+1) + 7 = -3(a-9)$ _____ **6.** $6x^2 - 7x - 5 = 0$

_____ **7.** $25z^2 = 16$

_____ **8.** $10(x^2 - 2x) = -3(x+1)$

Section 7.2

Reduce the fractions. Leave the answers in factored form.

_____ **9.** $\dfrac{8}{24}$ _____ **10.** $\dfrac{-35}{21}$

_____ **11.** $\dfrac{12x^7}{6x^2}$ _____ **12.** $\dfrac{45x^4y^3}{36x^2y}$

_____ **13.** $\dfrac{5a+15b}{5}$ _____ **14.** $\dfrac{21x}{27x+3x^2}$

_____ **15.** $\dfrac{64x^5}{8x^3+8x}$ _____ **16.** $\dfrac{-7a-7b}{35a+35b}$

_____ **17.** $\dfrac{x^2-y^2}{5x+5y}$ _____ **18.** $\dfrac{x^2-121}{x^2-11x}$

_____ **19.** $\dfrac{5a+6b}{25a^2-36b^2}$ _____ **20.** $\dfrac{9x-9y}{3x^2-3y^2}$

_____ **21.** $\dfrac{x^2-10x+25}{2x^2-13x+15}$ _____ **22.** $\dfrac{12x^2+8xy-4y^2}{12x-4y}$

Section 7.3

Simplify.

_____ **23.** $\dfrac{-3}{4} \cdot \dfrac{16}{21}$ _____ **24.** $\dfrac{2}{3} \div \dfrac{4}{15}$

_____ **25.** $-\dfrac{3}{5} \cdot 35$ _____ **26.** $\dfrac{3x}{7y} \cdot \dfrac{49y^2}{x^3} \cdot \dfrac{x^2y}{18}$

_____ **27.** $\dfrac{a^3b^2}{3c} \div \dfrac{a^5b}{27c^2}$ _____ **28.** $\left(\dfrac{56x}{49x^5} \div \dfrac{12x^2y}{21x^3}\right) \cdot \dfrac{15}{y^3}$

_____ 29. $\dfrac{4x + 4}{x + 4} \cdot \dfrac{1}{x + 1}$

_____ 30. $\dfrac{10a + 20b}{a^3 b^2} \cdot \dfrac{a^4 b^5}{a^2 + 2ab}$

_____ 31. $\dfrac{x^2 - 25y^2}{x^2 + 25y^2} \div \dfrac{3x^2 - 15xy}{x^2 + 25y^2}$

_____ 32. $\dfrac{(x + 4)^2}{64yz^2} \cdot \dfrac{24y^3 z}{3x^2 - 48}$

_____ 33. $\dfrac{x^2 - 12x + 35}{54x} \div \dfrac{2x - 14}{9}$

_____ 34. $\dfrac{10x^2 - 11x - 6}{2x^2 + 5x - 12} \cdot \dfrac{3x^2 + 11x - 4}{15x^2 - 11x + 2}$

_____ 35. $\left[\dfrac{x - 5y}{5} \cdot \dfrac{25y}{(x - 5y)^2} \right] \div \dfrac{5x + 25y}{x^2 - 25y^2}$

_____ 36. $\dfrac{4x^2 - 8xy}{x^2 - 4y^2} \cdot \dfrac{2x^2 + xy - 6y^2}{16x^2 - 24xy}$

8 Working More with Rational Expressions

In Chapter 7 we learned how to recognize, reduce, multiply, and divide rational expressions. In this chapter we learn how to

1. Add and subtract rational expressions.
2. Solve equations containing rational algebraic expressions.

8.1 Building Rational Expressions

We spent much time in Chapter 7 reducing fractions to simplest form by dividing numerator and denominator by common factors. For the work that lies ahead in Chapter 8, we must learn how to "build up" a fraction. Building a fraction is a process by which we rewrite a given fraction as a new fraction with a new denominator (and numerator) that are some multiple of the original denominator (and numerator).

8.1-1 Building Numerical Fractions

Suppose that we wished to write the fraction $\frac{1}{2}$ in an equivalent form with a denominator of, say, 10. We are interested in finding ? so that

$$\frac{1}{2} = \frac{?}{10}$$

You will surely agree that ? must be 5, because you know that

$$\frac{1}{2} = \frac{5}{10}$$

Notice that the new denominator of 10 was just 5 times the old denominator of 2, so the new numerator had to be 5 times the old numerator of 1. Hence we concluded that $5 \cdot 1 = 5$ was the new numerator. Let's try another.

$$\frac{4}{13} = \frac{?}{39}$$

Here we see that the new denominator of 39 is 3 times the old denominator of 13. So the new numerator must be 3 times the old numerator of 4. Hence we conclude that $3 \cdot 4 = 12$ is the new numerator and we write

$$\frac{4}{13} = \frac{12}{39}$$

What we are doing here is just the reverse of what we did in reducing fractions. To reduce fractions, we *divided* numerator and denominator by a common factor.

> To build fractions, we may *multiply* numerator and denominator by the same (nonzero) factor.

Notice that the resulting fraction will always be equal to the original fraction, because when we multiply the numerator and denominator by the same factor, we are actually multiplying the entire fraction by *one*. For instance, in changing $\frac{4}{13}$ to $\frac{12}{39}$:

$$\frac{4}{13} = \frac{4}{13} \cdot 1$$

$$= \frac{4}{13} \cdot \frac{3}{3}$$

$$= \frac{12}{39}$$

Example 1. $\dfrac{2}{3} = \dfrac{?}{21}.$

Solution. $\dfrac{2}{3} = \dfrac{?}{21} = \dfrac{?}{3 \cdot 7} = \dfrac{2 \cdot 7}{3 \cdot 7}$, so $\dfrac{2}{3} = \dfrac{14}{21}.$

Example 2. $\dfrac{-5}{9} = \dfrac{?}{180}.$

Solution. $\dfrac{-5}{9} = \dfrac{?}{180} = \dfrac{?}{9 \cdot 20} = \dfrac{-5 \cdot 20}{9 \cdot 20}$, so $\dfrac{-5}{9} = \dfrac{-100}{180}.$

Example 3. $\dfrac{6}{-11} = \dfrac{?}{121}.$

Solution. $\dfrac{6}{-11} = \dfrac{?}{121} = \dfrac{?}{(-11)(-11)} = \dfrac{6(-11)}{(-11)(-11)}$, so $\dfrac{6}{-11} = \dfrac{-66}{121}.$

8.1-2   Building Algebraic Fractions Containing Monomials

To build a new fraction with a certain denominator from an old fraction, we agreed that we must

> **1.** Decide by what factor the old denominator was multiplied to arrive at the new denominator.
> **2.** Then multiply the old numerator by the same factor to arrive at the new numerator.

This method also works when we wish to build fractions containing monomials. For instance, suppose that we consider

$$\frac{6x}{3y} = \frac{?}{12y^4}$$

By what factor was the old denominator of $3y$ multiplied to arrive at $12y^4$? The factor must be $4y^3$, since $3y(4y^3) = 12y^4$. If we multiplied the old denominator by $4y^3$, we must also multiply the old numerator by $4y^3$. We write this problem as

$$\frac{6x}{3y} = \frac{?}{12y^4}$$

$$= \frac{?}{3y(4y^3)}$$

$$= \frac{6x(4y^3)}{3y(4y^3)}$$

$$= \frac{24xy^3}{12y^4}$$

Notice again that when we multiply numerator and denominator by $4y^3$, we are actually multiplying the entire fraction by $\dfrac{4y^3}{4y^3}$ or 1. Therefore, our new fraction will be equal to our old fraction.

Example 4. $\dfrac{5}{xy} = \dfrac{?}{3x^2y^3}$.

Solution

$$\dfrac{5}{xy} = \dfrac{?}{3x^2y^3}$$

$$= \dfrac{?}{xy(3xy^2)}$$

$$= \dfrac{5(3xy^2)}{xy(3xy^2)}$$

$$= \dfrac{15xy^2}{3x^2y^3}$$

Example 5. $\dfrac{-7a}{10bc} = \dfrac{?}{10abc}$.

Solution

$$\dfrac{-7a}{10bc} = \dfrac{?}{10abc}$$

$$= \dfrac{?}{10bc(a)}$$

$$= \dfrac{-7a(a)}{10bc(a)}$$

$$= \dfrac{-7a^2}{10abc}$$

Example 6. $\dfrac{3}{ab} = \dfrac{?}{5(ab)^2}$.

Solution

$$\dfrac{3}{ab} = \dfrac{?}{5(ab)^2}$$

$$= \dfrac{?}{ab(5ab)}$$

$$= \dfrac{3(5ab)}{ab(5ab)}$$

$$= \dfrac{15ab}{5(ab)^2}$$

Trial Run   *Find the numerator needed to replace ? so that the fractions will be equal.*

———— **1.** $\dfrac{4}{9} = \dfrac{?}{36}$

———— **2.** $\dfrac{-3}{8} = \dfrac{?}{24}$

———— **3.** $\dfrac{5}{-12} = \dfrac{?}{84}$

———— **4.** $\dfrac{-2}{7} = \dfrac{?}{91}$

_____ 5. $\dfrac{5}{-7x} = \dfrac{?}{14x^2}$

_____ 6. $\dfrac{8x}{5y} = \dfrac{?}{15y^4}$

_____ 7. $\dfrac{2a}{9bc} = \dfrac{?}{27abc}$

_____ 8. $\dfrac{ax}{by} = \dfrac{?}{a^2b^2y}$

ANSWERS
1. 16.   **2.** -9.   **3.** -35.   **4.** -26.   **5.** $-10x$.   **6.** $24xy^3$.   **7.** $6a^2$.   **8.** a^3bx.

8.1-3   Building Algebraic Fractions Containing Polynomials

Building fractions containing polynomials requires a bit more care because the building factor may not be obvious. If the old and new denominators are in factored form, however, the procedure used is exactly the same as we have already discussed.

For instance, consider the building problem

$$\frac{3}{x+2} = \frac{?}{x(x+2)}$$

By what factor was the old denominator of $x+2$ multiplied to arrive at the new denominator of $x(x+2)$? The building factor must be x, and we must multiply the old numerator 3 by x also.

$$\frac{3}{x+2} = \frac{?}{x(x+2)}$$
$$= \frac{3x}{x(x+2)}$$

Let's build another fraction,

$$\frac{x}{2x+1} = \frac{?}{(2x+1)(x-3)}$$

It should be clear that the old denominator of $2x+1$ was multiplied by the factor $(x-3)$ to arrive at the new denominator of $(2x+1)(x-3)$. Therefore, we must multiply the old numerator x by $x-3$ also.

$$\frac{x}{2x+1} = \frac{?}{(2x+1)(x-3)}$$
$$= \frac{x(x-3)}{(2x+1)(x-3)}$$

Notice again that the numerator and denominator were both multiplied by $x-3$. In other words, the entire fraction was multiplied by $\dfrac{x-3}{x-3}$ or 1. We can be sure that our new fraction is equal to our old fraction.

Consider the problem

$$\frac{7x}{x-3} = \frac{?}{x^2-9}$$

Before we can decide by what factor the old denominator was multiplied, we must rewrite the new denominator in factored form.

$$\frac{7x}{x-3} = \frac{?}{(x-3)(x+3)}$$

Now it becomes clear that the old denominator, $x - 3$, was multiplied by the factor $x + 3$. Therefore, we must multiply the old numerator by $x + 3$ also.

$$\frac{7x}{x - 3} = \frac{7x(x + 3)}{(x - 3)(x + 3)}$$

Notice that we did not find the product of the factors in the numerator. It is not incorrect to carry this procedure one step further and write

$$\frac{7x}{x - 3} = \frac{7x(x + 3)}{(x - 3)(x + 3)}$$
$$= \frac{7x^2 + 21x}{(x - 3)(x + 3)}$$

but this is not our primary concern here. Notice, however, that it is preferable to leave the denominator in factored form.

Example 7. $\dfrac{x + 2}{x + 3} = \dfrac{?}{x^3 + 3x^2}$.

Solution

$$\frac{x + 2}{x + 3} = \frac{?}{x^3 + 3x^2}$$
$$= \frac{?}{x^2(x + 3)}$$
$$= \frac{x^2(x + 2)}{x^2(x + 3)}$$
$$= \frac{x^3 + 2x^2}{x^2(x + 3)}$$

Example 8. $\dfrac{2x - 5}{x - 1} = \dfrac{?}{x^2 - 1}$.

Solution

$$\frac{2x - 5}{x - 1} = \frac{?}{x^2 - 1}$$
$$= \frac{?}{(x - 1)(x + 1)}$$
$$= \frac{(2x - 5)(x + 1)}{(x - 1)(x + 1)}$$
$$= \frac{2x^2 - 3x - 5}{(x - 1)(x + 1)}$$

Example 9. $\dfrac{x + 6}{2x + 3} = \dfrac{?}{2x^2 - 7x - 15}$.

Solution

$$\frac{x + 6}{2x + 3} = \frac{?}{2x^2 - 7x - 15}$$
$$= \frac{?}{(2x + 3)(x - 5)}$$
$$= \frac{(x + 6)(x - 5)}{(2x + 3)(x - 5)}$$
$$= \frac{x^2 + x - 30}{(2x + 3)(x - 5)}$$

Example 10. $\dfrac{a-8}{a+2} = \dfrac{?}{3a(a^2-4)}$.

Solution

$$\frac{a-8}{a+2} = \frac{?}{3a(a^2-4)}$$

$$= \frac{?}{3a(a-2)(a+2)}$$

$$= \frac{3a(a-2)(a-8)}{3a(a-2)(a+2)}$$

$$= \frac{3a(a^2-10a+16)}{3a(a-2)(a+2)}$$

$$= \frac{3a^3-30a^2+48a}{3a(a-2)(a+2)}$$

Trial Run   *Find the numerator needed to replace ? so that the fractions will be equal.*

_____ 1. $\dfrac{1}{3} = \dfrac{?}{3(a-2)}$

_____ 2. $\dfrac{-4}{x+5} = \dfrac{?}{x(x+5)}$

_____ 3. $\dfrac{x}{2x-y} = \dfrac{?}{(2x-y)(2x+y)}$

_____ 4. $\dfrac{5x}{x-4} = \dfrac{?}{x^2-16}$

_____ 5. $\dfrac{x+1}{x+4} = \dfrac{?}{x^3+4x^2}$

_____ 6. $\dfrac{x-3}{2x-1} = \dfrac{?}{2x^2+9x-5}$

ANSWERS
1. $a-2$.   **2.** $-4x$.   **3.** $2x^2+xy$.   **4.** $5x^2+20x$.   **5.** x^3+x^2.   **6.** $x^2+2x-15$.

Building fractions will be an important tool in Section 8.3. It is not a difficult skill to master, but it does require practice.

EXERCISE SET 8.1

Find the numerator needed to replace ? so that the fractions will be equal.

_____ 1. $\dfrac{6}{7} = \dfrac{?}{49}$

_____ 2. $\dfrac{3}{5} = \dfrac{?}{25}$

_____ 3. $\dfrac{8}{3} = \dfrac{?}{36}$

_____ 4. $\dfrac{9}{4} = \dfrac{?}{48}$

_____ 5. $\dfrac{-4}{11} = \dfrac{?}{99}$

_____ 6. $\dfrac{-2}{13} = \dfrac{?}{65}$

_____ 7. $\dfrac{17}{-15} = \dfrac{?}{45}$

_____ 8. $\dfrac{13}{-12} = \dfrac{?}{60}$

_____ 9. $\dfrac{7}{8} = \dfrac{?}{72}$

_____ 10. $\dfrac{7}{10} = \dfrac{?}{120}$

_____ 11. $9 = \dfrac{?}{5}$

_____ 12. $7 = \dfrac{?}{4}$

_____ 13. $\dfrac{31}{84} = \dfrac{?}{252}$

_____ 14. $\dfrac{29}{132} = \dfrac{?}{396}$

_____ 15. $\dfrac{2}{3} = \dfrac{?}{3x}$

_____ 16. $\dfrac{4}{5} = \dfrac{?}{5x}$

_____ 17. $\dfrac{-1}{5x} = \dfrac{?}{15x^2}$

_____ 18. $\dfrac{-1}{6x} = \dfrac{?}{24x^2}$

_____ 19. $\dfrac{4}{xy} = \dfrac{?}{2x^2y^2}$

_____ 20. $\dfrac{5}{xy} = \dfrac{?}{4x^2y^2}$

_____ 21. $\dfrac{3x}{7y} = \dfrac{?}{21y^3}$

_____ 22. $\dfrac{4x}{9y} = \dfrac{?}{36y^3}$

_____ 23. $\dfrac{2a}{-bc} = \dfrac{?}{15abc}$

_____ 24. $\dfrac{5a}{-xy} = \dfrac{?}{10axy}$

_____ 25. $x = \dfrac{?}{3y}$

_____ 26. $a = \dfrac{?}{2y}$

_____ 27. $\dfrac{-x^2}{y^2} = \dfrac{?}{3y^3}$

_____ 28. $\dfrac{-a^2}{b^2} = \dfrac{?}{12b^4}$

_____ 29. $-2b^2 = \dfrac{?}{4a^2b^2}$

_____ 30. $-3x^2 = \dfrac{?}{9x^2y^2}$

_____ 31. $\dfrac{2}{3} = \dfrac{?}{12(x-y)}$ _____ 32. $\dfrac{3}{4} = \dfrac{?}{20(a-b)}$

_____ 33. $\dfrac{5}{2x} = \dfrac{?}{10x^3}$ _____ 34. $\dfrac{3}{5x} = \dfrac{?}{15x^4}$

_____ 35. $\dfrac{-4}{a} = \dfrac{?}{a(a+b)}$ _____ 36. $\dfrac{-7}{x} = \dfrac{?}{x(x+y)}$

_____ 37. $\dfrac{3x}{x+y} = \dfrac{?}{x^2+xy}$ _____ 38. $\dfrac{2y}{x-y} = \dfrac{?}{xy-y^2}$

_____ 39. $\dfrac{2a}{b+1} = \dfrac{?}{b^2-1}$ _____ 40. $\dfrac{3a}{b-2} = \dfrac{?}{b^2-4}$

_____ 41. $\dfrac{6x^2}{x-5} = \dfrac{?}{x^2-3x-10}$ _____ 42. $\dfrac{5x}{x-7} = \dfrac{?}{x^2-9x+14}$

_____ 43. $\dfrac{x+5}{x+3} = \dfrac{?}{x^3+3x^2}$ _____ 44. $\dfrac{a+3}{a+2} = \dfrac{?}{a^4+2a^3}$

_____ 45. $\dfrac{a-1}{a-3} = \dfrac{?}{a^2-7a+12}$ _____ 46. $\dfrac{x-2}{x-5} = \dfrac{?}{x^2-8x+15}$

_____ 47. $\dfrac{-3y}{x^2-x} = \dfrac{?}{x^3-2x^2+x}$ _____ 48. $\dfrac{-2x}{y^3-2y^2} = \dfrac{?}{y^4-4y^2}$

_____ 49. $\dfrac{x-y}{2x+2y} = \dfrac{?}{2x^2-4xy-6y^2}$ _____ 50. $\dfrac{x+2y}{3x-3y} = \dfrac{?}{3x^2+9xy-12y^2}$

Name _____ Date _____

CHECKUP 8.1

Find the numerator needed to replace ? so that the fractions will be equal.

_____ 1. $\dfrac{5}{7} = \dfrac{?}{56}$

_____ 2. $\dfrac{25}{-12} = \dfrac{?}{48}$

_____ 3. $\dfrac{-x}{6y^2} = \dfrac{?}{18y^4}$

_____ 4. $a^2 = \dfrac{?}{16a^2b^2}$

_____ 5. $\dfrac{6a}{-bc} = \dfrac{?}{5abc}$

_____ 6. $\dfrac{7}{3x} = \dfrac{?}{12x^3}$

_____ 7. $\dfrac{2a}{a+b} = \dfrac{?}{a^2b + ab^2}$

_____ 8. $\dfrac{5x}{x+6} = \dfrac{?}{x^2 - 36}$

_____ 9. $\dfrac{x-2}{x-1} = \dfrac{?}{x^2 + 3x - 4}$

_____ 10. $\dfrac{a-b}{5a+5b} = \dfrac{?}{5a^2 + 10ab + 5b^2}$

8.2 Adding and Subtracting Fractions with the Same Denominator

We must now turn our attention to the last two operations with fractions: addition and subtraction. From our earlier work with addition and subtraction, we recall that sums and differences can only be combined into a single term if the quantities being added or subtracted are alike. The same requirement holds in adding and subtracting fractions.

8.2-1 Adding and Subtracting Numerical Fractions

Suppose that we wished to find the sum

$$\frac{1}{7} + \frac{2}{7} + \frac{3}{7}$$

Reading this problem aloud, we wish to find the sum of one-seventh and two-sevenths and three-sevenths. We agree that these are like terms and conclude that their sum is six-sevenths.

$$\frac{1}{7} + \frac{2}{7} + \frac{3}{7} = \frac{6}{7}$$

Another approach would be to use the distributive property to factor $\frac{1}{7}$ from each term.

$$\frac{1}{7} + \frac{2}{7} + \frac{3}{7} = 1 \cdot \frac{1}{7} + 2 \cdot \frac{1}{7} + 3 \cdot \frac{1}{7}$$

$$= (1 + 2 + 3) \frac{1}{7}$$

$$= 6 \left(\frac{1}{7} \right)$$

$$= \frac{6}{7}$$

Both of these approaches should provide us with a procedure for adding or subtracting fractions that have the *same* denominator. *We simply combine the numerators and keep the same denominator.* We would write such a problem as

$$\frac{1}{7} + \frac{2}{7} + \frac{3}{7} = \frac{1 + 2 + 3}{7} = \frac{6}{7}$$

Example 1. Find $\frac{1}{9} - \frac{2}{9} + \frac{3}{9}$.

Solution

$$\frac{1}{9} - \frac{2}{9} + \frac{3}{9} = \frac{1 - 2 + 3}{9}$$

$$= \frac{2}{9}$$

Example 2. Find $\frac{-5}{11} - \frac{1}{11} - \frac{12}{11}$.

Solution

$$\frac{-5}{11} - \frac{1}{11} - \frac{12}{11} = \frac{-5 - 1 - 12}{11}$$

$$= \frac{-18}{11}$$

Example 3. Find $\dfrac{2}{3} - \dfrac{4}{3} - \dfrac{1}{3}$.

Solution

$$\dfrac{2}{3} - \dfrac{4}{3} - \dfrac{1}{3} = \dfrac{2 - 4 - 1}{3}$$
$$= \dfrac{-3}{3}$$
$$= 1$$

Example 4. Find $\dfrac{-7}{45} + \dfrac{8}{45} + \dfrac{14}{45}$.

Solution

$$\dfrac{-7}{45} + \dfrac{8}{45} + \dfrac{14}{45} = \dfrac{-7 + 8 + 14}{45}$$
$$= \dfrac{15}{45}$$
$$= \dfrac{1}{3}$$

Notice that we must remember to reduce our answers whenever possible.

Example 5. Find $\dfrac{12}{29} - \dfrac{3}{29} - \dfrac{9}{29}$.

Solution

$$\dfrac{12}{29} - \dfrac{3}{29} - \dfrac{9}{29} = \dfrac{12 - 3 - 9}{29}$$
$$= \dfrac{0}{29}$$
$$= 0$$

8.2-2 Adding and Subtracting Algebraic Fractions

The procedure for combining algebraic fractions having the same denominator is exactly the same as the procedure for combining numerical fractions.

To combine fractions with the same denominator:

1. Combine the numerators.
2. Keep the same denominator.
3. Simplify the numerator and reduce if possible.

For instance, suppose that we wished to find the sum

$$\dfrac{2}{x} + \dfrac{7}{x}$$

Since the fractions have the same denominator, we may write

$$\dfrac{2}{x} + \dfrac{7}{x} = \dfrac{2 + 7}{x}$$
$$= \dfrac{9}{x}$$

Similarly, we may find the sum

$$\frac{x}{3y} + \frac{2x}{3y} = \frac{x + 2x}{3y}$$

$$= \frac{3x}{3y}$$

$$= \frac{x}{y}$$

And once again,

$$\frac{x}{y} + \frac{3}{y} = \frac{x + 3}{y}$$

The rules for combining the terms in the new numerator are the same rules we have always used for *combining like terms*.

Example 6. Find $\frac{a}{5} + \frac{2a}{5}$.

Solution

$$\frac{a}{5} + \frac{2a}{5} = \frac{a + 2a}{5}$$

$$= \frac{3a}{5}$$

Example 7. Find $\frac{3}{7x} + \frac{4}{7x}$.

Solution

$$\frac{3}{7x} + \frac{4}{7x} = \frac{3 + 4}{7x}$$

$$= \frac{7}{7x}$$

$$= \frac{1}{x}$$

Example 8. Find $\frac{-2y}{9x} + \frac{y}{9x} - \frac{8y}{9x}$.

Solution

$$\frac{-2y}{9x} + \frac{y}{9x} - \frac{8y}{9x} = \frac{-2y + y - 8y}{9x}$$

$$= \frac{-9y}{9x}$$

$$= -\frac{y}{x}$$

Example 9. Find $\frac{5x}{4y} - \frac{x}{4y} + \frac{3}{4y} + \frac{5}{4y}$.

Solution

$$\frac{5x}{4y} - \frac{x}{4y} + \frac{3}{4y} + \frac{5}{4y} = \frac{5x - x + 3 + 5}{4y}$$

$$= \frac{4x + 8}{4y}$$

$$= \frac{4(x + 2)}{4y}$$

$$= \frac{x + 2}{y}$$

Example 10. Find $\dfrac{3x + 1}{7} - \dfrac{x + 2}{7}$.

Solution. Be careful with the minus sign before the second fraction. It shows that the entire second numerator is being subtracted. Parentheses may help.

$$\frac{3x + 1}{7} - \frac{x + 2}{7} = \frac{3x + 1 - (x + 2)}{7}$$

$$= \frac{3x + 1 - x - 2}{7}$$

$$= \frac{3x - x + 1 - 2}{7}$$

$$= \frac{2x - 1}{7}$$

Example 11. Find $\dfrac{2a - b}{4a} - \dfrac{2a - 3b}{4a}$.

Solution

$$\frac{2a - b}{4a} - \frac{2a - 3b}{4a} = \frac{2a - b - (2a - 3b)}{4a}$$

$$= \frac{2a - b - 2a + 3b}{4a}$$

$$= \frac{2a - 2a - b + 3b}{4a}$$

$$= \frac{2b}{4a}$$

$$= \frac{b}{2a}$$

Keep in mind that we must reduce whenever possible.

Trial Run   *Combine the following fractions. Write answers in reduced form.*

_____ 1. $\dfrac{3}{8} + \dfrac{2}{8}$

_____ 2. $\dfrac{7}{15} - \dfrac{4}{15}$

_____ 3. $\dfrac{16}{21} - \dfrac{5}{21} - \dfrac{11}{21}$

_____ 4. $\dfrac{7}{20} - \dfrac{21}{20} + \dfrac{3}{20}$

_____ 5. $\dfrac{4x}{5y} + \dfrac{6x}{5y}$

_____ 6. $\dfrac{a}{7} - \dfrac{5}{7}$

_____ 7. $\dfrac{7x}{3y} + \dfrac{2}{3y} - \dfrac{x}{3y} + \dfrac{1}{3y}$

8. $\dfrac{7a - 3b}{10b} - \dfrac{a - 3b}{10b}$

ANSWERS

1. $\dfrac{5}{8}$. **2.** $\dfrac{1}{5}$. **3.** 0. **4.** $\dfrac{-11}{20}$. **5.** $\dfrac{2x}{y}$. **6.** $\dfrac{a - 5}{7}$. **7.** $\dfrac{2x + 1}{y}$. **8.** $\dfrac{3a}{5b}$.

Let's try adding fractions with more than one term in the denominators. As long as the denominators all match, our method will be the same.

For instance, let's find the sum

$$\frac{2}{x + 3} + \frac{5}{x + 3}$$
$$= \frac{2 + 5}{x + 3}$$
$$= \frac{7}{x + 3}$$

Once again, we combine the numerators, keep the same denominator, and reduce if possible.

Example 12. Find $\dfrac{2x}{x + 5} + \dfrac{10}{x + 5}$.

Solution

$$\frac{2x}{x + 5} + \frac{10}{x + 5} = \frac{2x + 10}{x + 5}$$
$$= \frac{2(x + 5)}{x + 5}$$
$$= 2$$

Example 13. Find $\dfrac{3x}{2x - 1} - \dfrac{x}{2x - 1} + \dfrac{3}{2x - 1}$.

Solution

$$\frac{3x}{2x - 1} - \frac{x}{2x - 1} + \frac{3}{2x - 1} = \frac{3x - x + 3}{2x - 1}$$
$$= \frac{2x + 3}{2x - 1}$$

Example 14. Find $\dfrac{x^2}{x + 1} - \dfrac{2x}{x + 1} - \dfrac{3}{x + 1}$.

Solution

$$\frac{x^2}{x + 1} - \frac{2x}{x + 1} - \frac{3}{x + 1} = \frac{x^2 - 2x - 3}{x + 1}$$
$$= \frac{(x - 3)(x + 1)}{x + 1}$$
$$= x - 3$$

Notice that reducing occurs *after* the numerators have been combined and then factored.

Example 15. Find $\dfrac{2x^2}{x^2 + 4x} - \dfrac{x^2}{x^2 + 4x} - \dfrac{16}{x^2 + 4x}$.

Solution

$$\frac{2x^2}{x^2 + 4x} - \frac{x^2}{x^2 + 4x} - \frac{16}{x^2 + 4x} = \frac{2x^2 - x^2 - 16}{x^2 + 4x}$$

$$= \frac{x^2 - 16}{x^2 + 4x}$$

$$= \frac{(x - 4)(x + 4)}{x(x + 4)}$$

$$= \frac{x - 4}{x}$$

Example 16. Find $\dfrac{2x - 3}{x^2 - 1} - \dfrac{x - 4}{x^2 - 1}$.

Solution

$$\frac{2x - 3}{x^2 - 1} - \frac{x - 4}{x^2 - 1} = \frac{2x - 3 - (x - 4)}{x^2 - 1}$$

$$= \frac{2x - 3 - x + 4}{x^2 - 1}$$

$$= \frac{x + 1}{x^2 - 1}$$

$$= \frac{x + 1}{(x + 1)(x - 1)}$$

$$= \frac{1}{x - 1}$$

Example 17. Find $\dfrac{5x - 4}{x^2 + 5x + 6} - \dfrac{2(x - 5)}{x^2 + 5x + 6}$.

Solution

$$\frac{5x - 4}{x^2 + 5x + 6} - \frac{2(x - 5)}{x^2 + 5x + 6} = \frac{5x - 4 - 2(x - 5)}{x^2 + 5x + 6}$$

$$= \frac{5x - 4 - 2x + 10}{x^2 + 5x + 6}$$

$$= \frac{3x + 6}{x^2 + 5x + 6}$$

$$= \frac{3(x + 2)}{(x + 3)(x + 2)}$$

$$= \frac{3}{x + 3}$$

Trial Run   *Combine the following fractions. Write answers in reduced form.*

———— **1.** $\dfrac{3}{x + 7} + \dfrac{5}{x + 7}$

———— **2.** $\dfrac{3a}{a - 3} - \dfrac{9}{a - 3}$

———— **3.** $\dfrac{4x}{x - 2y} - \dfrac{3x}{x - 2y} + \dfrac{2y}{x - 2y}$

_____ **4.** $\dfrac{x^2}{x+2} - \dfrac{13x}{x+2} - \dfrac{30}{x+2}$

_____ **5.** $\dfrac{2a^2}{a^2-8a} - \dfrac{17a}{a^2-8a} + \dfrac{8}{a^2-8a}$

_____ **6.** $\dfrac{5x-3}{x^2-4} - \dfrac{4x-1}{x^2-4}$

ANSWERS

1. $\dfrac{8}{x+7}$.   **2.** 3.   **3.** $\dfrac{x+2y}{x-2y}$.   **4.** $x-15$.   **5.** $\dfrac{2a-1}{a}$.   **6.** $\dfrac{1}{x+2}$.

EXERCISE SET 8.2

Combine the following fractions. Write the answers in reduced form.

_____ 1. $\dfrac{3}{5} + \dfrac{1}{5}$ _____ 2. $\dfrac{4}{5} - \dfrac{1}{5}$

_____ 3. $\dfrac{2}{27} + \dfrac{7}{27}$ _____ 4. $\dfrac{3}{8} + \dfrac{1}{8}$

_____ 5. $\dfrac{2}{13} - \dfrac{7}{13}$ _____ 6. $\dfrac{4}{15} - \dfrac{11}{15}$

_____ 7. $\dfrac{3}{11} + \dfrac{2}{11} - \dfrac{4}{11}$ _____ 8. $\dfrac{9}{13} + \dfrac{2}{13} - \dfrac{10}{13}$

_____ 9. $\left(\dfrac{9}{11} - \dfrac{2}{11}\right) + \dfrac{4}{11}$ _____ 10. $\left(\dfrac{10}{17} - \dfrac{5}{17}\right) + \dfrac{12}{17}$

_____ 11. $\left(\dfrac{4}{5} - \dfrac{3}{5}\right) + \left(\dfrac{2}{5} - \dfrac{3}{5}\right)$ _____ 12. $\left(\dfrac{4}{7} - \dfrac{2}{7}\right) + \left(\dfrac{3}{7} - \dfrac{5}{7}\right)$

_____ 13. $\dfrac{5x}{3} + \dfrac{4x}{3}$ _____ 14. $\dfrac{9x}{7} + \dfrac{5x}{7}$

_____ 15. $\dfrac{2x}{3} + \dfrac{5}{3}$ _____ 16. $\dfrac{4x}{5} - \dfrac{1}{5}$

_____ 17. $\dfrac{2a}{7} + \dfrac{3a}{7} - \dfrac{6}{7}$ _____ 18. $\dfrac{5a}{9} + \dfrac{2a}{9} - \dfrac{4}{9}$

_____ 19. $\dfrac{2}{x} + \dfrac{3}{x}$ _____ 20. $\dfrac{4}{x} + \dfrac{7}{x}$

_____ 21. $\dfrac{7}{5x} - \dfrac{2}{5x}$ _____ 22. $\dfrac{15}{11x} - \dfrac{4}{11x}$

_____ 23. $\dfrac{12}{xy} + \dfrac{7}{xy}$ _____ 24. $\dfrac{5}{xy} + \dfrac{9}{xy}$

_____ 25. $\dfrac{5x}{2y} - \dfrac{3}{2y} - \dfrac{x}{2y} + \dfrac{1}{2y}$ _____ 26. $\dfrac{8x}{5y} - \dfrac{3}{5y} + \dfrac{2x}{5y} - \dfrac{2}{5y}$

_____ 27. $\dfrac{2a - 5}{3b} - \dfrac{2a + 3}{3b}$ _____ 28. $\dfrac{5a - 2}{7b} - \dfrac{a - 2}{7b}$

_____ 29. $\dfrac{7}{x + y} - \dfrac{3}{x + y}$ _____ 30. $\dfrac{3}{x + y} - \dfrac{1}{x + y}$

_____ 31. $\dfrac{4y}{x + 3y} + \dfrac{5y}{x + 3y}$ _____ 32. $\dfrac{6y}{x + 2y} - \dfrac{4y}{x + 2y}$

_____ 33. $\dfrac{x}{x+7} + \dfrac{7}{x+7}$

_____ 34. $\dfrac{x}{x-5} - \dfrac{5}{x-5}$

_____ 35. $\dfrac{3x}{x-1} - \dfrac{3}{x-1}$

_____ 36. $\dfrac{5x}{x-2} - \dfrac{10}{x-2}$

_____ 37. $\dfrac{3x}{2x+5} - \dfrac{x}{2x+5}$

_____ 38. $\dfrac{7x}{3x-1} - \dfrac{4x}{3x-1}$

_____ 39. $\dfrac{a}{a^2-16} - \dfrac{4}{a^2-16}$

_____ 40. $\dfrac{a}{a^2-25} - \dfrac{5}{a^2-25}$

_____ 41. $\dfrac{x^2}{x^2-4} + \dfrac{2x}{x^2-4}$

_____ 42. $\dfrac{x^2}{x^2-9} + \dfrac{3x}{x^2-9}$

_____ 43. $\dfrac{x^2}{x+2} + \dfrac{4}{x+2}$

_____ 44. $\dfrac{x^2}{x+9} + \dfrac{81}{x+9}$

_____ 45. $\dfrac{3x^2}{x+5} - \dfrac{75}{x+5}$

_____ 46. $\dfrac{2x^2}{x+4} - \dfrac{32}{x+4}$

_____ 47. $\dfrac{x^2}{x+3} + \dfrac{5x}{x+3} + \dfrac{6}{x+3}$

_____ 48. $\dfrac{x^2}{x+5} + \dfrac{3x}{x+5} - \dfrac{10}{x+5}$

_____ 49. $\dfrac{2a^2}{a^2+7a} + \dfrac{11a}{a^2+7a} - \dfrac{21}{a^2+7a}$

_____ 50. $\dfrac{2a^2}{a^2+3a} + \dfrac{a}{a^2+3a} - \dfrac{15}{a^2+3a}$

_____ 51. $\dfrac{6x^2}{4x^2-25} - \dfrac{13x+5}{4x^2-25}$

_____ 52. $\dfrac{3x^2}{9x^2-4} + \dfrac{17x+10}{9x^2-4}$

_____ 53. $\dfrac{7x-y}{3x^2-6xy} - \dfrac{4x-y}{3x^2-6xy}$

_____ 54. $\dfrac{6x-y}{4x^2-8xy} - \dfrac{2x-y}{4x^2-8xy}$

_____ 55. $\dfrac{2x-3}{x^2-5x+6} - \dfrac{1}{x^2-5x+6}$

_____ 56. $\dfrac{2x-5}{x^2+x-12} - \dfrac{1}{x^2+x-12}$

_____ 57. $\dfrac{2x^2-3x}{2x^2-9x+4} + \dfrac{4x-1}{2x^2-9x+4}$

_____ 58. $\dfrac{x^2-11x}{x^2-3x-4} + \dfrac{x+24}{x^2-3x-4}$

_____ 59. $\dfrac{3x(x-2)}{3x^2-13x+4} - \dfrac{x-2}{3x^2-13x+4}$

_____ 60. $\dfrac{2x(x-4)}{2x^2+9x-35} - \dfrac{3(x-5)}{2x^2+9x-35}$

Name _____ Date _____

CHECKUP 8.2

Combine the following fractions and reduce if possible.

_____ 1. $\dfrac{7}{9} - \dfrac{4}{9}$

_____ 2. $\dfrac{7x}{4} - \dfrac{9x}{4}$

_____ 3. $\dfrac{4x}{5} + \dfrac{1}{5}$

_____ 4. $\dfrac{10}{7x} - \dfrac{3}{7x}$

_____ 5. $\dfrac{6a}{5b} - \dfrac{4}{5b} + \dfrac{4a}{5b} - \dfrac{1}{5b}$

_____ 6. $\dfrac{x}{x + 3} + \dfrac{3}{x + 3}$

_____ 7. $\dfrac{3x}{x - 2} - \dfrac{6}{x - 2}$

_____ 8. $\dfrac{2x}{4x^2 - 9} - \dfrac{3}{4x^2 - 9}$

_____ 9. $\dfrac{7x - 10}{2x^2 - 6x} - \dfrac{2x + 5}{2x^2 - 6x}$

_____ 10. $\dfrac{x^2}{x - 1} - \dfrac{9x}{x - 1} + \dfrac{8}{x - 1}$

8.3 Adding and Subtracting Fractions with Different Denominators

What happens when we must add or subtract fractions with denominators which are *not* the same? Problems such as

$$\frac{2}{3} + \frac{1}{2} + \frac{5}{6}$$

$$\frac{3}{x} + \frac{7}{x^2} + \frac{5}{3x}$$

$$\frac{2}{x+1} - \frac{x}{x-3}$$

$$\frac{7x}{x^2-1} - \frac{4}{x-1} + \frac{2}{3x-3}$$

are examples of sums and differences of fractions having *different* denominators. Somehow, we must find a way to make all the denominators *match* in an addition or subtraction problem. Here is where we must use the tool of building fractions. We must decide on a good choice for a new denominator and then build up each of the old fractions to a new fraction having the new denominator.

8.3-1 Adding and Subtracting Numerical Fractions

Consider the first problem in our list:

$$\frac{2}{3} + \frac{1}{2} + \frac{5}{6}$$

What would be a good choice for a new denominator? Remember that each old denominator must be multiplied by some factor to obtain the new denominator. We may choose any number that is a multiple of 3 and 2 and 6. The numbers 6, 12, 18, 24, 30, and so on, are all good choices, but the *best* choice is the *smallest* number in that list. We shall use 6.

Now we must build a new fraction for each of our old fractions, and each new fraction must have a denominator of 6.

$$\frac{2}{3} + \frac{1}{2} + \frac{5}{6}$$

$$= \frac{?}{6} + \frac{?}{6} + \frac{?}{6}$$

Using the same procedure as earlier, we should agree that

$$\frac{2 \cdot 2}{6} + \frac{1 \cdot 3}{6} + \frac{5 \cdot 1}{6}$$

is the correct sum, so our problem becomes

$$\frac{4}{6} + \frac{3}{6} + \frac{5}{6}$$

$$= \frac{4+3+5}{6} = \frac{12}{6} = 2$$

The search that we performed in this problem is called **finding a common denominator**.

A common denominator for a set of fractions is a quantity which is a multiple of each of the denominators of the given fractions.

The smallest such common denominator is called the **lowest common denominator**, which we shall abbreviate by LCD.

To find the LCD, we must consider the factors of each of the given denominators. If the LCD is to be a multiple of each of the old denominators, the LCD must contain any factor that appears in any of the old denominators.

For instance, consider the problem

$$\frac{6}{7} + \frac{4}{5} - \frac{3}{10}$$

Factoring each of the denominators, we have

$$\frac{6}{7} + \frac{4}{5} - \frac{3}{2 \cdot 5}$$

In order for the LCD to be a multiple of the first denominator, the LCD must contain the factor 7.

In order for the LCD to be a multiple of the second denominator, the LCD must contain the factor 5.

In order for the LCD to be a multiple of the third denominator, the LCD must contain the factor 5 (which we already had from the second denominator) and the factor 2.

Therefore,

$$\text{LCD: } 7 \cdot 5 \cdot 2 = 70$$

Now we must build new fractions from the original fractions, being sure that each new fraction has the LCD as its denominator.

$$\frac{6}{7} + \frac{4}{5} - \frac{3}{2 \cdot 5}$$
$$= \frac{?}{7 \cdot 5 \cdot 2} + \frac{?}{7 \cdot 5 \cdot 2} - \frac{?}{7 \cdot 5 \cdot 2}$$

With the LCD written in factored form, it is easy to decide by what factors each of the old denominators was multiplied to arrive at the new denominator. Then we multiply each of the old numerators by those same factors.

$$\frac{6}{7} + \frac{4}{5} - \frac{3}{2 \cdot 5}$$
$$= \frac{6 \cdot 5 \cdot 2}{7 \cdot 5 \cdot 2} + \frac{4 \cdot 7 \cdot 2}{7 \cdot 5 \cdot 2} - \frac{3 \cdot 7}{7 \cdot 5 \cdot 2}$$
$$= \frac{60}{70} + \frac{56}{70} - \frac{21}{70}$$
$$= \frac{60 + 56 - 21}{70}$$
$$= \frac{95}{70} = \frac{5 \cdot 19}{5 \cdot 14} = \frac{19}{14}$$

Consider another problem:

$$\frac{7}{3} - \frac{5}{9} + \frac{1}{12}$$
$$= \frac{7}{3} - \frac{5}{3 \cdot 3} + \frac{1}{3 \cdot 2 \cdot 2}$$

The only different factors we see are 2 and 3. Does this mean that our LCD is $2 \cdot 3 = 6$? Obviously, 6 is not a whole multiple of the second denominator (9) nor of the third denominator (12). The second denominator contains 3 used as a factor *twice;* the third denominator contains 2 used as a factor *twice.* To form the LCD, we must use any factor *the most times that it occurs in any single denominator.*

Here, we must use the factor 2 twice and the factor 3 twice, so for the fractions

$$\frac{7}{3} - \frac{5}{3 \cdot 3} + \frac{1}{3 \cdot 2 \cdot 2}$$

we choose

$$\text{LCD: } 3 \cdot 3 \cdot 2 \cdot 2 = 36$$

Again we build new fractions, each having the LCD of $3 \cdot 3 \cdot 2 \cdot 2$.

$$\frac{7}{3} - \frac{5}{3 \cdot 3} + \frac{1}{3 \cdot 2 \cdot 2}$$

$$= \frac{7 \cdot 3 \cdot 2 \cdot 2}{3 \cdot 3 \cdot 2 \cdot 2} - \frac{5 \cdot 2 \cdot 2}{3 \cdot 3 \cdot 2 \cdot 2} + \frac{1 \cdot 3}{3 \cdot 3 \cdot 2 \cdot 2}$$

$$= \frac{84}{36} - \frac{20}{36} + \frac{3}{36}$$

$$= \frac{84 - 20 + 3}{36} = \frac{67}{36}$$

Let's summarize the steps we have used to find the LCD.

1. Factor the denominators of the given fractions.
2. Decide what different factors appear in the denominators.
3. Notice the most times each factor appears in any single denominator.
4. The LCD is the product of all the different factors, with each one used the most times it appears in any single denominator.

Example 1. Find $\dfrac{7}{25} + \dfrac{3}{2} - \dfrac{1}{8}$.

Solution. $\dfrac{7}{25} + \dfrac{3}{2} - \dfrac{1}{8} = \dfrac{7}{5 \cdot 5} + \dfrac{3}{2} - \dfrac{1}{2 \cdot 2 \cdot 2}$, so

$$\text{LCD: } 5 \cdot 5 \cdot 2 \cdot 2 \cdot 2 = 200$$

$$\frac{7}{5 \cdot 5} + \frac{3}{2} - \frac{1}{2 \cdot 2 \cdot 2}$$

$$= \frac{7 \cdot 2 \cdot 2 \cdot 2}{5 \cdot 5 \cdot 2 \cdot 2 \cdot 2} + \frac{3 \cdot 5 \cdot 5 \cdot 2 \cdot 2}{5 \cdot 5 \cdot 2 \cdot 2 \cdot 2} - \frac{1 \cdot 5 \cdot 5}{5 \cdot 5 \cdot 2 \cdot 2 \cdot 2}$$

$$= \frac{56}{200} + \frac{300}{200} - \frac{25}{200}$$

$$= \frac{56 + 300 - 25}{200}$$

$$= \frac{331}{200}$$

If you prefer, you may use exponents to indicate the number of times that a factor appears in a product.

Example 2. Find $\dfrac{3}{49} - \dfrac{5}{14} - \dfrac{1}{16}$.

Solution

$$\frac{3}{49} - \frac{5}{14} - \frac{1}{16} = \frac{3}{7 \cdot 7} - \frac{5}{2 \cdot 7} - \frac{1}{2 \cdot 2 \cdot 2 \cdot 2}$$

$$= \frac{3}{7^2} - \frac{5}{2 \cdot 7} - \frac{1}{2^4}$$

so

$$\text{LCD: } 7^2 \cdot 2^4 = 784$$

$$\frac{3}{7^2} - \frac{5}{2 \cdot 7} - \frac{1}{2^4}$$

$$= \frac{3 \cdot 2^4}{7^2 \cdot 2^4} - \frac{5 \cdot 7 \cdot 2^3}{7^2 \cdot 2^4} - \frac{1 \cdot 7^2}{7^2 \cdot 2^4}$$

$$= \frac{3 \cdot 16}{784} - \frac{5 \cdot 7 \cdot 8}{784} - \frac{1 \cdot 49}{784}$$

$$= \frac{48}{784} - \frac{280}{784} - \frac{49}{784}$$

$$= \frac{-281}{784}$$

Since the only factors of the denominator are 7 and 2, and neither of these is a factor of 281, we know that this fraction cannot be reduced.

Trial Run *Combine the fractions. Reduce if possible.*

———— **1.** $\dfrac{3}{10} + \dfrac{5}{14}$

———— **2.** $\dfrac{3}{4} + \dfrac{2}{5} - \dfrac{5}{6}$

———— **3.** $\dfrac{5}{2} - \dfrac{1}{4} - \dfrac{11}{18}$

———— **4.** $\dfrac{9}{16} + \dfrac{1}{36} - \dfrac{2}{3}$

———— **5.** $\dfrac{-11}{45} - \dfrac{7}{30} + 1$

———— **6.** $\dfrac{7}{6} - \dfrac{1}{3} - \dfrac{5}{7}$

ANSWERS

1. $\dfrac{23}{35}$.　　**2.** $\dfrac{19}{60}$.　　**3.** $\dfrac{59}{36}$.　　**4.** $\dfrac{-11}{144}$.　　**5.** $\dfrac{47}{90}$.　　**6.** $\dfrac{5}{49}$.

8.3-2　Adding and Subtracting Algebraic Fractions

The method for adding and subtracting numerical fractions with different denominators works well for algebraic fractions also. Remember that we must

1. Factor the denominators of the original fractions.
2. Choose the LCD as the product of all the different factors appearing in the original denominators, with each factor used the most times it appears in any single denominator.
3. Build new fractions from the old fractions, with each new fraction having the LCD.
4. Simplify each new numerator.
5. Combine the numerators to form a new fraction with the LCD as its denominator.
6. Simplify and reduce if possible.

Let us try an example with variables in the numerators.

$$\frac{x}{3} + \frac{2x}{5} - \frac{x}{9}$$

$$= \frac{x}{3} + \frac{2x}{5} - \frac{x}{3 \cdot 3} \qquad \text{LCD: } 3 \cdot 3 \cdot 5 = 45$$

$$= \frac{x \cdot 3 \cdot 5}{3 \cdot 3 \cdot 5} + \frac{2x \cdot 3 \cdot 3}{3 \cdot 3 \cdot 5} - \frac{x \cdot 5}{3 \cdot 3 \cdot 5}$$

$$= \frac{15x}{45} + \frac{18x}{45} - \frac{5x}{45}$$

$$= \frac{15x + 18x - 5x}{45}$$

$$= \frac{28x}{45}$$

Now try an example with monomials in the denominators. The procedure is still the same.

$$\frac{3}{x} + \frac{7}{x^2} + \frac{5}{3x}$$

$$= \frac{3}{x} + \frac{7}{x \cdot x} + \frac{5}{3 \cdot x} \qquad \text{LCD: } 3 \cdot x \cdot x = 3x^2$$

$$= \frac{3 \cdot 3 \cdot x}{3 \cdot x \cdot x} + \frac{7 \cdot 3}{3 \cdot x \cdot x} + \frac{5 \cdot x}{3 \cdot x \cdot x}$$

$$= \frac{9x}{3x^2} + \frac{21}{3x^2} + \frac{5x}{3x^2}$$

$$= \frac{9x + 21 + 5x}{3x^2}$$

$$= \frac{14x + 21}{3x^2} = \frac{7(2x + 3)}{3x^2}$$

Notice that we factored the numerator in the final step to see whether a factor common to the numerator and denominator might appear. Since no common factor appeared, the fractions cannot be reduced.

Let's consider a problem with binomial denominators. We are still looking for factors to make up the LCD.

$$\frac{2}{x + 1} - \frac{x}{x + 3}$$

$$= \frac{2}{1(x + 1)} - \frac{x}{1(x + 3)} \qquad \text{LCD: } (x + 1)(x + 3)$$

$$= \frac{2(x + 3)}{(x + 1)(x + 3)} - \frac{x(x + 1)}{(x + 1)(x + 3)}$$

$$= \frac{2x + 6}{(x + 1)(x + 3)} - \frac{x^2 + x}{(x + 1)(x + 3)}$$

$$= \frac{2x + 6 - (x^2 + x)}{(x + 1)(x + 3)}$$

$$= \frac{2x + 6 - x^2 - x}{(x + 1)(x + 3)}$$

$$= \frac{6 + x - x^2}{(x + 1)(x + 3)}$$

$$= \frac{(3 - x)(2 + x)}{(x + 1)(x + 3)}$$

We should always try to factor the final numerator, in hopes of finding a factor common to the numerator and denominator. Unfortunately, no common factors occur in this problem, so our fraction cannot be reduced.

Example 3. Find $\dfrac{1}{x+2} + \dfrac{1}{5x}$.

Solution

$$\frac{1}{x+2} + \frac{1}{5x}$$

$$= \frac{1}{1(x+2)} + \frac{1}{5 \cdot x} \qquad \text{LCD: } 5x(x+2)$$

$$= \frac{1 \cdot 5x}{5x(x+2)} + \frac{1 \cdot (x+2)}{5x(x+2)}$$

$$= \frac{5x}{5x(x+2)} + \frac{x+2}{5x(x+2)}$$

$$= \frac{5x + x + 2}{5x(x+2)}$$

$$= \frac{6x + 2}{5x(x+2)}$$

$$= \frac{2(3x+1)}{5x(x+2)}$$

Example 4. Find $\dfrac{4}{2x-1} - \dfrac{2}{2x^2-x}$.

Solution

$$\frac{4}{2x-1} - \frac{2}{2x^2-x}$$

$$= \frac{4}{2x-1} - \frac{2}{x(2x-1)} \qquad \text{LCD: } x(2x-1)$$

$$= \frac{4x}{x(2x-1)} - \frac{2}{x(2x-1)}$$

$$= \frac{4x-2}{x(2x-1)}$$

$$= \frac{2(2x-1)}{x(2x-1)}$$

$$= \frac{2}{x}$$

Example 5. Find $\dfrac{5x^2}{(x+7)^2} + \dfrac{2x}{x+7}$.

Solution

$$\frac{5x^2}{(x+7)^2} + \frac{2x}{x+7} \qquad \text{LCD: } (x+7)^2$$

$$= \frac{5x^2}{(x+7)^2} + \frac{2x(x+7)}{(x+7)^2}$$

$$= \frac{5x^2}{(x+7)^2} + \frac{2x^2 + 14x}{(x+7)^2}$$

$$= \frac{5x^2 + 2x^2 + 14x}{(x + 7)^2}$$

$$= \frac{7x^2 + 14x}{(x + 7)^2}$$

$$= \frac{7x(x + 2)}{(x + 7)^2}$$

Example 6. Find $\dfrac{7x}{x^2 - 1} - \dfrac{4}{x + 1} + \dfrac{2}{3x - 3}$.

Solution

$$\frac{7x}{x^2 - 1} - \frac{4}{x + 1} + \frac{2}{3x - 3}$$

$$= \frac{7x}{(x + 1)(x - 1)} - \frac{4}{1(x + 1)} + \frac{2}{3(x - 1)} \qquad \text{LCD: } 3(x + 1)(x - 1)$$

$$= \frac{7x \cdot 3}{3(x + 1)(x - 1)} - \frac{4 \cdot 3(x - 1)}{3(x + 1)(x - 1)} + \frac{2(x + 1)}{3(x + 1)(x - 1)}$$

$$= \frac{21x}{3(x + 1)(x - 1)} - \frac{12x - 12}{3(x + 1)(x - 1)} + \frac{2x + 2}{3(x + 1)(x - 1)}$$

$$= \frac{21x - (12x - 12) + 2x + 2}{3(x + 1)(x - 1)}$$

$$= \frac{21x - 12x + 12 + 2x + 2}{3(x + 1)(x - 1)}$$

$$= \frac{11x + 14}{3(x + 1)(x - 1)}$$

Example 7. Find $\dfrac{x}{x^2 + 6x + 9} + \dfrac{2x}{x^2 - 9}$.

Solution

$$\frac{x}{x^2 + 6x + 9} + \frac{2x}{x^2 - 9}$$

$$= \frac{x}{(x + 3)(x + 3)} + \frac{2x}{(x + 3)(x - 3)}$$

Notice that $x + 3$ appears as a factor twice in a single denominator.

$$\frac{x}{(x + 3)(x + 3)} + \frac{2x}{(x + 3)(x - 3)} \qquad \text{LCD: } (x + 3)(x + 3)(x - 3)$$

$$= \frac{x(x - 3)}{(x + 3)(x + 3)(x - 3)} + \frac{2x(x + 3)}{(x + 3)(x + 3)(x - 3)}$$

$$= \frac{x^2 - 3x}{(x + 3)^2(x - 3)} + \frac{2x^2 + 6x}{(x + 3)^2(x - 3)}$$

$$= \frac{x^2 - 3x + 2x^2 + 6x}{(x + 3)^2(x - 3)}$$

$$= \frac{3x^2 + 3x}{(x + 3)^2(x - 3)}$$

$$= \frac{3x(x + 1)}{(x + 3)^2(x - 3)}$$

Trial Run   *Combine the fractions. Reduce if possible.*

_____ 1. $\dfrac{x}{4} + \dfrac{3x}{5}$

_____ 2. $\dfrac{5x}{9} - \dfrac{x}{6} + \dfrac{1}{4}$

_____ 3. $\dfrac{3}{x} + \dfrac{2}{x^2} - \dfrac{1}{3x}$

_____ 4. $\dfrac{5}{2x} - \dfrac{4}{x + 3}$

_____ 5. $\dfrac{1}{x - 1} + \dfrac{3}{x + 2}$

_____ 6. $\dfrac{x}{x - 9} - \dfrac{2x - 4}{x^2 - 11x + 18}$

ANSWERS

1. $\dfrac{17x}{20}$.   2. $\dfrac{14x + 9}{36}$.   3. $\dfrac{2(4x + 3)}{3x^2}$.   4. $\dfrac{x + 15}{2x(x + 3)}$.   5. $\dfrac{4x - 1}{(x - 1)(x + 2)}$.

6. $\dfrac{x - 2}{x - 9}$.

EXERCISE SET 8.3

Combine the fractions and, if possible, reduce.

_____ 1. $\dfrac{1}{3} + \dfrac{1}{5}$

_____ 2. $\dfrac{1}{7} + \dfrac{1}{9}$

_____ 3. $\dfrac{2}{3} - \dfrac{4}{9}$

_____ 4. $\dfrac{3}{5} - \dfrac{11}{25}$

_____ 5. $\dfrac{2}{3} + \dfrac{5}{21} - \dfrac{5}{7}$

_____ 6. $\dfrac{1}{2} - \dfrac{7}{10} + \dfrac{2}{5}$

_____ 7. $\dfrac{4}{5} - 3$

_____ 8. $\dfrac{3}{7} - 2$

_____ 9. $\dfrac{7}{12} - \dfrac{11}{28} + \dfrac{17}{21}$

_____ 10. $\dfrac{3}{10} + \dfrac{12}{35} + \dfrac{5}{14}$

_____ 11. $\dfrac{29}{45} - \dfrac{17}{36} - \dfrac{7}{20}$

_____ 12. $\dfrac{37}{63} - \dfrac{11}{18} - \dfrac{3}{14}$

_____ 13. $\dfrac{3x}{4} + \dfrac{7x}{4}$

_____ 14. $\dfrac{11x}{6} - \dfrac{2x}{6}$

_____ 15. $\dfrac{4x}{7} + \dfrac{9}{10}$

_____ 16. $\dfrac{3x}{5} + \dfrac{7}{8}$

_____ 17. $\dfrac{x}{6} - \dfrac{y}{10}$

_____ 18. $\dfrac{x}{9} - \dfrac{y}{6}$

_____ 19. $\dfrac{9}{a} + \dfrac{3}{a^2}$

_____ 20. $\dfrac{25}{a} + \dfrac{5}{a^2}$

_____ 21. $\dfrac{3}{x} + \dfrac{5}{x^2} - \dfrac{2}{5x^2}$

_____ 22. $\dfrac{5}{x} + \dfrac{6}{x^2} - \dfrac{7}{3x^2}$

_____ 23. $7 - \dfrac{1}{y}$

_____ 24. $9 - \dfrac{1}{y}$

_____ 25. $\dfrac{4}{x} - 2x$

_____ 26. $\dfrac{6}{x} - 3x$

_____ 27. $\dfrac{5}{x} + \dfrac{2}{y}$

_____ 28. $\dfrac{7}{x} - \dfrac{11}{y}$

_____ 29. $\dfrac{x+1}{2x} - \dfrac{2x+3}{3x}$

_____ 30. $\dfrac{x+2}{5x} - \dfrac{3x+2}{4x}$

_____ 31. $\dfrac{2}{x} - \dfrac{3}{y} + \dfrac{2}{5x} - \dfrac{1}{3y}$

_____ 32. $\dfrac{3}{x} - \dfrac{2}{y} + \dfrac{1}{4x} - \dfrac{4}{3y}$

_____ 33. $\dfrac{x + 1}{6x^2} - \dfrac{x + 3}{8x^2}$

_____ 34. $\dfrac{x + 3}{5x^2} - \dfrac{5x + 7}{35x^2}$

_____ 35. $\dfrac{4}{x^2y} - \dfrac{5}{xy^2}$

_____ 36. $\dfrac{3}{x^2y} - \dfrac{7}{xy^2}$

_____ 37. $\dfrac{x - 2y}{3x^2y} - \dfrac{x + 2y}{4xy^2} + \dfrac{1}{6xy}$

_____ 38. $\dfrac{2x - y}{2x^2y} - \dfrac{2x + y}{5xy^2} - \dfrac{8}{5xy}$

_____ 39. $\dfrac{3}{a + 2} + \dfrac{5}{a + 2}$

_____ 40. $\dfrac{4}{a + 3} + \dfrac{2}{a + 3}$

_____ 41. $\dfrac{7}{x + 2} - \dfrac{5}{(x + 2)^2}$

_____ 42. $\dfrac{11}{x + 3} - \dfrac{4}{(x + 3)^2}$

_____ 43. $\dfrac{3}{x + 4} - \dfrac{5}{x - 4}$

_____ 44. $\dfrac{4}{x + 5} - \dfrac{3}{x - 5}$

_____ 45. $\dfrac{x}{3} - \dfrac{2}{x + 1}$

_____ 46. $\dfrac{x}{4} - \dfrac{3}{x - 1}$

_____ 47. $\dfrac{4}{3x + 3} - \dfrac{5}{x + 1}$

_____ 48. $\dfrac{7}{2x + 2} - \dfrac{4}{x + 1}$

_____ 49. $\dfrac{x}{x^2 - 25} - \dfrac{1}{x^2 - 5x}$

_____ 50. $\dfrac{x}{x^2 - 49} - \dfrac{3}{x^2 - 7x}$

_____ 51. $\dfrac{1}{x^2 - 2x} - \dfrac{2}{x^3 - 2x^2}$

_____ 52. $\dfrac{1}{x^2 - 3x} - \dfrac{3}{x^3 - 3x^2}$

_____ 53. $\dfrac{1}{x^3 - 4x^2} - \dfrac{16}{x^5 - 4x^4}$

_____ 54. $\dfrac{1}{x^3 - 5x^2} - \dfrac{25}{x^5 - 5x^4}$

_____ 55. $\dfrac{4}{x^2 - 49} - \dfrac{1}{x^2 + 14x + 49}$

_____ 56. $\dfrac{5}{x^2 - 36} - \dfrac{1}{x^2 + 12x + 36}$

_____ 57. $\dfrac{x + 6}{x^2 + 2x - 8} + \dfrac{3}{x + 4}$

_____ 58. $\dfrac{x - 8}{x^2 - 11x + 18} - \dfrac{4}{x - 9}$

_____ 59. $\dfrac{2x - 6}{2x^2 - 7x + 3} - \dfrac{5}{2x^2 - x}$

_____ 60. $\dfrac{3x - 21}{3x^2 - 22x + 7} - \dfrac{4}{3x^2 - x}$

CHECKUP 8.3

Combine and, if possible, reduce.

_____ 1. $\dfrac{2}{3} - \dfrac{4}{21}$

_____ 2. $\dfrac{4x}{5} + \dfrac{2x}{3}$

_____ 3. $\dfrac{3x}{2} - \dfrac{y}{3}$

_____ 4. $\dfrac{3}{x} + \dfrac{2}{y}$

_____ 5. $\dfrac{4}{x-3} + \dfrac{7}{x}$

_____ 6. $\dfrac{3}{x-2} + \dfrac{4}{x+2}$

_____ 7. $\dfrac{7}{2x+2} + \dfrac{4}{x+1}$

_____ 8. $\dfrac{4}{x^2-25} - \dfrac{1}{x^2+10x+25}$

_____ 9. $\dfrac{3}{3x^2-2x} - \dfrac{2}{3x^3-2x^2}$

_____ 10. $\dfrac{x+8}{x^2-2x-63} - \dfrac{1}{x+7}$

8.4 Solving Fractional Equations

Fractional equations are equations that contain rational expressions. From our earlier work with equations we recall that we can

> 1. Add (or subtract) the same quantity to (or from) both sides of an equation.
> 2. Multiply (or divide) both sides of an equation by the same nonzero quantity.

To solve a fractional equation, our first goal is to get rid of all the denominators in the equation. To solve an equation such as

$$\frac{x}{3} + \frac{5}{6} = \frac{9x}{2}$$

we would like to eliminate the denominator on both sides of the equation. We can accomplish this by multiplying both sides of the equation by the LCD for all the fractions in the equation. Here, the LCD is clearly 6. Let's multiply both sides by 6.

$$6\left(\frac{x}{3} + \frac{5}{6}\right) = 6\left(\frac{9x}{2}\right)$$

Using the distributive property on the left-hand side, we have

$$6\left(\frac{x}{3}\right) + 6\left(\frac{5}{6}\right) = 6\left(\frac{9x}{2}\right)$$

Performing each multiplication by the usual methods, we have

$$\overset{2}{\cancel{6}}{\over 1} \cdot \frac{x}{\underset{1}{\cancel{3}}} + \frac{\overset{1}{\cancel{6}}}{1} \cdot \frac{5}{\underset{1}{\cancel{6}}} = \frac{\overset{3}{\cancel{6}}}{1} \cdot \frac{9x}{\underset{1}{\cancel{2}}}$$

$$2x + 5 = 27x$$

Notice that we are left with a typical first-degree equation in which we must isolate the variable and solve.

$$2x + 5 - 2x = 27x - 2x$$

$$5 = 25x$$

$$\frac{5}{25} = \frac{25x}{25}$$

$$\frac{5}{25} = x$$

$$\frac{1}{5} = x$$

Example 1. Solve $\frac{3a}{4} = 10 + \frac{a}{3}$.

Solution. LCD is 12.

$$12\left(\frac{3a}{4}\right) = 12(10) + 12\left(\frac{a}{3}\right)$$

$$3(3a) = 12(10) + 4a$$

$$9a = 120 + 4a$$

$$9a - 4a = 120 + 4a - 4a$$

$$5a = 120$$

$$\frac{5a}{5} = \frac{120}{5}$$

$$a = 24$$

Example 2. Solve $\dfrac{7}{x^2} - \dfrac{3}{x} = \dfrac{5}{2x}$.

Solution. LCD is $2x^2$.

$$2x^2\left(\frac{7}{x^2}\right) - 2x^2\left(\frac{3}{x}\right) = 2x^2\left(\frac{5}{2x}\right)$$

$$2 \cdot 7 - 2x(3) = x(5)$$

$$14 - 6x = 5x$$

$$14 - 6x + 6x = 5x + 6x$$

$$14 = 11x$$

$$\frac{14}{11} = \frac{11x}{11}$$

$$\frac{14}{11} = x$$

Trial Run *Solve.*

———— 1. $\dfrac{x}{5} + \dfrac{7}{10} = \dfrac{3x}{2}$

———— 2. $\dfrac{2a}{7} = 5 + \dfrac{a}{4}$

———— 3. $\dfrac{4}{x^2} - \dfrac{2}{x} = \dfrac{3}{2x}$

———— 4. $\dfrac{y}{5} = 4 - y$

———— 5. $\dfrac{3}{x} - \dfrac{5}{2x} = \dfrac{7}{2}$

———— 6. $\dfrac{3x-2}{x^2} - \dfrac{4}{3x} = 0$

ANSWERS

1. $x = \dfrac{7}{13}$.   2. $a = 140$.   3. $x = \dfrac{8}{7}$.   4. $y = \dfrac{10}{3}$.   5. $x = \dfrac{1}{7}$.   6. $x = \dfrac{6}{5}$.

Let's try another problem. Solve

$$\frac{x}{x+3} + \frac{5}{4} = \frac{2}{x+3}$$

Noting the factors of all the denominators, we choose

$$\text{LCD: } 4(x+3)$$

Now we multiply each term in our equation by that LCD and solve.

$$4(x+3)\left(\frac{x}{x+3}\right) + 4(x+3)\left(\frac{5}{4}\right) = 4(x+3)\left(\frac{2}{x+3}\right)$$

$$4x + 5(x+3) = 4 \cdot 2$$

$$4x + 5x + 15 = 8$$

$$9x + 15 = 8$$

$$9x + 15 - 15 = 8 - 15$$
$$9x = -7$$
$$\frac{9x}{9} = \frac{-7}{9}$$
$$x = \frac{-7}{9}$$

Remember that we agreed to multiply both sides of equations by *nonzero* quantities only. In some of our problems, we multiplied both sides of the equation by factors containing variables. How can we be sure that those factors are not zero? We must **restrict the variable** so that the LCD we use is never equal to zero. In the problem where the LCD was

$$4(x + 3)$$

we must avoid any x-value that makes

$$4(x + 3) = 0$$

Since 4 can never be zero, we must avoid x-values for which

$$x + 3 = 0$$
$$x = -3$$

We therefore restrict our variable by saying that $x \neq -3$.

In the problem where the LCD was $2x^2$, we wished to avoid any x-value that would make

$$2x^2 = 0$$

The only value of x for which $2x^2 = 0$ is

$$x = 0$$

We therefore restrict our variable by saying that $x \neq 0$.

From now on, whenever we solve a fractional equation, we shall always restrict our variable as soon as we have chosen the LCD. Let's work another problem. Solve

$$\frac{3}{x - 5} - \frac{7}{x + 1} = \frac{5}{2(x + 1)}$$

The factors we need for the LCD are

$$2(x - 5)(x + 1)$$

We must restrict the variable to avoid x-values for which

$$2(x - 5)(x + 1) = 0$$

From the zero product rule, we conclude that

$$x - 5 = 0 \quad \text{or} \quad x + 1 = 0$$
$$x = 5 \quad \text{or} \quad x = -1$$

We therefore restrict our variable by saying that

$$x \neq 5$$
$$x \neq -1$$

Now we solve the equation by multiplying each term by the LCD: $2(x - 5)(x + 1)$.

$$2(x - 5)(x + 1)\left(\frac{3}{x - 5}\right) - 2(x - 5)(x + 1)\left(\frac{7}{x + 1}\right) = 2(x - 5)(x + 1)\left[\frac{5}{2(x + 1)}\right]$$

$$2(x + 1)(3) - 2(x - 5)(7) = (x - 5)(5)$$
$$6(x + 1) - 14(x - 5) = 5(x - 5)$$
$$6x + 6 - 14x + 70 = 5x - 25$$

$$-8x + 76 = 5x - 25$$
$$-8x + 76 - 5x = 5x - 25 - 5x$$
$$-13x + 76 = -25$$
$$-13x + 76 - 76 = -25 - 76$$
$$-13x = -101$$
$$\frac{-13x}{-13} = \frac{-101}{-13}$$
$$x = \frac{101}{13}$$

In general, to solve fractional equations, we must

1. Choose the LCD for the entire equation.
2. Restrict the variable to avoid any value that makes the LCD zero.
3. Multiply each term in the equation by the LCD.
4. Remove parentheses and combine like terms.
5. Solve the equation and check the solution in the original equation.

Example 3. Solve $\dfrac{5}{2x} = \dfrac{2}{x + 7}$.

Solution. LCD is $2x(x + 7)$. Restrictions are $x \neq 0$, $x \neq -7$.

$$2x(x + 7)\left(\frac{5}{2x}\right) = 2x(x + 7)\left(\frac{2}{x + 7}\right)$$
$$(x + 7)5 = 2x(2)$$
$$5x + 35 = 4x$$
$$5x + 35 - 4x = 4x - 4x$$
$$x + 35 = 0$$
$$x + 35 - 35 = 0 - 35$$
$$x = -35$$

Example 4. Solve $\dfrac{x}{x - 3} + \dfrac{x - 6}{x - 3} = 1$.

Solution. LCD is $x - 3$. Restriction is $x \neq 3$.

$$(x - 3)\left(\frac{x}{x - 3}\right) + (x - 3)\left(\frac{x - 6}{x - 3}\right) = (x - 3)(1)$$
$$x + x - 6 = x - 3$$
$$2x - 6 = x - 3$$
$$2x - 6 - x = x - 3 - x$$
$$x - 6 = -3$$
$$x - 6 + 6 = -3 + 6$$
$$x = 3$$

But our restriction was $x \neq 3$. Therefore, this problem has *no solution*. Notice what happens if we try to check by substituting $x = 3$ into our original equation.

$$\frac{x}{x - 3} + \frac{x - 6}{x - 3} = 1$$

$$\frac{3}{3-3} + \frac{3-6}{3-3} \overset{?}{=} 1$$

$$\frac{3}{0} + \frac{-3}{0} \overset{?}{=} 1$$

Since division by zero is undefined, this equation is *not* a true statement. Our conclusion that there was *no* solution must be correct.

Example 5. Solve $\dfrac{5}{x+2} = \dfrac{10}{x^2-4}$.

Solution. To find the LCD, we must factor our denominators:

$$\frac{5}{x+2} = \frac{10}{(x+2)(x-2)}$$

LCD is $(x+2)(x-2)$. Restrictions are $x \neq -2$, $x \neq 2$.

$$(x+2)(x-2)\left(\frac{5}{x+2}\right) = (x+2)(x-2)\left[\frac{10}{(x+2)(x-2)}\right]$$

$$5(x-2) = 10$$

$$5x - 10 = 10$$

$$5x - 10 + 10 = 10 + 10$$

$$5x = 20$$

$$\frac{5x}{5} = \frac{20}{5}$$

$$x = 4$$

Since our answer does not disagree with our restrictions, we accept it. Perhaps a check would be a good idea.

CHECK: $x = 4$

$$\frac{5}{x+2} = \frac{10}{x^2-4}$$

$$\frac{5}{4+2} \overset{?}{=} \frac{10}{4^2-4}$$

$$\frac{5}{6} \overset{?}{=} \frac{10}{16-4}$$

$$\frac{5}{6} \overset{?}{=} \frac{10}{12}$$

$$\frac{5}{6} = \frac{5}{6}$$

Example 6. Solve $\dfrac{4}{x^2+x-2} - \dfrac{2}{x+2} = \dfrac{1}{3x-3}$.

Solution. To find the LCD, we must factor the denominators

$$\frac{4}{(x+2)(x-1)} - \frac{2}{x+2} = \frac{1}{3(x-1)}$$

LCD is $3(x+2)(x-1)$. Restrictions are $x \neq -2$, $x \neq 1$.

$$3(x+2)(x-1)\left[\frac{4}{(x+2)(x-1)}\right] - 3(x+2)(x-1)\left(\frac{2}{x+2}\right)$$

$$= 3(x+2)(x-1)\left[\frac{1}{3(x-1)}\right]$$

$$3 \cdot 4 - 3(x - 1)(2) = (x + 2)(1)$$
$$12 - 6(x - 1) = x + 2$$
$$12 - 6x + 6 = x + 2$$
$$18 - 6x = x + 2$$
$$18 - 6x + 6x = x + 2 + 6x$$
$$18 = 7x + 2$$
$$18 - 2 = 7x + 2 - 2$$
$$16 = 7x$$
$$\frac{16}{7} = \frac{7x}{7}$$
$$\frac{16}{7} = x$$

Trial Run   *Solve.*

_____ 1. $\dfrac{x}{x - 6} + \dfrac{3}{2} = \dfrac{4}{x - 6}$

_____ 2. $\dfrac{4}{x - 5} - \dfrac{5}{x + 1} = \dfrac{3}{2x + 2}$

_____ 3. $\dfrac{10}{3x} = \dfrac{2}{x + 9}$

_____ 4. $\dfrac{x}{x - 2} + \dfrac{7}{3} = \dfrac{2}{x - 2}$

_____ 5. $\dfrac{x}{x - 4} - \dfrac{8}{x - 4} = 2$

_____ 6. $\dfrac{5}{x + 3} = \dfrac{7}{x^2 - 9}$

ANSWERS

1. $x = \dfrac{26}{5}$.   2. $x = \dfrac{-73}{5}$.   3. $x = \dfrac{-45}{2}$.   4. No solution.   5. $x = 0$.   6. $x = \dfrac{22}{5}$.

EXERCISE SET 8.4

Solve.

_____ 1. $\dfrac{x}{5} = \dfrac{7}{10}$

_____ 2. $\dfrac{x}{8} = \dfrac{9}{20}$

_____ 3. $\dfrac{x}{4} = \dfrac{x-2}{2}$

_____ 4. $\dfrac{y}{10} = \dfrac{y-10}{5}$

_____ 5. $\dfrac{2x}{7} = \dfrac{x-2}{6}$

_____ 6. $\dfrac{3x}{5} = \dfrac{x-1}{4}$

_____ 7. $\dfrac{y+5}{5} = \dfrac{y-5}{10}$

_____ 8. $\dfrac{y+6}{6} = \dfrac{y-6}{18}$

_____ 9. $\dfrac{x}{9} + \dfrac{x}{2} = 5$

_____ 10. $\dfrac{x}{3} - \dfrac{x}{9} = 2$

_____ 11. $\dfrac{x+1}{3} - \dfrac{x}{9} = 4$

_____ 12. $\dfrac{x+3}{5} - \dfrac{x}{25} = 1$

_____ 13. $\dfrac{3}{y} - \dfrac{7}{y} = 4$

_____ 14. $\dfrac{5}{y} - \dfrac{11}{y} = 6$

_____ 15. $\dfrac{7}{x} - \dfrac{3}{2} = \dfrac{5}{2x}$

_____ 16. $\dfrac{11}{x} - \dfrac{4}{3} = \dfrac{1}{3x}$

_____ 17. $\dfrac{3}{x-4} = \dfrac{2}{x-1}$

_____ 18. $\dfrac{-1}{x+2} = \dfrac{2}{x+5}$

_____ 19. $\dfrac{z-1}{z+1} = 2$

_____ 20. $\dfrac{z+4}{z-4} = 3$

_____ 21. $\dfrac{2x-1}{x^2} - \dfrac{5}{3x} = 0$

_____ 22. $\dfrac{3x+1}{x^2} - \dfrac{7}{5x} = 0$

_____ 23. $\dfrac{x}{x-7} + \dfrac{5}{4} = \dfrac{1}{x-7}$

_____ 24. $\dfrac{x}{x-9} + \dfrac{1}{2} = \dfrac{5}{x-9}$

_____ 25. $\dfrac{8}{x-3} - \dfrac{2}{x+4} = \dfrac{5}{2x-6}$

_____ 26. $\dfrac{7}{x-5} - \dfrac{3}{x+1} = \dfrac{2}{3x+3}$

_____ 27. $\dfrac{11}{4x} = \dfrac{5}{x+4}$

_____ 28. $\dfrac{3}{7x} = \dfrac{2}{x+7}$

_____ 29. $\dfrac{a}{a-4} + \dfrac{a-8}{a-4} = 3$

_____ 30. $\dfrac{2a}{a-3} + \dfrac{a-9}{a-3} = 1$

_____ 31. $\dfrac{4}{x-7} = \dfrac{5}{x^2-49}$

_____ 32. $\dfrac{3}{x-5} = \dfrac{2}{x^2-25}$

_____ 33. $\dfrac{4}{2x-1} + \dfrac{1}{2x+1} = \dfrac{2}{4x^2-1}$

_____ 34. $\dfrac{5}{3x+1} - \dfrac{2}{3x-1} = \dfrac{1}{9x^2-1}$

_____ 35. $\dfrac{7}{a-2} = \dfrac{3}{a^2-7a+10}$

_____ 36. $\dfrac{3}{a-9} = \dfrac{2}{a^2-6a-27}$

_____ 37. $\dfrac{3b}{4b^2-1} = \dfrac{-1}{2b-1}$

_____ 38. $\dfrac{4b}{9b^2-4} = \dfrac{-2}{3b+2}$

_____ 39. $\dfrac{3}{x^2-x-12} - \dfrac{2}{x-4} = \dfrac{1}{2x+6}$

_____ 40. $\dfrac{1}{x^2-6x-7} - \dfrac{3}{x-7} = \dfrac{2}{3x+3}$

_____ 41. $\dfrac{5}{x-1} = \dfrac{3x-2}{x^2+15x-16}$

_____ 42. $\dfrac{3}{x-2} = \dfrac{7x-4}{x^2-8x+12}$

_____ 43. $\dfrac{5x+3}{x-7} = 2$

_____ 44. $\dfrac{6x+5}{x-9} = 3$

_____ 45. $\dfrac{a-2}{a+5} = \dfrac{a-1}{a+1}$

_____ 46. $\dfrac{a-7}{a-2} = \dfrac{a+3}{a+2}$

_____ 47. $\dfrac{x}{x-5} + \dfrac{2}{5} = \dfrac{5}{x-5}$

_____ 48. $\dfrac{x}{x-4} + \dfrac{3}{2} = \dfrac{4}{x-4}$

_____ 49. $\dfrac{2x}{2x-3} - \dfrac{6}{2x-3} = 2$

_____ 50. $\dfrac{3x}{3x-1} - \dfrac{4}{3x-1} = 4$

CHECKUP 8.4

Solve.

_____ 1. $\dfrac{14}{y} = \dfrac{7}{9}$

_____ 2. $\dfrac{x}{2} = \dfrac{x-3}{5}$

_____ 3. $\dfrac{x-4}{5} = x + 8$

_____ 4. $\dfrac{y+5}{2} = \dfrac{y-4}{3}$

_____ 5. $\dfrac{6}{a} - \dfrac{4}{9} = \dfrac{8}{9a}$

_____ 6. $\dfrac{5}{11x} = \dfrac{2}{x+11}$

_____ 7. $\dfrac{5}{2b+1} = \dfrac{-4}{b-1}$

_____ 8. $\dfrac{6x}{x-3} - \dfrac{x+15}{x-3} = 6$

_____ 9. $\dfrac{3}{a-5} = \dfrac{-4a}{a^2-a-20}$

_____ 10. $\dfrac{10x-3}{25x^2-16} = \dfrac{5}{5x+4}$

8.5 Switching from Word Statements to Fractional Equations

Suppose that we practice again switching from word statements to variable statements. Remember that in each problem we must identify the variable and translate the word phrases into variable expressions before coming up with an equation to solve.

Problem 1

Maria and Tony earn the same monthly salary. Maria saves $\frac{1}{5}$ of her pay and Tony saves $\frac{1}{6}$ of his pay. Together, they save $550 each month. What is their monthly salary?

$$\text{let } x = \text{monthly salary of each}$$

$$\frac{1}{5}x = \text{Maria's savings}$$

$$\frac{1}{6}x = \text{Tony's savings}$$

$$\frac{1}{5}x + \frac{1}{6}x = \text{total savings}$$

Then

$$\frac{1}{5}x + \frac{1}{6}x = \$550$$

$$\frac{x}{5} + \frac{x}{6} = 550 \qquad \text{LCD: } 30$$

$$30\left(\frac{x}{5}\right) + 30\left(\frac{x}{6}\right) = 30(550)$$

$$6x + 5x = 16,500$$

$$11x = 16,500$$

$$\frac{11x}{11} = \frac{16,500}{11}$$

$$x = 1500$$

Their monthly salary is $1500 each.

Problem 2

On long trips, Mary drives her car 5 miles per hour faster than her sister Jo. If Mary can drive 312 miles in the same length of time that Jo drives 282 miles, how fast does each sister drive?

$$\text{let } r = \text{Jo's rate}$$
$$r + 5 = \text{Mary's rate}$$
$$282 = \text{Jo's distance}$$
$$312 = \text{Mary's distance}$$

We know that

$$\text{distance} = \text{rate} \cdot \text{time}$$

so

$$\text{time} = \frac{\text{distance}}{\text{rate}}$$

and we can represent driving time for each sister as her distance divided by her rate.

$$\frac{282}{r} = \text{Jo's time}$$

$$\frac{312}{r + 5} = \text{Mary's time}$$

Since we are told that the driving times are the same, we may write

$$\frac{282}{r} = \frac{312}{r + 5} \qquad \text{LCD: } r(r + 5)$$

$$r(r + 5)\left(\frac{282}{r}\right) = r(r + 5)\left(\frac{312}{r + 5}\right)$$

$$282(r + 5) = 312r$$

$$282r + 1410 = 312r$$

$$1410 = 312r - 282r$$

$$1410 = 30r$$

$$\frac{1410}{30} = r$$

$$47 = r$$

So the sisters' rates are

$$\text{Jo's rate} = r = 47 \text{ miles per hour}$$

$$\text{Mary's rate} = r + 5 = 52 \text{ miles per hour}$$

Problem 3

Ratios are handy tools to use in solving everyday mathematical problems. A **ratio** is a way of comparing two numbers by means of a fraction.

For instance, if you earn $1500 per month and your brother earns $1200 per month, the ratio of your salary to your brother's salary is $1500 to $1200, which we may write as

$$\frac{1500}{1200} = \frac{5 \cdot 3 \cdot 100}{4 \cdot 3 \cdot 100} = \frac{5}{4}$$

Your salaries are in the ratio of 5 to 4; you earn $5 for every $4 your brother earns.

Suppose that your salary is raised to $1800. What would your brother's new salary have to be in order for your new salaries to remain in the ratio of 5 to 4? If we let S represent your brother's new salary, we are wanting the ratio of $1800 to S to be equal to the ratio of 5 to 4. In fractional equation form, we want

$$\frac{1800}{S} = \frac{5}{4}$$

Using $4S$ as the LCD, we may solve

$$4S\left(\frac{1800}{S}\right) = 4S\left(\frac{5}{4}\right)$$

$$4(1800) = 5S$$

$$7200 = 5S$$

$$\frac{7200}{5} = \frac{5S}{5}$$

$$1440 = S$$

Your brother's new salary must be $1440.

An equation which says that two ratios are equal is called a **proportion**. In our salary problem, we wrote a proportion when we wrote

$$\frac{1800}{S} = \frac{5}{4}$$

Let's solve another problem using a proportion.

Problem 4

In a class of 95 students, the ratio of females to males is 3 to 2. How many females and how many males are in this class?

If we let x be the number of females in the class, how many males are in the class? If we said there were 80 females, how would you figure out how many males there were? You would subtract: $95 - 80$. Or if we said there were 43 females, you would say that there must be $95 - 43$ males. So if there are x females, there must be $95 - x$ males.

$$\text{let } x = \text{number of females}$$
$$95 - x = \text{number of males}$$
$$\frac{x}{95 - x} = \text{ratio of females to males}$$

Then

$$\frac{x}{95 - x} = \frac{3}{2} \qquad \text{LCD: } 2(95 - x)$$
$$2(95 - x)\left(\frac{x}{95 - x}\right) = 2(95 - x)\left(\frac{3}{2}\right)$$
$$2x = 3(95 - x)$$
$$2x = 285 - 3x$$
$$2x + 3x = 285 - 3x + 3x$$
$$5x = 285$$
$$\frac{5x}{5} = \frac{285}{5}$$
$$x = 57$$

So there are 57 females and $95 - 57 = 38$ males in the class.

Problem 5

Suppose that the denominator of a fraction is 3 more than twice the numerator. Write a rational expression for this fraction.

$$\text{let } x = \text{numerator}$$
$$2x = \text{twice the numerator}$$
$$2x + 3 = \text{3 more than twice the numerator}$$
$$\frac{x}{2x + 3} = \text{the fraction}$$

Suppose we know that when the numerator and denominator of the above fraction are each decreased by 5, the new fraction is equivalent to $\frac{1}{4}$. Let's find the original fraction.

$$\text{let } x - 5 = \text{new numerator}$$
$$(2x + 3) - 5 = 2x - 2 = \text{new denominator}$$
$$\frac{x - 5}{2x - 2} = \text{new fraction}$$

Then

$$\frac{x - 5}{2x - 2} = \frac{1}{4}$$
$$\frac{x - 5}{2(x - 1)} = \frac{1}{2 \cdot 2} \qquad \text{LCD: } 2 \cdot 2(x - 1)$$
$$2 \cdot 2(x - 1)\left[\frac{x - 5}{2(x - 1)}\right] = 2 \cdot 2(x - 1)\left(\frac{1}{2 \cdot 2}\right)$$
$$2(x - 5) = x - 1$$
$$2x - 10 = x - 1$$
$$2x - 10 - x = x - 1 - x$$

$$x - 10 = -1$$
$$x - 10 + 10 = -1 + 10$$
$$x = 9$$

So we know the original numerator was 9 and the original denominator was

$$2x + 3 = 2(9) + 3 = 21$$

Therefore, the original fraction was $\dfrac{9}{21}$.

EXERCISE SET 8.5

For each problem, write an equation and solve it.

_____ **1.** Eric and Kate each own interest in the Specialty Shop. Each month Eric receives $\frac{1}{10}$ of the profits and Kate receives $\frac{1}{8}$ of the profits. Together one month they received $630. How much profit did the Specialty Shop make that month?

_____ **2.** A number is added to the numerator and twice the same number is subtracted from the denominator of $\frac{4}{15}$. The resulting fraction is $\frac{7}{9}$. Find the number.

_____ **3.** Driving 3 mph faster than his normal rate, Henry can travel 638 miles in the same time it takes him to travel 605 miles at his normal rate. What is Henry's normal rate?

_____ **4.** José and Rick can assemble 36 parts of a carburetor in an hour. The ratio of the number Rick did to the number José did is 4 to 5. Find how many parts each assembled during the hour.

_____ **5.** On a road map, 1 inch represents 15 miles. If the distance between Birmingham and Chattanooga is 11 inches on the map, how far apart are the two cities?

_____ **6.** The ratio of students to teachers at Benton High School is 3 teachers for every 70 students. If the school presently has 24 teachers, what is the enrollment of the student body?

_____ **7.** The width of a rectangle is $\frac{2}{3}$ of its length. If the perimeter is 60 feet, find the dimensions. (Remember: $P = 2l + 2w$.)

_____ **8.** Sidney owns $\frac{5}{8}$ interest in a farm that was sold at auction. His share was $150,000. What was the total selling price of the farm?

_____ **9.** In a certain math course the ratio of success to failure is 20 to 3. If there are 115 students enrolled in the course, how many can be expected to succeed?

_____ **10.** If the denominator of $\frac{x}{12}$ is decreased by x, the new fraction is equal to $\frac{1}{2}$. Find the value of x.

Summary

In this chapter we learned that to add and subtract fractions we must make certain that they have the same denominator. If they do not, we must find their lowest common denominator (LCD). The LCD is the product of all factors in the original denominators, with each factor used the most times that it appears in any single denominator. We agreed on the following procedure for adding and subtracting fractions.

1. Factor all the denominators.
2. Choose the LCD.
3. Build new fractions from the old fractions, with each new fraction having the LCD.
4. Combine the numerators to form a new fraction with the LCD as its denominator.
5. Simplify and reduce the fraction if possible.

We learned to solve fractional equations by multiplying each term of the equation by the LCD for all the fractions in the equation. This allowed us to get rid of all the denominators and we were left with familiar equations to solve by addition, subtraction, multiplication, and division.

Finally, we practiced switching from word statements to equations that contained rational algebraic expressions.

REVIEW EXERCISES

Find the numerator to replace ? so that the fractions will be equivalent.

SECTION 8.1

_____ 1. $\dfrac{8}{9} = \dfrac{?}{54}$ _____ 2. $\dfrac{5}{3x} = \dfrac{?}{9x^3}$

_____ 3. $\dfrac{3x}{7y} = \dfrac{?}{35y^5}$ _____ 4. $\dfrac{2a}{a + b} = \dfrac{?}{a^2 + ab}$

_____ 5. $\dfrac{3y}{y - 2} = \dfrac{?}{y^2 - 4}$ _____ 6. $\dfrac{x - 3}{x - 6} = \dfrac{?}{x^2 - 11x + 30}$

Combine the following fractions. Write your answers in reduced form.

SECTION 8.2

_____ 7. $\dfrac{11}{3x} - \dfrac{8}{3x}$ _____ 8. $\dfrac{7x}{2y} + \dfrac{5}{2y} - \dfrac{3x}{2y} + \dfrac{1}{2y}$

_____ 9. $\dfrac{a}{a - b} - \dfrac{b}{a - b}$ _____ 10. $\dfrac{2x}{4x^2 - 25} - \dfrac{5}{4x^2 - 25}$

_____ 11. $\dfrac{2a^2}{a^2 - 7a} - \dfrac{13a}{a^2 - 7a} - \dfrac{7}{a^2 - 7a}$

_____ 12. $\dfrac{2x^2 + 5x}{6x^2 + 7x - 3} + \dfrac{6x - 5}{6x^2 + 7x - 3}$

SECTION 8.3

_____ 13. $\dfrac{11}{20} - \dfrac{13}{12} + \dfrac{4}{15}$ _____ 14. $\dfrac{3x}{7} + \dfrac{2x}{5}$

_____ 15. $\dfrac{8}{x} + \dfrac{1}{x^2} - \dfrac{3}{4x^2}$ _____ 16. $\dfrac{x + 6}{3x} - \dfrac{3x - 4}{2x}$

_____ 17. $\dfrac{x + y}{4x^2y} - \dfrac{x - 2y}{3xy^2} - \dfrac{21}{12xy}$ _____ 18. $\dfrac{5}{a - 1} - \dfrac{4}{a + 1}$

_____ 19. $\dfrac{5x}{6x - 18} - \dfrac{3x}{2x - 6}$ _____ 20. $\dfrac{2a}{a^2 - 4} - \dfrac{3}{a^2 - a - 2}$

_____ 21. $\dfrac{x + 5}{x^2 + 3x - 18} - \dfrac{4}{x + 6}$

_____ 22. $\dfrac{3}{y^2 + y - 2} - \dfrac{2}{y^3 - y}$

Solve.

SECTION 8.4

_____ 23. $\dfrac{x - 3}{5} - \dfrac{x}{10} = 1$ _____ 24. $\dfrac{2x - 5}{x^2} - \dfrac{7}{3x} = 0$

_____ **25.** $\dfrac{y}{y-3} + \dfrac{7}{3} = \dfrac{3}{y-3}$ _____ **26.** $\dfrac{9}{4x} = \dfrac{3}{x+6}$

_____ **27.** $\dfrac{a+3}{a+1} = \dfrac{a-3}{a-2}$ _____ **28.** $\dfrac{8}{x-7} = \dfrac{3x-5}{3x^2 - 19x - 14}$

SECTION 8.5

_____ **29.** If 4 gallons of paint will cover 600 square feet, how many gallons will be needed to paint 1800 square feet?

_____ **30.** The number of games the Braves lost last year was 5 less than the number they won. If the ratio of the games won to the games lost is 3 to 2, find how many games the Braves played last year.

9 Working with Inequalities

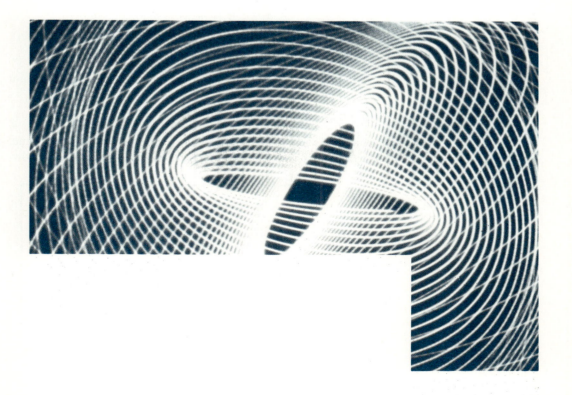

In everyday conversation we constantly compare numbers. Remarks such as

"It's warmer today than it was yesterday."
"Marie is taller than Todd."
"This bike is less expensive than that bike."

are all familiar to us. In other terms, we could translate those remarks to the following:

"Today's temperature is *greater than* yesterday's temperature."
"Marie's height is *greater than* Todd's height."
"The price of this bike is *less than* the price of that bike."

In Chapter 1 we agreed that the number line was a useful tool for picturing numbers. We discovered that two numbers can be easily compared by describing their locations on the number line. Statements that compare two quantities using phrases such as

"is less than"
"is greater than"
"is less than or equal to"
"is greater than or equal to"

are called **inequalities**.

In this chapter we learn how to

1. Write and graph numerical inequalities and inequalities containing variables.
2. Simplify and solve first-degree inequalities.
3. Combine inequalities containing variables.
4. Switch from word statements to variable inequalities.

9.1 Writing and Graphing Inequalities

Mathematicians have invented some very handy symbols to be used in writing inequalities.

$2 < 3$ is read "2 is less than 3"
$5 > -1$ is read "5 is greater than -1"

On the number line,

$2 < 3$ means "the point 2 lies to the *left* of the point 3"
$5 > -1$ means "the point 5 lies to the *right* of the point -1"

9.1-1 Writing Numerical Inequalities

The rational numbers are very orderly. If we are given any two rational numbers, *a* and *b*, it is always possible to compare them using an inequality. We know

> $a < b$ if a lies to the left of b
> on the number line.
>
> $a > b$ if a lies to the right of b
> on the number line.
>
> $a = b$ if a and b correspond to the
> same point on the number line.

Using a number line to picture each of the possible inequalities, we see

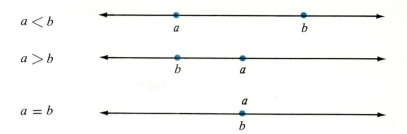

For instance, suppose that we were given the following number line with some labeled points.

Comparing B and A, we may write

$$B < A \quad \text{or} \quad A > B$$

Comparing A and C, we may write

$$A = C$$

Comparing F and B, we may write

$$F > B \quad \text{or} \quad B < F$$

Comparing F and E and D, we may write

$$F < E < D \quad \text{or} \quad D > E > F$$

Notice that when one number, such as E, lies between two other numbers, such as F and D, we may use a three-part inequality to express that fact. For instance, to express the fact that 5 is between 2 and 9, we may write

$$2 < 5 < 9$$

Let's practice comparing some numbers using our inequality symbols to fill in the blanks.

Examples	*Solutions*
2 _____ 10	$2 < 10$
-5 _____ 0	$-5 < 0$
-3 _____ -7	$-3 > -7$
0 _____ 1	$0 < 1$
$\dfrac{1}{2}$ _____ $\dfrac{1}{3}$	$\dfrac{1}{2} > \dfrac{1}{3}$
$\dfrac{5}{2}$ _____ $\dfrac{10}{4}$	$\dfrac{5}{2} = \dfrac{10}{4}$
-127 _____ -126	$-127 < -126$
$\dfrac{1001}{1000}$ _____ 1	$\dfrac{1001}{1000} > 1$
$\dfrac{-750}{250}$ _____ -3	$\dfrac{-750}{250} = -3$

Example 1. Write an inequality to say that 0 is between -1 and 3.

    Solution. $-1 < 0 < 3$.

Example 2. Write an inequality to say that $-\frac{1}{2}$ is between $-\frac{3}{4}$ and $-\frac{1}{4}$.

    Solution. $-\frac{3}{4} < -\frac{1}{2} < -\frac{1}{4}$.

9.1-2   Writing Variable Inequalities

Suppose that we were asked to find a number which is less than 7. One student might respond with the answer 6. Another might suggest 0. Still others might offer $\frac{5}{2}$ or -3 or -659. Which of these answers is correct? All of them are correct, of course. There is, in fact, an endless list of numbers less than 7. We sometimes say there is an *infinite* number of correct answers to such a problem.

    The problem we have just described was actually asking us to find all the x-values that satisfy the inequality

$$x < 7$$

We discovered that there was no practical way to list all the solutions to such a variable inequality. The only way that we might get a better idea of what numbers are in this infinite list would be to use the number line.

    On the number line, all the numbers that satisfy the inequality

$$x < 7$$

are the numbers that lie to the *left* of 7. We can picture those numbers by

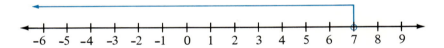

Such a picture is called a **graph** of the inequality. Notice the open dot at the point 7, which shows that 7 itself is *not* a solution to the inequality but that any number up to 7 (such a 6.9 or 6.999 and so on) *does* satisfy the inequality. Notice also that the arrow pointing to the left on the graph shows that it extends forever in that direction.

    Let's graph the solution to the inequality

$$a > -2$$

Since this is a greater-than inequality, our solutions are all the numbers to the *right* of -2. On the number line, we shall graph an open dot at the point -2 and an arrow extending forever to the right.

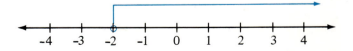

Once again, we see that there are infinitely many solutions to our inequality.

    Consider the three-part inequality

$$-1 < x < 3$$

Remember that such an inequality demands that the middle quantity, x, be located *be-*

tween the other quantities. We are looking for x-values between -1 and 3. On a number line, we can picture all such x-values by

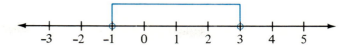

Suppose that we are asked to find all numbers which are either less than *or* equal to 5. Then suppose that we were asked to find all numbers which are either greater than *or* equal to -2. Mathematicians use the following symbols to stand for such phrases.

> $a \leq b$ is read "a is less than *or* equal to b"
>
> $a \geq b$ is read "a is greater than *or* equal to b"

On the number line,

> $a \leq b$ means that a lies to the left of b,
>   *or* that a and b are located at the same point
>
> $a \geq b$ means that a lies to the right of b,
>   *or* that a and b are located at the same point

For instance, to find all numbers less than or equal to 5, we must solve the inequality

$$a \leq 5$$

On the number line, we need all the numbers that lie to the left of 5 or at the point 5. We graph our solutions as

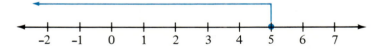

Notice the solid dot at 5 to show that 5 itself is a solution to the inequality.

Now let's find all the numbers greater than or equal to -2. Here we must solve the inequality

$$x \geq -2$$

On the number line, we must graph all the numbers that lie to the right of -2 or at the point -2.

Once again, a solid dot at -2 shows that -2 itself is a solution to the inequality.

To solve a three-part inequality such as

$$0 \leq x \leq 4$$

our graph must include all the numbers between 0 and 4, together with the points 0 and 4.

Example 3. Graph the solutions of $x < -1$.

Solution.

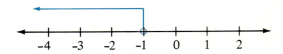

Example 4. Graph the solutions of $a \geq 2$.

Solution.

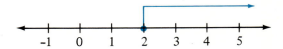

Example 5. Graph the solutions of $-1 \leq x < 7$.

Solution.

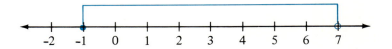

Example 6. Graph the solutions of $0 < x \leq \dfrac{3}{2}$.

Solution.

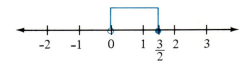

Trial Run   *Graph the solutions on a number line.*

1. $a > 5$

2. $x \leq -2$

3. $y \geq -1$

4. $0 \leq x \leq \dfrac{5}{2}$

5. $-2 < x \leq 1$

6. $-5 \leq a < -\dfrac{3}{2}$

ANSWERS

1.

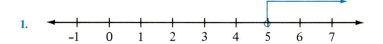

2.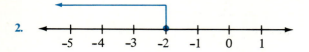

3.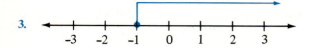

4.

5.

6.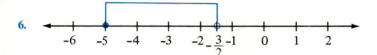

EXERCISE SET 9.1

Fill each blank with the correct symbol ($<$, $>$, $=$).

1. -11 _____ 0

2. -9 _____ -1

3. 3 _____ -19

4. 5 _____ -23

5. $\dfrac{1}{3}$ _____ $\dfrac{4}{12}$

6. $\dfrac{10}{15}$ _____ $\dfrac{2}{3}$

7. -23 _____ -2

8. -45 _____ 0

9. $\dfrac{-800}{50}$ _____ -15

10. $\dfrac{-600}{30}$ _____ -19

11. -2 _____ -10

12. 5 _____ -35

13. -15 _____ -59

14. -17 _____ -38

Write an inequality for each of the following.

_____ 15. -5 is between -11 and 0.

_____ 16. -6 is between -9 and -1.

_____ 17. 3 is between -2 and 7.

_____ 18. 4 is between 0 and 10.

_____ 19. $-\dfrac{2}{5}$ is between 0 and $-\dfrac{4}{5}$.

_____ 20. $-\dfrac{3}{8}$ is between $-\dfrac{7}{8}$ and $-\dfrac{1}{8}$.

Graph the solutions on a number line.

21. $x > 1$

22. $x > 3$

23. $a \leq 3$

24. $a \leq 0$

25. $y \geq -2$

26. $y \geq -3$

27. $b < 0$

28. $b < -6$

29. $0 < x < 6$

30. $-2 < x < 3$

31. $-1 \leq x < \dfrac{7}{2}$

32. $0 \leq x < \dfrac{5}{2}$

33. $-6 \leq a \leq -2$

34. $-5 \leq a \leq -3$

35. $-\dfrac{3}{2} < y \leq 0$

36. $-1 < y \leq \dfrac{1}{2}$

37. $3 < x < 4$

38. $-1 < x < 0$

39. $-5 \leq x \leq 1$

40. $-4 \leq x \leq 2$

CHECKUP 9.1

Fill each blank with the correct symbol ($<$, $>$, $=$).

_____ 1. -7 _____ 1

_____ 2. $\dfrac{6}{10}$ _____ $\dfrac{15}{25}$

_____ 3. -18 _____ -22

_____ 4. 0 _____ $\dfrac{-4}{5}$

Write an inequality for each of the following.

_____ 5. -2 is between -5 and 0.

_____ 6. 3 is between $\dfrac{3}{2}$ and $\dfrac{7}{2}$.

Graph the solution on a number line.

_____ 7. $a > -1$

_____ 8. $x \le 0$

_____ 9. $-3 \le y \le \dfrac{9}{2}$

_____ 10. $-2 < x \le 2$

9.2 Operating on Inequalities

We have already learned how to use the operations of addition, subtraction, multiplication, and division on both sides of an equation. Can we perform those same operations on both sides of an inequality? Is an inequality still true after performing such operations? We must investigate before we can decide.

9.2-1 Operating on Numerical Inequalities

Suppose that we consider the numerical inequality

$$6 < 20$$

What happens if we add the same number to both sides of this inequality? Let's add 5.

$$6 + 5 \overset{?}{<} 20 + 5$$
$$11 \overset{?}{<} 25$$

We certainly can agree that $11 < 25$, so the inequality is still true. Let's subtract 29 from both sides of the original inequality.

$$6 < 20$$
$$6 - 29 \overset{?}{<} 20 - 29$$
$$-23 \overset{?}{<} -9$$

Yes, $-23 < -9$, so the inequality is again true. Our conclusion?

> If we add (or subtract) the same quantity to (or from) both sides of an inequality, the inequality will still be true.

Returning to our original inequality,

$$6 < 20$$

let's now try multiplying and dividing both sides by some numbers.
First, let's multiply both sides by 7.

$$7 \cdot 6 \overset{?}{<} 7 \cdot 20$$
$$42 \overset{?}{<} 140$$

We agree that $42 < 140$ and our inequality is still true. Let's divide both sides by 2.

$$6 < 20$$
$$\frac{6}{2} \overset{?}{<} \frac{20}{2}$$
$$3 \overset{?}{<} 10$$

Of course, $3 < 10$ is still a true statement.
Now let's multiply both sides by -4.

$$6 < 20$$
$$-4 \cdot 6 \overset{?}{<} -4 \cdot 20$$
$$-24 \overset{?}{<} -80$$

Is this true? Does -24 lie to the *left* of -80 on the number line? Definitely *not*! In fact,

$$-24 > -80$$

In the process of multiplying by -4, the direction of the inequality was *reversed*. What happens if we divide both sides by -2?

$$6 < 20$$
$$\frac{6}{-2} \overset{?}{<} \frac{20}{-2}$$
$$-3 \overset{?}{<} -10$$

Once again, this is *not* a true statement. In fact,

$$-3 > -10$$

In the process of dividing both sides of our inequality by a *negative* number, the direction of the inequality was *reversed*. Our conclusion?

If we multiply (or divide) both sides of an inequality by the same *positive* number, the *direction* of the resulting inequality will be the same.

If we multiply (or divide) both sides of an inequality by the same *negative* number, the *direction* of the resulting inequality must be reversed.

Example	*Operation on both sides*	*Result*
$-1 < 5$	add 3	$-1 + 3 < 5 + 3$ $2 < 8$
$12 > 7$	add -5	$12 + (-5) > 7 + (-5)$ $7 > 2$
$0 < 3$	subtract 4	$0 - 4 < 3 - 4$ $-4 < -1$
$6 > 4$	multiply by 3	$3 \cdot 6 > 3 \cdot 4$ $18 > 12$
$-2 < -1$	multiply by 5	$5(-2) < 5(-1)$ $-10 < -5$
$10 < 15$	divide by 5	$\frac{10}{5} < \frac{15}{5}$ $2 < 3$
$18 > 9$	divide by -3	$\frac{18}{-3} < \frac{9}{-3}$ $-6 < -3$
$16 > -8$	multiply by -1	$(-1)(16) < (-1)(-8)$ $-16 < 8$

9.2-2 Operating on Variable Inequalities

Now that we know how to operate on both sides of an inequality, we may solve variable inequalities such as

$$x + 3 \le 2$$
$$a - 4 > 7$$
$$3x - 1 \ge 17$$
$$5 - 2y < -3$$
$$-y + 2 \ge 2$$

These inequalities are similar to first-degree equations because the variable terms have exponents of 1. As with first-degree equations, our goal is to isolate the variable.

In the inequality

$$x + 3 \leq 2$$

we must subtract 3 from both sides.

$$x + 3 - 3 \leq 2 - 3$$
$$x \leq -1$$

Graphing our solutions, we have

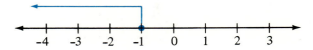

To solve the inequality

$$a - 4 > 7$$

we must add 4 to both sides.

$$a - 4 + 4 > 7 + 4$$
$$a > 11$$

The graph of our solutions is

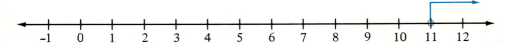

For the inequality

$$3x - 4 \geq 17$$

we first add 4 to both sides.

$$3x - 4 + 4 \geq 17 + 4$$
$$3x \geq 21$$

Then we divide by 3.

$$\frac{3x}{3} \geq \frac{21}{3}$$

$$x \geq 7$$

Graphing the solutions gives

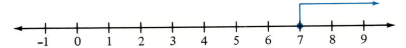

To solve the inequality

$$5 - 2y < -3$$

we first subtract 5 from both sides.

$$5 - 2y - 5 < -3 - 5$$
$$-2y < -8$$

Now divide by -2, remembering to reverse the direction of the inequality.

$$-2y < -8$$
$$\frac{-2y}{-2} > \frac{-8}{-2}$$
$$y > 4$$

Our graph becomes

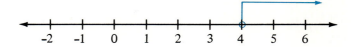

For the inequality

$$-y + 2 \geq 2$$

we first subtract 2 from both sides.

$$-y + 2 - 2 \geq 2 - 2$$
$$-y \geq 0$$

Now divide both sides by -1, remembering to reverse the direction of the inequality.

$$-y \geq 0$$
$$\frac{-y}{-1} \leq \frac{0}{-1}$$
$$y \leq 0$$

We graph our solutions as

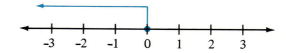

To solve first-degree inequalities, we must

1. Perform any necessary addition or subtractions on both sides of the inequality.
2. Perform any necessary multiplications and/or divisions on both sides of the inequality, remembering to reverse the direction of the inequality whenever multiplying or dividing by a negative number.

Example 1. Solve $2x + 3 > -4$.

Solution

$$2x + 3 > -4$$
$$2x + 3 - 3 > -4 - 3$$
$$2x > -7$$
$$\frac{2x}{2} > \frac{-7}{2}$$
$$x > \frac{-7}{2}$$

Example 2. Solve $\frac{x}{2} - 4 \leq 1$.

Solution

$$\frac{x}{2} - 4 \leq 1$$
$$\frac{x}{2} - 4 + 4 \leq 1 + 4$$

$$\frac{x}{2} \leq 5$$

$$2\left(\frac{x}{2}\right) \leq 2(5)$$

$$x \leq 10$$

Example 3. Solve $5 - x < 9$.

    Solution

$$5 - x < 9$$

$$5 - x - 5 < 9 - 5$$

$$-x < 4$$

$$\frac{-x}{-1} > \frac{4}{-1}$$

$$x > -4$$

Example 4. Solve $\dfrac{-3x}{2} + 5 \geq 2$.

    Solution

$$\frac{-3x}{2} + 5 \geq 2$$

$$\frac{-3x}{2} + 5 - 5 \geq 2 - 5$$

$$\frac{-3x}{2} \geq -3$$

$$2\left(\frac{-3x}{2}\right) \geq 2(-3)$$

$$-3x \geq -6$$

$$\frac{-3x}{-3} \leq \frac{-6}{-3}$$

$$x \leq 2$$

To solve inequalities containing parentheses or variables on both sides, we use the same methods we used for equations. We must remove parentheses and then get all variable terms on one side and constant terms on the other side. A few examples are worked here.

Example 5. Solve $5(x + 1) - 3 \leq 12$.

    Solution

$$5(x + 1) - 3 \leq 12$$

$$5x + 5 - 3 \leq 12$$

$$5x + 2 \leq 12$$

$$5x + 2 - 2 \leq 12 - 2$$

$$5x \leq 10$$

$$\frac{5x}{5} \leq \frac{10}{5}$$

$$x \leq 2$$

Example 6. Solve $3x + 7 > 2x - 6$.

Solution

$$3x + 7 > 2x - 6$$
$$3x + 7 - 2x > 2x - 6 - 2x$$
$$x + 7 > -6$$
$$x + 7 - 7 > -6 - 7$$
$$x > -13$$

Example 7. Solve $4(x - 1) - 10 \geq 7x + 1$.

Solution

$$4(x - 1) - 10 \geq 7x + 1$$
$$4x - 4 - 10 \geq 7x + 1$$
$$4x - 14 \geq 7x + 1$$
$$4x - 14 - 7x \geq 7x + 1 - 7x$$
$$-3x - 14 \geq 1$$
$$-3x - 14 + 14 \geq 1 + 14$$
$$-3x \geq 15$$
$$\frac{-3x}{-3} \leq \frac{15}{-3}$$
$$x \leq -5$$

Trial Run   *Solve each inequality.*

———— **1.** $-3x > 15$

———— **2.** $a + 2 < -7$

———— **3.** $\dfrac{2a}{5} - 3 \geq 1$

———— **4.** $9 - 6y \leq -6$

———— **5.** $2y - 3(y + 1) > 5$

———— **6.** $5y - 7 \leq 2y + 5$

ANSWERS

1. $x < -5$.   **2.** $a < -9$.   **3.** $a \geq 10$.   **4.** $y \geq \dfrac{5}{2}$.   **5.** $y < -8$.   **6.** $y \leq 4$.

Name _____ **Date** _____

EXERCISE SET 9.2

Perform the indicated operation on both sides of the inequality.

_____ **1.** $-3 < 5$, add 4   _____ **2.** $2 < 7$, add -2

_____ **3.** $13 > 2$, add -7 _____ **4.** $3 > -2$, add 6

_____ **5.** $2 > -3$, multiply by 4   _____ **6.** $4 > -5$, multiply by 3

_____ **7.** $-18 < -6$, divide by 3   _____ **8.** $-21 < -14$, divide by 7

_____ **9.** $13 > 6$, subtract 20   _____ **10.** $12 > 3$, subtract 15

_____ **11.** $10 > -5$, divide by -5 _____ **12.** $12 > -16$, divide by -4

_____ **13.** $-13 < -1$, multiply by -1 _____ **14.** $-19 < -3$, divide by -1

_____ **15.** $-8 < 16$, divide by -2 _____ **16.** $16 > 10$, multiply by -3

Solve and graph each inequality.

_____ **17.** $x + 3 \leq 7$ _____ **18.** $x + 5 \leq 7$

_____ **19.** $3x > -15$ _____ **20.** $4x > -16$

_____ **21.** $-2a > 7$ _____ **22.** $-3a > 9$

_____ **23.** $3a - 3 \geq 12$ _____ **24.** $4a - 5 \geq 15$

_____ **25.** $\frac{x}{3} - 1 < -2$ _____ **26.** $\frac{x}{5} - 3 < -4$

_____ **27.** $-5z + 12 \leq 17$ _____ **28.** $-6z + 14 \leq 26$

_____ **29.** $3 - y > 1$ _____ **30.** $4 - y > 1$

_____ **31.** $\frac{5x}{2} + 1 \geq 6$ _____ **32.** $\frac{4x}{3} + 5 \geq 9$

Solve each inequality.

_____ **33.** $4(x - 5) \geq -2$ _____ **34.** $3(x - 6) \geq -4$

_____ **35.** $4y - 2(y - 5) < 10$ _____ **36.** $7y - 3(y - 4) < 12$

_____ **37.** $\frac{2x - 9}{3} > 5$ _____ **38.** $\frac{5x + 1}{4} > 9$

_____ **39.** $2a - 4 > 5a - 1$ _____ **40.** $5a - 8 > 7a - 12$

_____ **41.** $5x - (x - 4) \leq 2x + 5$ _____ **42.** $3x - (x - 5) \leq x + 7$

_____ **43.** $\frac{x - 4}{3} > \frac{2x - 1}{2}$ _____ **44.** $\frac{3x - 1}{5} > \frac{4x - 3}{2}$

_____ **45.** $2(z - 2) - 3 < 5(z + 5) - 23$ _____ **46.** $3(z + 1) + 5 < 9(2z + 3) + 1$

_____ **47.** $-\frac{x}{4} + 2 \geq \frac{x}{3} - 5$ _____ **48.** $-\frac{x}{3} + 2 \geq \frac{x}{5} - 6$

_____ **49.** $\frac{x}{4} - \frac{2x}{3} \geq \frac{1}{12}$ _____ **50.** $\frac{x}{5} - \frac{5x}{3} \geq \frac{4}{15}$

CHECKUP 9.2

Perform the indicated operation on both sides of the inequality.

_____ 1. $-2 > -9$, add 5

_____ 2. $-5 < 0$, multiply by -3

_____ 3. $4 > -6$, divide by -2

Solve and graph each inequality.

_____ 4. $x - 5 \geq -7$

_____ 5. $-5a \leq 15$

_____ 6. $8 - y > 7$

_____ 7. $-\dfrac{4x}{3} + 7 < 3$

Solve each inequality.

_____ 8. $9(a - 2) \leq -3$

_____ 9. $\dfrac{4x - 1}{2} > -3$

_____ 10. $5y - 2 < 9y + 4$

9.3 Combining Variable Inequalities (Optional)

Sometimes it is necessary to find the values of a variable that satisfy more than one inequality at the same time.

9.3-1 Satisfying Three-Part Inequalities

We have already looked at three-part inequalities as a way of saying that some number is located between two other numbers. We wrote

$$3 < 5 < 9$$

to say that 5 is located between 3 and 9. We wrote

$$-1 < x < 3$$

to say that x stands for any number between -1 and 3.

Actually, the three-part inequality

$$-1 < x < 3$$

is saying that x must satisfy two conditions:

$$x > -1 \quad \text{and} \quad x < 3$$

Only numbers that are greater than -1 and less than 3 *at the same time* will satisfy the original inequality. Let's consider each inequality separately and graph each on the number line.

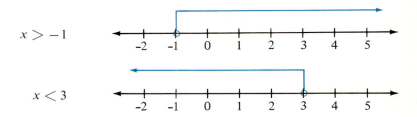

In order to see where both inequalities are satisfied at the same time, let us put both graphs on the same number line.

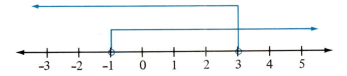

Can you see where both inequalities are satisfied?

This result should not be surprising, since we agreed earlier that graphs of three-part inequalities always look like this.

Suppose that we consider three-part inequalities in which the x is not completely isolated in the middle. For instance, let's solve

$$-1 < 3x + 5 < 8$$

How can we isolate the x in the middle? Remember that this inequality is saying

$$3x + 5 > -1 \quad \text{and} \quad 3x + 5 < 8$$

We have already learned how to solve such inequalities. First we subtract 5.

$$3x + 5 - 5 > -1 - 5 \quad \text{and} \quad 3x + 5 - 5 < 8 - 5$$
$$3x > -6 \quad \text{and} \quad 3x < 3$$

Then we divide by 3.

$$\frac{3x}{3} > \frac{-6}{3} \quad \text{and} \quad \frac{3x}{3} < \frac{3}{3}$$
$$x < -2 \quad \text{and} \quad x > -4$$

Our result is that x must be greater than -2 *and* less than 1 *at the same time*. We may write

$$-2 < x < 1$$

A shorter way of solving the original inequality

$$-1 < 3x + 5 < 8$$

allows us to operate on all three parts at the same time. We decide what operation must be performed to isolate the variable in the middle, and then perform that operation on all three parts of the inequality.

$$-1 < 3x + 5 < 8$$

Subtract 5:
$$-1 - 5 < 3x + 5 - 5 < 8 - 5$$
$$-6 < 3x < 3$$

Divide by 3:
$$\frac{-6}{3} < \frac{3x}{3} < \frac{3}{3}$$
$$-2 < x < 1$$

Notice this is the same result as before. Our graph is

To check our solutions, we might choose any number between -2 and 1 to see if it really does satisfy the original inequality. For instance, suppose that we choose -1 and substitute it for x in the original inequality.

$$-1 < 3x + 5 \quad < 8$$
$$-1 \overset{?}{<} 3(-1) + 5 \overset{?}{<} 8$$
$$-1 \overset{?}{<} -3 + 5 \quad \overset{?}{<} 8$$
$$-1 \overset{?}{<} \quad 2 \quad \overset{?}{<} 8$$

Since 2 does lie between -1 and 8, the inequality is satisfied. Of course, we have not checked *all* the numbers between -2 and 1, but that would be impossible. Checking just one such number is helpful; we can be pretty sure that our work is correct.

On the other hand, suppose that we check a number somewhere else on the number line, *not* between -2 and 1. Let's try letting x be 4 in our original inequality.

$$-1 < 3x + 5 \quad < 8$$
$$-1 \overset{?}{<} 3(4) + 5 \overset{?}{<} 8$$
$$-1 \overset{?}{<} 12 + 5 \quad \overset{?}{<} 8$$
$$-1 \overset{?}{<} \quad 17 \quad \overset{?}{<} 8 \quad \text{No!}$$

As you can see, x-values that do not satisfy $-2 < x < 1$ will not satisfy the original inequality.

Example 1. Solve $5 \leq 2x - 1 \leq 19$.

Solution

$$5 \leq 2x - 1 \leq 19$$

Add 1:

$$5 + 1 \leq 2x - 1 + 1 \leq 19 + 1$$
$$6 \leq \quad 2x \quad \leq 20$$

Divide by 2:

$$\frac{6}{2} \leq \frac{2x}{2} \leq \frac{20}{2}$$

$$3 \leq \; x \; \leq 10$$

Example 2. Solve $-5 < \frac{1}{2}x + 3 \leq 0$.

Solution

$$-5 < \frac{1}{2}x + 3 \leq 0$$

Subtract 3:

$$-5 - 3 < \frac{1}{2}x + 3 - 3 \leq 0 - 3$$

$$-8 < \quad \frac{1}{2}x \quad \leq -3$$

Multiply by 2:

$$2(-8) < 2\left(\frac{1}{2}x\right) \leq 2(-3)$$

$$-16 < \quad x \quad \leq -6$$

You should check this result by substituting some x-value between -16 and -6 in the original inequality.

When the coefficient of the variable term is negative, we know that we must be careful because solving will require multiplying or dividing our inequality by a negative number. Consider the problem

$$3 < 1 - x < 5$$

which can also be written as

$$1 - x > 3 \quad \text{and} \quad 1 - x < 5$$

First we subtract 1.

$$1 - x - 1 > 3 - 1 \qquad \text{and} \qquad 1 - x - 1 < 5 - 1$$
$$-x > 2 \qquad \text{and} \qquad -x < 4$$

Now divide by -1, remembering to reverse the direction of each inequality.

$$\frac{-x}{-1} < \frac{2}{-1} \qquad \text{and} \qquad \frac{-x}{-1} > \frac{4}{-1}$$

$$x < -2 \qquad \text{and} \qquad x > -4$$

On the same number line we graph our solutions

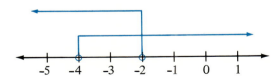

and we see that both inequalities are satisfied between -4 and -2. We write the solution

$$-4 < x < -2$$

What would have happened if we had operated on all three parts of this inequality at the same time? Let's see.

$$3 < 1 - x < 5$$

Subtract 1:
$$3 - 1 < 1 - x - 1 < 5 - 1$$
$$2 < \quad -x \quad < 4$$

Divide by -1:
$$\frac{2}{-1} > \frac{-x}{-1} > \frac{4}{-1}$$
$$-2 > \quad x \quad > -4$$

Putting our answer in proper order (with the smaller number on the left), we may write

$$-4 < x < -2$$

which matches our earlier solution.

Example 3. Solve $-3 \leq -5x - 3 < 2$.

Solution

$$-3 \leq -5x - 3 < 2$$

Add 3:
$$-3 + 3 \leq -5x - 3 + 3 < 2 + 3$$
$$0 \leq \quad -5x \quad < 5$$

Divide by -5:
$$\frac{0}{-5} \geq \frac{-5x}{-5} > \frac{5}{-5}$$
$$0 \geq \quad x \quad > -1$$
$$-1 < \quad x \quad \leq 0$$

Example 4. Solve $4 \leq 6 - \frac{2}{3}x \leq 10$.

Solution

$$4 \leq 6 - \frac{2}{3}x \leq 10$$

Subtract 6:
$$4 - 6 \leq 6 - \frac{2}{3}x - 6 \leq 10 - 6$$
$$-2 \leq \quad -\frac{2}{3}x \quad \leq 4$$

Multiply by 3:
$$3(-2) \leq 3\left(-\frac{2}{3}x\right) \leq 3(4)$$
$$-6 \leq \quad -2x \quad \leq 12$$

Divide by -2:
$$\frac{-6}{-2} \geq \frac{-2x}{-2} \geq \frac{12}{-2}$$
$$3 \geq \quad x \quad \geq -6$$
$$-6 \leq \quad x \quad \leq 3$$

Trial Run *Solve.*

_____ 1. $3 \leq x - 5 \leq 5$

_____ 2. $-6 < 3x \leq 9$

_____ 3. $1 \leq 2x + 1 \leq 7$

_____ **4.** $-3 < \frac{1}{2}x - 4 < 1$

_____ **5.** $1 \le 2 - x < 4$

_____ **6.** $-1 \le 2 - \frac{3}{4}x \le 8$

ANSWERS
1. $8 \le x \le 10$.   **2.** $-2 < x \le 3$.   **3.** $0 \le x \le 3$.   **4.** $2 < x < 10$.   **5.** $-2 < x \le 1$.
6. $-8 \le x \le 4$.

9.3-2   Satisfying One Inequality *and* Another Inequality

From our work with three-part inequalities, we should have a pretty good idea how to satisfy two inequalities when the problem requires that both conditions be met at the same time. For instance, to satisfy

$$x \le 3 \quad \text{and} \quad x > -1$$

we must consider the graphs of the two inequalities on the same number line.

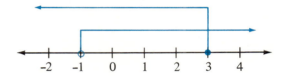

Then we check to see where *both* inequalities are satisfied at the same time. Where? Between -1 and 3 (including the point 3, not including the point -1).

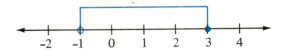

The inequality $-1 < x \le 3$ describes all the numbers satisfying both of the original inequalities.

Consider satisfying

$$x > 2 \quad \text{and} \quad x > 5$$

On the same number line, we graph our solutions.

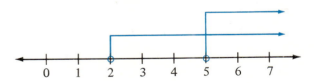

Where are *both* inequalities satisfied at the same time? Where do both lines appear at the same time? Only to the right of 5. Our conclusion? The inequality that describes the numbers satisfying *both* of the original inequalities is

$$x > 5$$

To check our solution, we might choose an x-value to the right of 5, say 7, and substitute it into the original inequalities to see if *both* are satisfied. For $x = 7$ we must check

$$x > 2 \quad \text{and} \quad x > 5$$
$$7 \overset{?}{>} 2 \quad \text{and} \quad 7 \overset{?}{>} 5$$

Yes, both inequalities are satisfied.

Now try some x-value that is *not* to the right of 5. Let's try $x = 4$ and see if *both* inequalities are satisfied.

$$x > 2 \quad \text{and} \quad x > 5$$
$$4 \overset{?}{>} 2 \quad \text{and} \quad 4 \overset{?}{>} 5$$

Although one of the inequalities is true, the other one is *not* true, so we cannot say that *both* inequalities are satisfied. Our solution $x > 5$ seems to be correct.

Let's see where we find numbers that satisfy

$$x < 7 \quad \text{and} \quad x \le -1$$

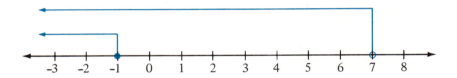

Both conditions are met from -1 to the left. The inequality that describes the numbers satisfying both of the original inequalities is

$$x \le -1$$

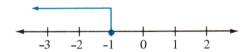

Checking the solutions is left to you.

Let's find the numbers that satisfy

$$x > 2 \quad \text{and} \quad x \le 0$$

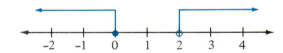

Where do both graphs appear at the same time? Nowhere! There are *no* numbers that satisfy both inequalities. We say that there is *no solution* to this problem.

To satisfy a pair of inequalities such as

$$2x + 1 \le 3 \quad \text{and} \quad 5 - x < 7$$

requires that we isolate x in each inequality, then graph and look for solutions satisfying both inequalities.

$$
\begin{array}{ccc}
2x + 1 \le 3 & \text{and} & 5 - x < 7 \\
2x + 1 - 1 \le 3 - 1 & & 5 - x - 5 < 7 - 5 \\
2x \le 2 & & -x < 2 \\
\dfrac{2x}{2} \le \dfrac{2}{2} & & \dfrac{-x}{-1} > \dfrac{2}{-1} \\
x \le 1 & \text{and} & x > -2
\end{array}
$$

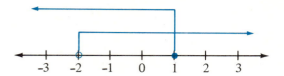

The inequality $-2 < x \leq 1$ describes the numbers that satisfy *both* of the original inequalities.

Let's check our final solutions $-2 < x \leq 1$ by choosing an x-value between -2 and 1. Suppose that we check $x = 0$ to see if *both* of the original inequalities are satisfied.

$$2x + 1 \leq 3 \qquad \text{and} \qquad 5 - x < 7$$
$$2(0) + 1 \overset{?}{\leq} 3 \qquad \qquad 5 - 0 \overset{?}{<} 7$$
$$0 + 1 \overset{?}{\leq} 3$$
$$1 \overset{?}{\leq} 3 \qquad \text{and} \qquad 5 \overset{?}{<} 7$$

Both inequalities are true, so our solution $-2 < x \leq 1$ seems correct.

Now try an x-value *not* between -2 and 1, say 3. Substituting $x = 3$ into the original inequalities, we have

$$2x + 1 \leq 3 \qquad \text{and} \qquad 5 - x < 7$$
$$2(3) + 1 \overset{?}{\leq} 3 \qquad \qquad 5 - 3 \overset{?}{<} 7$$
$$6 + 1 \overset{?}{\leq} 3$$
$$7 \overset{?}{\leq} 3 \qquad \text{and} \qquad 2 \overset{?}{<} 7$$

Although one of the inequalities is true, the other one is *not* true, so we cannot say that *both* inequalities are satisfied. Our checking suggests that the solution $-2 < x \leq 1$ is correct.

Example 5. Satisfy $2 + 3x \geq 11$ and $5x - 1 > 2$.

   Solution

$$2 + 3x \geq 11 \qquad \text{and} \qquad 5x - 1 > 2$$
$$2 + 3x - 2 \geq 11 - 2 \qquad \qquad 5x - 1 + 1 > 2 + 1$$
$$3x \geq 9 \qquad \qquad 5x > 3$$
$$\frac{3x}{3} \geq \frac{9}{3} \qquad \qquad \frac{5x}{5} > \frac{3}{5}$$
$$x \geq 3 \qquad \text{and} \qquad x > \frac{3}{5}$$

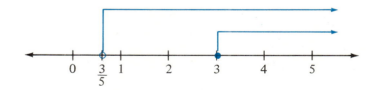

So $x \geq 3$ satisfied both inequalities. Checking is left to you.

Example 6. Satisfy $1 - x > 8$ and $\frac{2}{3}x \geq 0$.

   Solution

$$1 - x > 8 \qquad \text{and} \qquad \frac{2}{3}x \geq 0$$

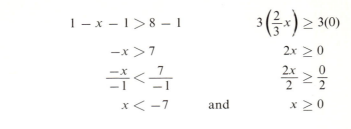

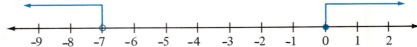

So there is no solution.

In general, to satisfy one inequality *and* another inequality, we must

1. Isolate the variable in each inequality.
2. Graph the solution for each inequality on the same number line.
3. Find the numbers that are in *both* graphs.
4. Write a single inequality that describes the numbers satisfying both of the original inequalities.

Trial Run *Find the solution for the pairs of inequalities.*

_____ 1. $x < -3$ and $x \leq 2$

_____ 2. $x \leq 7$ and $x > -1$

_____ 3. $x > 5$ and $x \geq 9$

_____ 4. $x < 0$ and $x > 4$

_____ 5. $x - 5 < 6$ and $2x - 1 \geq 9$

_____ 6. $5x + 6 > -4$ and $4 - x \leq 8$

_____ 7. $2x - 1 < 0$ and $2x - 1 \geq 1$

_____ 8. $4 - 3x \geq 13$ and $2x - 3 \leq -4$

ANSWERS
1. $x < -3$. 2. $-1 < x \leq 7$. 3. $x \geq 9$. 4. No solution. 5. $5 \leq x < 11$.
6. $x > -2$. 7. No solution. 8. $x \leq -3$.

9.3-3 Satisfying One Inequality *or* Another Inequality

Even in everyday English, the word *and* is much more demanding than the word *or*. If we are told that we are required to meet one condition *and* another, we shall be successful only when we have met *both* of those conditions at the same time. On the other hand, if we are told that we must meet one condition *or* another, we shall be successful if we meet either the first condition or the second condition or perhaps both.

The same is true for satisfying pairs of inequalities. If we must satisfy

$$x + 2 > 3 \qquad \text{and} \qquad x \geq 5$$

then only numbers satisfying *both* inequalities are acceptable solutions.

However, if we must satisfy

$$x + 2 > 3 \quad \text{or} \quad x \geq 5$$

then numbers that satisfy *either* of the inequalities (or both of them) are acceptable solutions. Let us solve these pairs of inequalities. First,

$$x + 2 > 3 \quad \text{and} \quad x \geq 5$$
$$x > 1 \quad \text{and} \quad x \geq 5$$

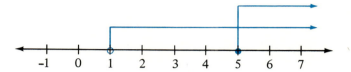

Our solution is $x \geq 5$.

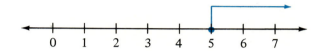

Now

$$x + 2 > 3 \quad \text{or} \quad x \geq 5$$
$$x > 1 \quad \text{or} \quad x \geq 5$$

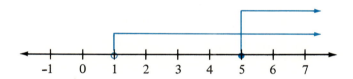

Where are all the numbers that satisfy *either* (or both) of the inequalities? They all lie to the right of 1. Our solution is $x > 1$.

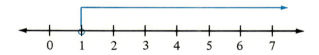

In looking at the number line, we accepted any numbers that appeared in the graph of either one of the inequalities.

Let us try to satisfy the pair of inequalities.

$$2x + 3 < 7 \qquad \text{or} \qquad 1 - x > 3$$
$$2x + 3 - 3 < 7 - 3 \qquad 1 - x - 1 > 3 - 1$$
$$2x < 4 \qquad -x > 2$$
$$\frac{2x}{2} < \frac{4}{2} \qquad \frac{-x}{-1} < \frac{2}{-1}$$
$$x < 2 \qquad \text{or} \qquad x < -2$$

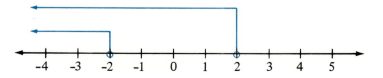

The numbers that appear in the graph of either one inequality or the other (or both) all lie to the left of 2. Our solution is

$$x < 2$$

Consider the inequalities

$$
\begin{array}{ccc}
5x - 1 > 14 & \text{or} & -2x + 5 \geq 0 \\
5x - 1 + 1 > 14 + 1 & & -2x + 5 - 5 \geq 0 - 5 \\
5x > 15 & & -2x \geq -5 \\
\dfrac{5x}{5} > \dfrac{15}{5} & & \dfrac{-2x}{-2} \leq \dfrac{-5}{-2} \\
x > 3 & \text{or} & x \leq \dfrac{5}{2}
\end{array}
$$

The numbers satisfying either one of the inequalities or the other lie to the right of 3 or to the left of (and including) $\dfrac{5}{2}$. Our solution cannot be written as a single inequality. We must write

$$x > 3 \quad \text{or} \quad x \leq \dfrac{5}{2}$$

In general, to satisfy one inequality *or* another inequality, we must

1. Isolate the variable in each inequality.
2. Graph the solutions for each inequality on the same number line.
3. Find the numbers that are in *either* graph.
4. Describe the numbers satisfying either of the original inequalities by writing a single inequality or a pair of inequalities.

Example 7. Satisfy $-3 + 7x > 4$ or $-\dfrac{5}{2}x \leq 0$.

Solution

$$
\begin{array}{ccc}
-3 + 7x > 4 & \text{or} & -\dfrac{5}{2}x \leq 0 \\
-3 + 7x + 3 > 4 + 3 & & 2\left(-\dfrac{5}{2}x\right) \leq 2(0) \\
7x > 7 & & -5x \leq 0 \\
\dfrac{7x}{7} > \dfrac{7}{7} & & \dfrac{-5x}{-5} \geq \dfrac{0}{-5} \\
x > 1 & \text{or} & x \geq 0
\end{array}
$$

Our solution is $x \geq 0$.

Example 8. Satisfy $2x - 1 \leq 0$ or $1 + 4x > 4$.

Solution

$$2x - 1 \leq 0 \qquad \text{or} \qquad 1 + 4x > 4$$
$$2x - 1 + 1 \leq 0 + 1 \qquad\qquad 1 + 4x - 1 > 4 - 1$$
$$2x \leq 1 \qquad\qquad 4x > 3$$
$$\frac{2x}{2} \leq \frac{1}{2} \qquad\qquad \frac{4x}{4} > \frac{3}{4}$$
$$x \leq \frac{1}{2} \qquad \text{or} \qquad x > \frac{3}{4}$$

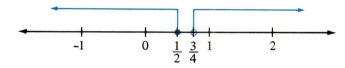

Our solution is $x \leq \dfrac{1}{2}$ or $x > \dfrac{3}{4}$.

Trial Run   *Find the solution for the pairs of inequalities.*

_____ **1.** $6x - 5 \geq 7$ or $3 - x < 8$

_____ **2.** $3x - 1 < 2$ or $x - 9 \leq -6$

_____ **3.** $2x + 13 < 9$ or $7x + 9 > 16$

_____ **4.** $\dfrac{x}{5} + 6 \leq 7$ or $\dfrac{2}{3}x < 0$

_____ **5.** $8 - x < 5$ or $2 - 3x < 8$

_____ **6.** $4x + 11 > 15$ or $-2x + 3 \geq 5$

ANSWERS
1. $x > -5$.   **2.** $x < 3$.   **3.** $x < -2$ or $x > 1$.   **4.** $x \leq 5$.   **5.** $x > -2$.
6. $x \leq -1$ or $x > 1$.

Remember, to satisfy one inequality *and* another inequality, we shall be happy with only those numbers that satisfy *both* inequalities at the same time. To satisfy one inequality *or* another inequality, we shall be happy with numbers that satisfy either the first inequality or the second inequality (or both).

EXERCISE SET 9.3

Solve each inequality.

_____ **1.** $4 < x + 6 < 9$

_____ **2.** $-5 \leq x - 3 < 1$

_____ **3.** $-1 \leq -2x \leq 8$

_____ **4.** $-6 \leq -4x \leq 4$

_____ **5.** $-5 \leq 3x + 4 \leq 16$

_____ **6.** $-7 \leq 5x + 3 \leq 28$

_____ **7.** $7 \leq \frac{1}{2}x + 9 < 17$

_____ **8.** $-7 \leq \frac{1}{3}x - 6 < -5$

_____ **9.** $4 \leq 6 - 3x \leq 9$

_____ **10.** $-5 \leq 7 - 2x \leq 5$

Find the solution for the pairs of inequalities.

_____ **11.** $x \geq -2$ and $x \leq 3$

_____ **12.** $x \geq -6$ and $x \leq 2$

_____ **13.** $x > -6$ and $x > 0$

_____ **14.** $x \geq -5$ and $x > -1$

_____ **15.** $x < 5$ or $x < -2$

_____ **16.** $x < -1$ or $x < 3$

_____ **17.** $x > 1$ or $x < -2$

_____ **18.** $x > 0$ or $x < -3$

_____ **19.** $x > 5$ and $x < -1$

_____ **20.** $x > -1$ and $x < -3$

_____ **21.** $x > \frac{1}{2}$ and $x < 4$

_____ **22.** $x > -2$ and $x < 3$

_____ **23.** $x > 0$ or $x > 4$

_____ **24.** $x > -3$ or $x > 2$

_____ **25.** $x + 6 < 8$ and $x + 2 > 0$

_____ **26.** $x + 5 < 6$ and $x + 5 > 1$

_____ **27.** $3x + 15 \geq 21$ and $5x - 2 < 13$

_____ **28.** $2x + 9 \geq 11$ and $3x - 27 < -15$

_____ **29.** $7x + 10 > 10$ or $9 - 2x \leq 19$

_____ **30.** $11x - 3 > 8$ or $5x - 7 \geq 3$

_____ **31.** $7 - 3x > 10$ or $\frac{2}{3}x > 0$

_____ **32.** $12 - 5x > 22$ or $\frac{4}{5}x > 0$

_____ **33.** $\frac{x}{2} + 3 > 5$ and $8 - \frac{x}{3} \leq 7$

_____ **34.** $\frac{x}{3} - 1 \geq 0$ and $5 - \frac{x}{2} < -6$

_____ **35.** $5x - 3 < 12$ or $2x - 3 < -5$

_____ **36.** $2x - 13 < -7$ or $3x + 11 < 5$

_____ 37. $3x + 1 < 7$ or $2x - 3 \leq 3$ _____ 38. $7x + 2 \leq 9$ or $3x - 1 < -4$

_____ 39. $-x - 6 < 3$ and $\dfrac{2x}{3} - 1 > 1$ _____ 40. $-x + 9 < 5$ and $\dfrac{5x}{2} - 4 > 1$

_____ 41. $3x - 5 < 2x + 6$ and $4x - 1 \geq 2x + 9$

_____ 42. $2x + 5 > -4x - 7$ and $2x - 11 \leq -6x + 13$

_____ 43. $2(x - 5) > 3(x - 2)$ or $3(x - 1) > 2(x + 5)$

_____ 44. $4(x - 3) > 5(x - 1)$ or $7(x - 5) > 2(3x - 5)$

_____ 45. $\dfrac{5}{6}x - \dfrac{2}{3}x \geq \dfrac{1}{2}$ and $\dfrac{1}{5}x - \dfrac{3}{10}x < \dfrac{1}{2}$ _____ 46. $\dfrac{2}{7}x - \dfrac{1}{2}x > \dfrac{3}{14}$ and $\dfrac{1}{5}x - \dfrac{2}{3}x \leq \dfrac{14}{15}$

_____ 47. $\dfrac{-x}{2} + \dfrac{x}{8} \leq 3$ or $\dfrac{2x}{3} - \dfrac{x}{4} \geq 5$ _____ 48. $\dfrac{-x}{5} + \dfrac{x}{10} \geq 3$ or $\dfrac{3x}{4} - \dfrac{x}{2} \geq 2$

CHECKUP 9.3

Solve each inequality.

_____ **1.** $-6 \leq 5y - 1 \leq 9$

_____ **2.** $-4 \leq 2 - x \leq 1$

_____ **3.** $-2 \leq \dfrac{3x}{2} - 5 \leq 4$

Find the solution for the pairs of inequalities.

_____ **4.** $x > -2$ or $x > 3$

_____ **5.** $a > -2$ and $a < -5$

_____ **6.** $x < -2$ or $x > 4$

_____ **7.** $z > -1$ and $z > 9$

_____ **8.** $4 - 3x > -8$ and $2x - 5 > -13$

_____ **9.** $\dfrac{x}{5} - 2 > -1$ or $7 - \dfrac{x}{3} < 1$

_____ **10.** $2x + 2 < 5$ and $5x - 7 > -12$

9.4 Switching from Word Statements to Variable Inequalities

Many everyday math problems involve inequalities rather than equations.

For example, suppose that a college student must have successfully completed at least 30 hours of course work to qualify as a sophomore. Letting

$$h = \text{number of hours}$$

our situation can be described by

$$h \geq 30$$

Miss America contestants must be between 18 and 25 years of age. Letting

$$a = \text{age}$$

we can describe the age qualification by

$$18 \leq a \leq 25$$

Suppose that we try solving some word problems using inequalities. We must continue with our usual methods: identify the variable, write any expressions containing the variable, write an inequality from the information in the problem, and solve.

Problem 1

Mr. Angelo, a retiree, will lose some retirement benefits if he earns over $8000 per year working part-time as a night watchman. If he has already earned $7200, how many more hours may he work at $5 per hour without losing any retirement benefits?

$$\text{let } h = \text{hours to be worked}$$
$$5h = \text{money to be earned}$$
$$7200 + 5h = \text{total earnings}$$

If Mr. Angelo does not wish to lose benefits, he must earn $8000 or less. We write the inequality

$$7200 + 5h \leq 8000$$
$$7200 + 5h - 7200 \leq 8000 - 7200$$
$$5h \leq 800$$
$$\frac{5h}{5} \leq \frac{800}{5}$$
$$h \leq 160$$

Mr. Angelo may work 160 hours or fewer.

Problem 2

In order to receive a B in her math course, Henrietta must have an average of at least 80 on six tests. If her grades on the first five tests were 73, 85, 80, 79, and 72, what grade must she receive on the sixth test in order to earn a B in the course?

$$\text{let } g = \text{grade on sixth test}$$
$$\frac{73 + 85 + 80 + 79 + 72 + g}{6} = \text{average grade for six tests}$$

Since Henrietta must have at least an 80 average, we write

$$\frac{73 + 85 + 80 + 79 + 72 + g}{6} \geq 80$$

$$\frac{389 + g}{6} \geq 80$$

$$6\left(\frac{389 + g}{6}\right) \geq 6(80)$$

$$389 + g \geq 480$$

$$389 + g - 389 \geq 480 - 389$$

$$g \geq 91$$

Henrietta must receive a 91 or better on her last test.

Problem 3

To qualify for federal funding, a special program must enroll between 150 and 210 students. Moreover, the number of rural students enrolled must be twice the number of nonrural students. How many of each type of student must be enrolled in the special program?

$$\text{let } x = \text{number of nonrural students}$$
$$2x = \text{number of rural students}$$
$$x + 2x = \text{total number of students}$$

Since this is an in-between statement, we shall write the three-part inequality

$$150 \leq x + 2x \leq 210$$
$$150 \leq \quad 3x \quad \leq 210$$
$$\frac{150}{3} \leq \quad \frac{3x}{3} \quad \leq \frac{210}{3}$$
$$50 \leq \quad x \quad \leq 70$$

Between 50 and 70 nonrural students must be enrolled. To find the number of rural students, we must look at $2x$. Since

$$50 \leq x \leq 70$$

we can multiply all parts of the inequality by 2 to obtain $2x$ in the middle.

$$2(50) \leq 2 \cdot x \leq 2(70)$$
$$100 \leq \quad 2x \quad \leq 140$$

Between 100 and 140 rural students must be enrolled in the special program.

EXERCISE SET 9.4

Solve.

_____ 1. In order for the Barminskis to have good television reception, the top of their antenna must be at least 24 feet above the ground. If the height of their house is 16 feet, what would be the shortest antenna that would assure good reception?

_____ 2. The total passenger weight of the persons riding the elevator at Holt's Department Store must not exceed 2175 pounds. If the average weight of an adult is 145 pounds, what would be the maximum number of adults allowed in the elevator?

_____ 3. In addition to her yearly salary of $12,000, Ms. Greenwell sells vacuum cleaners to supplement her income. She receives a commission of $50 on each vacuum cleaner she sells. If she wants to keep her total yearly income below $20,000 for tax purposes, what is the maximum number of cleaners she should sell each year?

_____ 4. Ella's diet requires her to have no more than 1100 calories per day. For lunch she can have 100 more calories than she has for breakfast, and for dinner she can have twice the number of calories she has for breakfast. Find the maximum number of calories that she can have for each meal.

_____ 5. The perimeter of a triangle can be no more than 35 feet. One side is 7 feet and the other two sides are the same. Find the longest possible lengths for the two equal sides.

_____ 6. If 5 times a number added to 9 is less than or equal to 4 times the number added to 12, find the largest value the number can have.

_____ 7. The sum of 7 times a number and 13 is greater than 27 and less than 48. Find the possible values for the number.

_____ 8. The Lovell Memorial Theatre has only 400 seats. If the theatre company charges admissions of $5 for adults and $3 for children, find the least number of adult tickets they can sell and still cover their production expenses of $1800.

_____ 9. The Quarterback Club is selling candy bars to raise money. Large bars sell for $2 and small bars for $1. They decide that they can sell 650 candy bars. Find the minimum number of $2 bars they should sell so that their total sales will be greater than or equal to $1000.

_____ 10. If the product of 8 and a number is added to 15, the result is greater than or equal to 11 times the number. Find the largest value for the number.

_____ 11. The length of a cattle-loading pen is to be 6 feet longer than the width. The perimeter must be less than or equal to 108 feet, which is the amount of fence available. Find the maximum width of the pen.

_____ 12. The perimeter of a triangle must be less than 75 centimeters. One side is 22 and the second side is to be 3 more than the third side. Find the largest possible length of the third side.

_____ 13. In order for the Associated Student Government to break even on a concert, they must take in at least $34,800. The price of a ticket sold in advance is $6 and the price is $8 at the door. If ASG sells 5000 tickets altogether, how many tickets must be sold at the door so that they do not lose money?

_____ 14. The length of a rectangle is to be two less than 5 times the width. If the difference between the two dimensions is not more than 4 meters, find the largest possible value for the width.

Summary

In this chapter we learned how to write and graph different kinds of inequalities.

$a < b$ (a is less than b)
$a > b$ (a is greater than b)
$a \leq b$ (a is less than or equal to b)
$a \geq b$ (a is greater than or equal to b)

We discovered that first-degree variable inequalities can be solved by the same methods used to solve first-degree equations, provided that we remember to *reverse* the direction of the inequality whenever we multiply or divide both sides by a *negative* number.

When solving pairs of inequalities, we learned that the numbers satisfying one inequality *and* another inequality must satisfy *both* inequalities at the same time, but the numbers satisfying one inequality *or* another inequality may satisfy either one inequality or the other (or both). The number line graphs of the inequalities helped us find the correct solutions.

Finally, we used our knowledge of inequalities to solve word problems involving phrases such as at least, no more than, fewer than, more than, and less than.

REVIEW EXERCISES

SECTION 9.1

Fill each blank with the correct symbols ($<$, $>$, $=$).

1. -27 _____ -3 **2.** 3 _____ -11

Write an inequality for each of the following.

_____ **3.** -7 is between -10 and 3.   _____ **4.** $\frac{5}{3}$ is between 0 and 4.

Graph the solutions on a number line.

_____ **5.** $y \geq -5$ _____ **6.** $a > 0$

_____ **7.** $-3 < x \leq 7$ _____ **8.** $-2 \leq y \leq \frac{5}{2}$

SECTION 9.2

Perform the indicated operation on both sides of the inequality.

_____ **9.** $-5 < 4$, add 9   _____ **10.** $7 < 10$, multiply by -3

_____ **11.** $10 > -4$, divide by -2 _____ **12.** $-3 > -9$, subtract 6

Solve and graph each inequality.

_____ **13.** $3x + 9 \leq 21$ _____ **14.** $8 - y > 3$

_____ **15.** $\frac{a}{2} + 2 \geq -2$ _____ **16.** $\frac{-3x}{2} + 6 < -3$

Solve each inequality.

_____ **17.** $2x - 5(x + 5) < -10$ _____ **18.** $8x - 9 \geq 3x + 1$

_____ **19.** $\frac{x - 7}{4} < \frac{3x + 1}{2}$

_____ **20.** $3(y - 1) - y \geq 4(y + 2) + 1$

_____ **21.** $\frac{-x}{3} + 2 \geq \frac{2x}{4} - 1$

SECTION 9.3

_____ **22.** $-3 \leq 3y - 1 \leq 5$ _____ **23.** $-8 \leq 4 - 2x < 8$

_____ **24.** $-3 \leq \frac{3}{5}x + 2 \leq 7$

Find the solution for the pairs of inequalities.

_____ **25.** $x > -9$ and $x \leq 3$

_____ **26.** $x + 4 > 7$ or $x - 6 \geq -1$

_____ **27.** $\frac{-x}{3} \leq -2$ and $\frac{3x}{2} - 5 > 1$

_____ **28.** $4(x - 2) > 2(x - 1)$ or $5(3 - x) < -2(3x - 7)$

Solve.

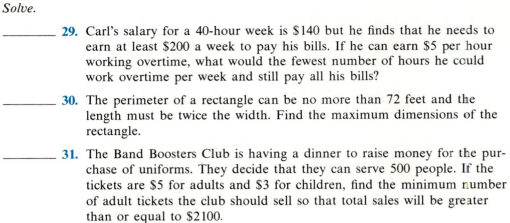

_____ **29.** Carl's salary for a 40-hour week is $140 but he finds that he needs to earn at least $200 a week to pay his bills. If he can earn $5 per hour working overtime, what would the fewest number of hours he could work overtime per week and still pay all his bills?

_____ **30.** The perimeter of a rectangle can be no more than 72 feet and the length must be twice the width. Find the maximum dimensions of the rectangle.

_____ **31.** The Band Boosters Club is having a dinner to raise money for the purchase of uniforms. They decide that they can serve 500 people. If the tickets are $5 for adults and $3 for children, find the minimum number of adult tickets the club should sell so that total sales will be greater than or equal to $2100.

10 Working with Two Variables

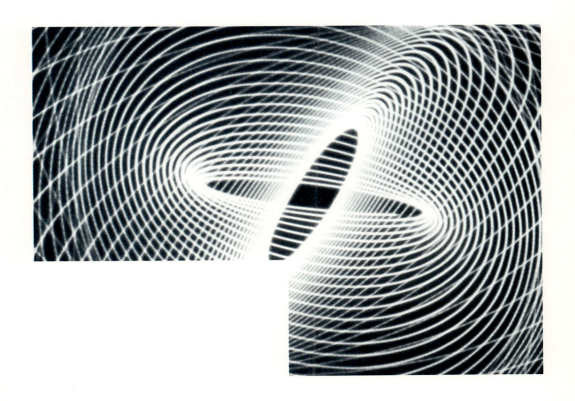

So far in our study of algebra we have concentrated on expressions and equations containing *one* variable. We have learned to solve first-degree and second-degree (quadratic) equations by finding the value of the variable that makes the equation a true statement.

In this chapter we deal with first-degree equations that contain *two* variables, and we learn how to

1. Find solutions of equations containing two variables.
2. Graph solutions of equations containing two variables.
3. Graph solutions of constant equations.

10.1 Solving Equations Containing Two Variables

Suppose that a student has two part-time jobs, one paying $3 per hour and one paying $4 per hour. Let's write an expression for the amount of money the student can earn in one week, working at both jobs. Since we cannot assume that the number of hours spent at one job is the same as the number of hours spent at the other job, we must use two variables.

$$\text{let } x = \text{hours spent per week at first job}$$
$$y = \text{hours spent per week at second job}$$
$$3x = \text{dollars earned per week at first job}$$
$$4y = \text{dollars earned per week at second job}$$

Then $3x + 4y = $ total earnings in one week.

If the student wishes to earn $72 per week, we could write

$$3x + 4y = 72$$

and we see that there are many possible values of x and y that would satisfy this equation.

For instance, if the student were to work 4 hours at the first job, how many hours must she spend at the second job? We can let $x = 4$ and find y.

$$3x + 4y = 72$$
$$3(4) + 4y = 72$$
$$12 + 4y = 72$$
$$4y = 60$$
$$\frac{4y}{4} = \frac{60}{4}$$
$$y = 15$$

So she could work 4 hours at the first job and 15 hours at the second job.

If she works 12 hours at the first job, how many hours must she work at the second job? We must let $x = 12$ and find y.

$$3x + 4y = 72$$
$$3(12) + 4y = 72$$
$$36 + 4y = 72$$
$$4y = 36$$
$$\frac{4y}{4} = \frac{36}{4}$$
$$y = 9$$

So she could work 12 hours at the first job and 9 hours at the second job.

Suppose that she works no hours at the first job. How many hours must she work at the second job? Here we let $x = 0$ and find y.

$$3x + 4y = 72$$

$$3(0) + 4y = 72$$
$$0 + 4y = 72$$
$$4y = 72$$
$$y = 18$$

So she could work 0 hours at the first job and 18 hours at the second job. Suppose that the student works no hours at the second job. How many hours must she work at the first job? Here we let $y = 0$ and find x.

$$3x + 4y = 72$$
$$3x + 4(0) = 72$$
$$3x + 0 = 72$$
$$3x = 72$$
$$x = 24$$

So she could work 24 hours at the first job and 0 hours at the second job.

10.1-1 Finding Ordered Pairs

In each part of the student-work-hours problem, we found an x-value *and* a y-value that would satisfy the equation $3x + 4y = 72$.

$$x = 4, \qquad y = 15$$
$$x = 12, \qquad y = 9$$
$$x = 0, \qquad y = 18$$
$$x = 24, \qquad y = 0$$

Sometimes we choose to write these corresponding x-values and y-values in the form of **ordered pairs** enclosed in parentheses with a comma in between.

$$x = 4, y = 15 \text{ is written } \quad (4, 15)$$
$$x = 12, y = 9 \text{ is written } \quad (12, 9)$$
$$x = 0, y = 18 \text{ is written } \quad (0, 18)$$
$$x = 24, y = 0 \text{ is written } \quad (24, 0)$$

$(8, 12)$ means when $x = 8$, then $y = 12$.
$(20, 3)$ means when $x = 20$, then $y = 3$.

The first number in the ordered pair is always the x-value and the second number in the ordered pair is always the corresponding y-value. An ordered pair **satisfies** an equation if we can make the equation *true* by substituting the x-value and y-value from the ordered pair into the equation.

Example 1. Does $(3, 1)$ satisfy $x + 2y = 5$?

Solution. From $(3, 1)$ we know that $x = 3$ and $y = 1$.

$$x + 2y = 5$$
$$3 + 2(1) \overset{?}{=} 5$$
$$3 + 2 \overset{?}{=} 5$$
$$5 = 5$$

Yes, $(3, 1)$ satisfies the equation $x + 2y = 5$.

Example 2. Does $(0, -3)$ satisfy $9x - 2y = 6$?

Solution. From $(0, -3)$ we know that $x = 0$ and $y = -3$.

$$9x - 2y = 6$$
$$9(0) - 2(-3) \overset{?}{=} 6$$
$$0 + 6 \overset{?}{=} 6$$
$$6 = 6$$

Yes, $(0, -3)$ satisfies the equation $9x - 2y = 6$.

Example 3. Does $(-2, -1)$ satisfy $y = 3 - 5x$?

Solution. From $(-2, -1)$ we know that $x = -2$ and $y = -1$.

$$y = 3 - 5x$$
$$-1 \stackrel{?}{=} 3 - 5(-2)$$
$$-1 \stackrel{?}{=} 3 + 10$$
$$-1 \stackrel{?}{=} 13$$

No, $(-2, -1)$ does not satisfy the equation $y = 3 - 5x$.

It is clear how we can check to see whether an ordered pair satisfies an equation. Suppose, however, that we are asked to *find* ordered pairs that satisfy a particular equation. For instance, suppose that we wish to find several ordered pairs which satisfy the equation

$$x + y = 7$$

Our goal is to find some x-values with their corresponding y-values that will make this equation a true statement. We may *choose any x-value* that we wish, and then we must find the y-value that belongs with it. Some possibilities might be

$(0, 7)$	because	$0 + 7 = 7$
$(3, 4)$	because	$3 + 4 = 7$
$(12, -5)$	because	$12 + (-5) = 7$
$(7, 0)$	because	$7 + 0 = 7$
$\left(\frac{1}{2}, 6\frac{1}{2}\right)$	because	$\frac{1}{2} + 6\frac{1}{2} = 7$

How many ordered pairs do you think we could find to satisfy the equation $x + y = 7$? If you guessed that there is an infinite number of such ordered pairs, you are correct.

Let's find some ordered pairs that satisfy

$$y = 2x + 3$$

Once again, we shall pick some x-value and try to find a corresponding y-value. Notice that the y-value is always equal to twice the x-value, plus 3.

if $x = 0$,	$y = 2(0) + 3 = 0 + 3 = 3$	$(0, 3)$
if $x = 4$,	$y = 2(4) + 3 = 8 + 3 = 11$	$(4, 11)$
if $x = -2$,	$y = 2(-2) + 3 = -4 + 3 = -1$	$(-2, -1)$
if $x = \frac{1}{2}$,	$y = 2\left(\frac{1}{2}\right) + 3 = 1 + 3 = 4$	$\left(\frac{1}{2}, 4\right)$

Since we could continue in this way forever, we agree that there is an infinite number of ordered pairs that will satisfy the equation $y = 2x + 3$.

Finding ordered pairs to satisfy the equation

$$5x + y = 16$$

looks a bit more difficult than our other two problems. One way to do this problem is to substitute an x-value into the equation and find the corresponding y-value each time. We used that method to solve the student-work-hours problem. Another approach to this problem, however, would be to *isolate y* right away, like this:

$$5x + y = 16$$
$$5x + y - 5x = 16 - 5x$$
$$y = 16 - 5x$$

Now we see that to find a y-value corresponding to some x-value, we must multiply the x-value by 5 and subtract that quantity from 16. Let's try it.

Let $x = 1$. Then $y = 16 - 5(1) = 16 - 5 = 11$ $(1, 11)$

Let $x = 0$. Then $y = 16 - 5(0) = 16 - 0 = 16$ $(0, 16)$

Let $x = -3$. Then $y = 16 - 5(-3) = 16 + 15 = 31$ $(-3, 31)$

Let $x = \dfrac{2}{3}$. Then $y = 16 - 5\left(\dfrac{2}{3}\right) = \dfrac{16}{1} - \dfrac{10}{3}$

$$= \dfrac{48}{3} - \dfrac{10}{3} = \dfrac{38}{3} \quad \left(\dfrac{2}{3}, \dfrac{38}{3}\right)$$

Remember that the x-value you choose is entirely up to you. Quite often, it is a good idea to choose some negative x-value, some positive x-value, and zero.

Example 4. Find three ordered pairs that satisfy

$$y = 3x - 1$$

Solution. Since y is isolated, we may substitute immediately.

Let $x = -2$. Then $y = 3(-2) - 1 = -6 - 1 = -7$ $(-2, -7)$

Let $x = 0$. Then $y = 3(0) - 1 = 0 - 1 = -1$ $(0, -1)$

Let $x = 4$. Then $y = 3(4) - 1 = 12 - 1 = 11$ $(4, 11)$

Example 5. Find three ordered pairs that satisfy

$$4x + 2y = 6$$

Solution. First we shall isolate y.

$$4x + 2y = 6$$
$$4x + 2y - 4x = 6 - 4x$$
$$2y = 6 - 4x$$
$$\frac{2y}{2} = \frac{6 - 4x}{2}$$
$$y = \frac{2(3 - 2x)}{2}$$
$$y = 3 - 2x$$

Let $x = -1$. Then $y = 3 - 2(-1) = 3 + 2 = 5$ $(-1, 5)$

Let $x = 0$. Then $y = 3 - 2(0) = 3 - 0 = 3$ $(0, 3)$

Let $x = 5$. Then $y = 3 - 2(5) = 3 - 10 = -7$ $(5, -7)$

Example 6. Find three ordered pairs that satisfy

$$3x - 5y = 1$$

Solution. First we shall isolate y. Let's add $5y$ to both sides so that the coefficient of y will be a positive number.

$$3x - 5y = 1$$
$$3x - 5y + 5y = 1 + 5y$$
$$3x = 1 + 5y$$
$$3x - 1 = 1 + 5y - 1$$
$$3x - 1 = 5y$$
$$\frac{3x - 1}{5} = \frac{5y}{5}$$
$$\frac{3x - 1}{5} = y$$

$$\text{Let } x = -2. \quad \text{Then } y = \frac{3(-2)-1}{5} = \frac{-6-1}{5} = \frac{-7}{5} \quad \left(-2, \frac{-7}{5}\right)$$

$$\text{Let } x = 0. \quad \text{Then } y = \frac{3(0)-1}{5} = \frac{0-1}{5} = \frac{-1}{5} \quad \left(0, \frac{-1}{5}\right)$$

$$\text{Let } x = 2. \quad \text{Then } y = \frac{3(2)-1}{5} = \frac{6-1}{5} = \frac{5}{5} = 1 \quad (2, 1)$$

Example 7. Find three ordered pairs that satisfy

$$2x + 3y = 0$$

Solution. First we must isolate y.

$$2x + 3y = 0$$
$$2x + 3y - 2x = 0 - 2x$$
$$3y = -2x$$
$$\frac{3y}{3} = \frac{-2x}{3}$$
$$y = \frac{-2x}{3}$$

$$\text{Let } x = -3. \quad \text{Then } y = \frac{-2(-3)}{3} = \frac{6}{3} = 2 \quad (-3, 2)$$

$$\text{Let } x = 0. \quad \text{Then } y = \frac{-2(0)}{3} = \frac{0}{3} = 0 \quad (0, 0)$$

$$\text{Let } x = 1. \quad \text{Then } y = \frac{-2(1)}{3} = \frac{-2}{3} \quad \left(1, \frac{-2}{3}\right)$$

Trial Run   *Decide if the given ordered pair satisfies the equation.*

_____ 1. $5x - 3y = 2$; $(1, 1)$

_____ 2. $x - 2y = 0$; $(4, -2)$

_____ 3. $y = \frac{1}{3}x - 4$; $(-3, -5)$

_____ 4. $-2x + 7y = -9$; $(-1, 1)$

Find three ordered pairs that satisfy each equation. There are lots of possible correct answers.

_____ 5. $y = 2x + 5$

_____ 6. $x - y = -2$

_____ 7. $3x + y = 4$

_____ 8. $5x - 2y = 1$

ANSWERS
1. Yes.   2. No.   3. Yes.   4. No.

You may have noticed that first-degree equations containing two variables can always be written in the form "coefficient times x plus coefficient times y equals constant." Using a, b, and c to represent constants,

---

General First-Degree Equation in Two Variables

$$ax + by = c$$

where a and b are not both zero.

To find several ordered pairs that satisfy a first-degree equation in two variables, we have used the following procedure.

---

1. Isolate y in the original equation.
2. Choose any x-values and find the corresponding y-values.
3. Write each solution as an ordered pair of the form (x, y).

Even though we find just a few ordered pairs for each equation, we realize that there is actually an infinite number of ordered pairs that will satisfy such an equation.

10-1.2 Switching from Words to Equations in Two Variables

The student-work-hours problem pointed out the everyday usefulness of equations in two variables. The methods we have just learned will help us find ordered pairs that satisfy such equations. As always when dealing with word problems, we must be sure to identify the variables in each situation.

Problem 1

Suppose that Aaron Gray wishes to fence in a rectangular portion of his yard for a dog run. What are three possible pairs of dimensions for the run if he has 60 feet of fencing to use?

We shall use a drawing to help us here.

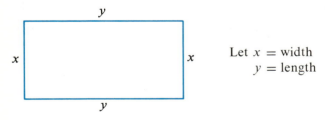

Let x = width
y = length

The amount of fencing is

$$x + y + x + y = 60$$
$$2x + 2y = 60$$

Isolating y, we see that

$$2x + 2y = 60$$
$$2x + 2y - 2x = 60 - 2x$$
$$2y = 60 - 2x$$
$$\frac{2y}{2} = \frac{60 - 2x}{2}$$
$$y = \frac{2(30 - x)}{2}$$
$$y = 30 - x$$

To find three ordered pairs, keep in mind that negative x-values would not make sense in a problem dealing with length and width.

$$\text{Let } x = 10. \qquad \text{Then } y = 30 - 10 = 20 \qquad (10, 20)$$

$$\text{Let } x = 15. \qquad \text{Then } y = 30 - 15 = 15 \qquad (15, 15)$$

$$\text{Let } x = 8\frac{1}{2}. \qquad \text{Then } y = 30 - 8\frac{1}{2} = 21\frac{1}{2} \qquad \left(8\frac{1}{2}, 21\frac{1}{2}\right)$$

These ordered pairs give the dimensions for just three possible rectangular dog runs.

$(10, 20)$ width = 10 feet and length = 20 feet

$(15, 15)$ width = 15 feet and length = 15 feet (a square)

$\left(8\frac{1}{2}, 21\frac{1}{2}\right)$ width = $8\frac{1}{2}$ feet and length = $21\frac{1}{2}$ feet

Keep in mind that there are infinitely many possibilities.

Problem 2

Suppose that a local restaurant sells chicken sandwiches at $1.50 each. Help the waitress make a chart of all possible ordered pairs (number of sandwiches, cost) for customers ordering between 1 and 5 chicken sandwiches.

$$\text{let } x = \text{number of sandwiches}$$

$$y = \text{cost of sandwiches (in dollars)}$$

$$y = 1.5x = \frac{15}{10}x = \frac{3}{2}x$$

So we must find ordered pairs to satisfy

$$y = \frac{3}{2}x$$

$$\text{Let } x = 1. \qquad \text{Then } y = \frac{3}{2}(1) = \frac{3}{2} \qquad \left(1, \frac{3}{2}\right)$$

$$\text{Let } x = 2. \qquad \text{Then } y = \frac{3}{2}(2) = 3 \qquad (2, 3)$$

$$\text{Let } x = 3. \qquad \text{Then } y = \frac{3}{2}(3) = \frac{9}{2} \qquad \left(3, \frac{9}{2}\right)$$

$$\text{Let } x = 4. \qquad \text{Then } y = \frac{3}{2}(4) = 6 \qquad (4, 6)$$

$$\text{Let } x = 5. \qquad \text{Then } y = \frac{3}{2}(5) = \frac{15}{2} \qquad \left(5, \frac{15}{2}\right)$$

We see that

1 sandwich costs $\$\frac{3}{2}$ (or $1.50)

2 sandwiches cost $3

3 sandwiches cost $\$\frac{9}{2}$ (or $4.50)

4 sandwiches cost $6

5 sandwiches cost $\$\frac{15}{2}$ (or $7.50)

Use your equation to find the cost when a busload of customers orders 44 sandwiches. Here $x = 44$.

$$y = \frac{3}{2}x$$

$$y = \frac{3}{2}(44) = \$66$$

Problem 3

Doug and Bob plan to take turns driving on a 600-mile trip. If Doug drives at 50 mph and Bob drives at 60 mph, write three ordered pairs which describe the number of hours that each might drive.

$$\text{let } x = \text{hours Doug drives}$$
$$y = \text{hours Bob drives}$$

Remembering that distance (miles) is equal to rate (miles per hour) multiplied by time (hours) we know that

$$50x = \text{miles Doug drives}$$
$$60y = \text{miles Bob drives}$$
$$50x + 60y = \text{total miles driven}$$

So

$$50x + 60y = 600$$

Isolating y, we have

$$50x + 60y = 600$$
$$50x + 60y - 50x = 600 - 50x$$
$$60y = 600 - 50x$$
$$y = \frac{600 - 50x}{60}$$
$$y = \frac{10(60 - 5x)}{60}$$
$$y = \frac{60 - 5x}{6}$$

Let $x = 12$.   Then $y = \dfrac{60 - 5(12)}{6} = \dfrac{60 - 60}{6} = 0$ $(12, 0)$

Let $x = 6$.   Then $y = \dfrac{60 - 5(6)}{6} = \dfrac{60 - 30}{6} = \dfrac{30}{6} = 5$ $(6, 5)$

Let $x = 4$.   Then $y = \dfrac{60 - 5(4)}{6} = \dfrac{60 - 20}{6} = \dfrac{40}{6}$ $\left(4, 6\frac{2}{3}\right)$

So

if Doug drives 12 hours, Bob drives no hours.

if Doug drives 6 hours, Bob drives 5 hours.

if Doug drives 4 hours, Bob drives $6\frac{2}{3}$ hours.

Trial Run   *Write each of the following word statements as an equation containing two variables and then list three ordered pairs that will satisfy the equation.*

_____ **1.** The sum of two numbers is 63.

_____ **2.** The length of a rectangle is 3 less than twice its width.

_____ **3.** A bank teller has a stack of both five- and ten-dollar bills. The total value of the money is $215.

_____ **4.** The perimeter of a rectangle is 36 feet.

_____ **5.** Denise drove 750 miles to see her mother. She averaged 60 miles per hour on the interstate highway and 50 miles per hour on two-lane roads.

_____ **6.** At French's Grocery, the price of tomatoes is 50 cents per pound. What equation would be used to make a chart showing the cost y of x pounds of tomatoes?

ANSWERS

1. $a + b = 63$.   **2.** $l = 2w - 3$.   **3.** $\$5x + \$10y = \$215$.   **4.** $2w + 2l = 36$.

5. $60x + 50y = 750$.   **6.** $y = 0.50x$ or $y = \dfrac{1}{2}x$.

EXERCISE SET 10.1

Decide if the given ordered pair satisfies the equation.

_____ 1. $6x - 5y = 2$; $(2, 2)$ _____ 2. $4x - 3y = 1$; $(1, 1)$

_____ 3. $2x - 4y = 0$; $(2, -1)$ _____ 4. $3x - 9y = 0$; $(3, -1)$

_____ 5. $y = \frac{1}{5}x - 6$; $(5, -5)$ _____ 6. $y = \frac{1}{2}x - 5$; $(2, -4)$

_____ 7. $-3x + 2y = 7$; $(1, 2)$ _____ 8. $-4x + 5y = 11$; $(-4, 5)$

_____ 9. $5x + 3y = 8$; $(0, 0)$ _____ 10. $7x + 11y = 18$; $(0, 0)$

_____ 11. $2x + 3y = -6$; $(3, -4)$ _____ 12. $3x + 8y = -21$; $(1, -3)$

Find three ordered pairs that satisfy each equation.

_____ 13. $x + y = 2$ _____ 14. $x - y = 3$

_____ 15. $2x + y = 4$ _____ 16. $x + 2y = 6$

_____ 17. $2x - 3y = 1$ _____ 18. $4x - 5y = 1$

_____ 19. $x - 2y = 0$ _____ 20. $2x + y = 0$

_____ 21. $x + 3y = 3$ _____ 22. $4x + y = 5$

_____ 23. $y = x + 6$ _____ 24. $y = x - 5$

_____ 25. $y = 3x - 5$ _____ 26. $y = 4x - 7$

_____ 27. $4x + 3y = -6$ _____ 28. $3x + 2y = -5$

Write each of the following word statements as an equation containing two variables and then list three ordered pairs that will satisfy the equation.

_____ 29. One number is 8 more than 4 times another.

_____ 30. One number is 3 less than 6 times another.

_____ 31. The sum of two numbers is 28.

_____ 32. The sum of one number and 10 is twice another number.

_____ 33. The length of a rectangle is 5 less than twice its width.

_____ 34. The length of a rectangle is 2 more than three times the width.

_____ 35. The Math Club treasurer has a stack of both $1 and $5 bills. The total value is $100.

_____ 36. Rich Yunkus has savings accounts at two different banks. The total amount of his savings is $12,500.

_____ 37. The perimeter of a rectangle is 48 feet.

_____ 38. For the basketball game, some student and some adult tickets were sold. Student tickets sell for $1 and adult tickets for $3. The total of gate receipts for the night was $1500.

_____ 39. Kelly walked 20 miles in the Walk-a-Thon to raise money for the Special Olympics. Part of the time she averaged 3 miles per hour, and the remainder she averaged 2 miles per hour.

_____ 40. In the Humane Society animal shelter the number of cats is 2 less than 5 times the number of dogs.

CHECKUP 10.1

Decide if the given ordered pair satisfies the equation.

_____ **1.** $2x - 6y = 4$; $(-1, -1)$

_____ **2.** $y = \frac{1}{3}x - 2$; $(6, -1)$

_____ **3.** $-4x + y = 0$; $(0, 0)$

_____ **4.** $5x - y = -4$; $(-2, 6)$

Find three ordered pairs that satisfy each equation.

_____ **5.** $x - y = -2$

_____ **6.** $2x - y = 0$

_____ **7.** $y = 3x - 2$

Write each of the following word statements as an equation containing two variables.

_____ **8.** The amount of money Mr. Brinkley has invested at Union Bank is $500 less than twice the amount invested at Farmers State Bank.

_____ **9.** Janyce bought a box of antique bottles. Some cost $2 each and some cost $3 each. The total cost was $17.

_____ **10.** The perimeter of a rectangular lot is 196 feet.

10.2 Graphing Equations in Two Variables

Earlier, we learned how to use the number line to graph solutions to first-degree equations and inequalities containing one variable. We would like to graph solutions to first-degree equations in two variables, but a number line will not work very well, since every solution is an ordered pair. Every solution pairs an *x*-value with a corresponding *y*-value.

Over three hundred years ago, a mathematician and philosopher named René Descartes thought of a way to graph ordered pairs by using *two* number lines. He placed those two number lines at right angles (perpendicular) to each other and labeled the horizontal line the **x-axis** and the vertical line the **y-axis**. Naming the point where the two axes intersect the **origin**, he marked off the units in the usual way.

> Points to the *right* of the origin on the *x*-axis correspond to *positive x*-values.
> Points to the *left* of the origin on the *x*-axis correspond to *negative x*-values.
> Points *above* the origin on the *y*-axis correspond to *positive y*-values.
> Points *below* the origin on the *y*-axis correspond to *negative y*-values.
> At the *origin*, the *x*-value is *zero* and the *y*-value is *zero*.

Descartes came up with the following diagram, sometimes called the **Cartesian coordinate plane** in his honor.

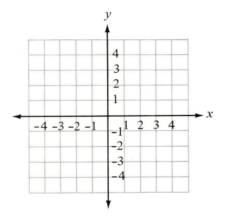

10.2-1 Graphing Ordered Pairs

Now we must discover how to use the Cartesian coordinate plane to graph ordered pairs. To graph an ordered pair such as (2, 3), we must keep in mind that

(2, 3) means "when *x* = 2, then *y* = 3"

On the Cartesian plane, we may use a dotted vertical line to show where *x* is 2 and a dashed horizontal line to show where *y* is 3.

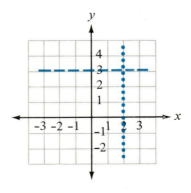

Since we wish x to be 2 and y to be 3 *at the same time,* we should agree that the point where the dotted line crosses the dashed line is the only sensible point for representing (2, 3). Indeed, that point is the graph of the ordered pair (2, 3).

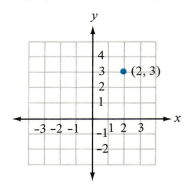

In plotting points in the Cartesian plane, we shall not sketch the dotted and dashed lines. You should do that in your head and then locate the point of intersection. All points should be labeled with an ordered pair after plotting. The point corresponding to an ordered pair contains two parts.

x-coordinate: the x-value in an ordered pair.
y-coordinate: the y-value in an ordered pair.

Example 1. Graph the ordered pairs and label the points: (5, 1), (1, 5).

Solution

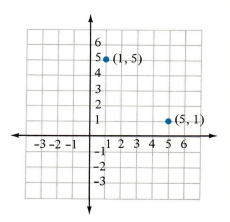

Notice that the ordered pairs (5, 1) and (1, 5) do not represent the same point. This means that (5, 1) $\neq$ (1, 5).

Example 2. Graph the ordered pairs and label the points: (1, -2), (-3, 4), (-5, -5).

Solution. For (1, -2) we must find the point where x is 1 and y is -2.

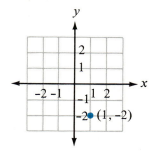

For $(-3, 4)$ we must find the point where x is -3 and y is 4.

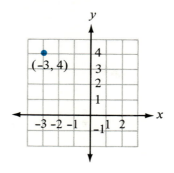

For $(-5, -5)$ we must find the point where x is -5 and y is -5.

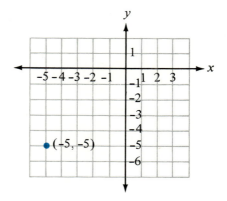

Suppose that we plot some points with zero for the x-coordinate or y-coordinate. First of all, where is the point $(0, 0)$? On the Cartesian plane, this point must be located at the origin.

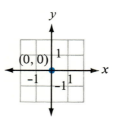

How about points such as $(1, 0)$ or $(3, 0)$ or $(-5, 0)$? In each ordered pair, the y-coordinate is zero. We graph these points as follows.

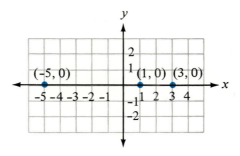

We notice that

If a point has a y-coordinate of 0, that point lies on the x-axis.

To graph points such as (0, 2), (0, 4), or (0, −3) we notice that each point has an *x*-coordinate of 0.

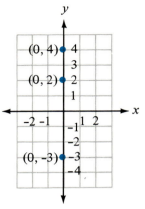

We observe that

> If a point has an *x*-coordinate of 0, that point lies on the *y*-axis.

Example 3. Graph and label the following points.

$$A(4, 1) \qquad B(-2, -3) \qquad C\left(0, \frac{3}{2}\right)$$

$$D\left(\frac{1}{2}, -\frac{1}{2}\right) \qquad E\left(-\frac{5}{2}, 0\right) \qquad F\left(-2, \frac{7}{3}\right)$$

Solution. When rational numbers occur as coordinates of points, we estimate their location as accurately as possible.

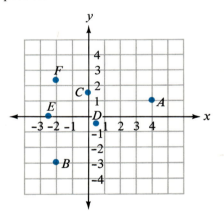

Example 4. Give the coordinates of each point.

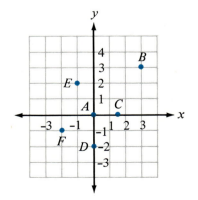

Solution

$$A(0,0) \qquad B(3,3) \qquad C\left(\frac{3}{2}, 0\right)$$

$$D(0, -2) \qquad E(-1, 2) \qquad F(-2, -1)$$

From our work in this section, we make the observation that each ordered pair of numbers corresponds to *one and only one* point in the Cartesian coordinate plane. Moreover, each point in the plane corresponds to *one and only one* ordered pair of numbers.

Trial Run

1. Graph the following points.

$$A(0, -2) \qquad B(-1, -3) \qquad C(-1, 2) \qquad D\left(1\frac{1}{2}, 2\right)$$

2. Give the coordinates of each point.

_____ *A.*
_____ *B.*
_____ *C.*
_____ *D.*

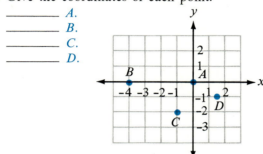

ANSWERS

1.

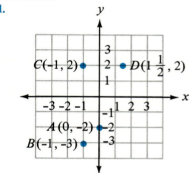

2. *A.* $(0, 0)$.   *B.* $(-4, 0)$.   *C.* $(-1, -2)$.   *D.* $\left(1\frac{1}{2}, -1\right)$.

10.2-2   Graphing Equations by "Choose an *x*, Find a *y*" Method

In Section 10.1 we learned how to find ordered pairs of numbers that would satisfy equations in two variables. We considered first-degree equations of the form

$$ax + by = c$$

Then, by choosing different *x*-values and finding corresponding *y*-values, we came up with several ordered pairs that satisfied such equations.

For instance, for the equation

$$x + y = 7$$

some solutions were

$$(7, 0), \quad (5, 2), \quad (10, -3), \quad \left(\frac{7}{2}, \frac{7}{2}\right), \quad (0, 7), \quad (-2, 9), \quad (1, 6)$$

and so on. We agreed that such an equation was satisfied by an infinite number of ordered pairs, each found by choosing an x-value and finding the corresponding y-value.

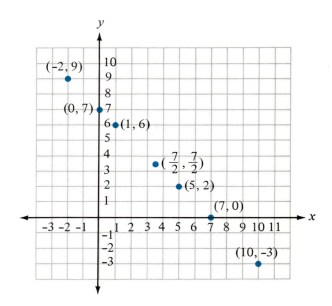

Suppose that we graph these ordered pairs in the Cartesian coordinate plane. remembering to label each point. It is not just a nice coincidence that all these points seem to lie on a straight line. You should see that if we were to choose more and more x-values and find their corresponding y-values, each ordered pair would lie on the same line. Indeed, any ordered pair that satisfies the equation

$$x + y = 7$$

lies on the line we have already observed.

The graph of an equation is the set of all points corresponding to the ordered pairs that satisfy the equation. Since it is impossible to find *all* such points, we shall settle for a reasonable number of ordered pairs, say 3 or 4. After finding 3 or 4 points, we hope that the appearance of the graph will be clear to us.

Consider graphing the first-degree equation.

$$2x - y = 6$$

In order to "choose an x and find a y," we discovered earlier that it is helpful to isolate y right away.

$$2x - y = 6$$
$$2x - y + y = 6 + y$$
$$2x = 6 + y$$
$$2x - 6 = 6 + y - 6$$
$$2x - 6 = y$$

Now we choose some x-values, remembering to include a negative value, a positive value, and zero.

Let $x = -2$.   Then $y = 2(-2) - 6 = -4 - 6 = -10$ $(-2, -10)$

Let $x = 0$.   Then $y = 2(0) - 6 = 0 - 6 = -6$ $(0, -6)$

Let $x = 3$.   Then $y = 2(3) - 6 = 6 - 6 = 0$ $(3, 0)$

Let $x = 5$.   Then $y = 2(5) - 6 = 10 - 6 = 4$ $(5, 4)$

For graphing purposes, it is useful to organize this work in a table with a column for the x-values, a column for finding the y-values, and a column for listing the ordered pairs.

x	$y = 2x - 6$	(x, y)
-2	$2(-2) - 6 = -4 - 6 = -10$	$(-2, -10)$
0	$2(0) - 6 = 0 - 6 = -6$	$(0, -6)$
3	$2(3) - 6 = 6 - 6 = 0$	$(3, 0)$
5	$2(5) - 6 = 10 - 6 = 4$	$(5, 4)$

Referring to the (x, y) column, we may plot the four points and notice that they lie on a straight line.

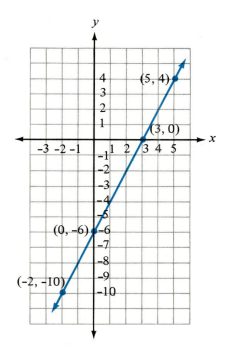

Once again we considered ordered pairs satisfying the equation $2x - y = 6$ and discovered that the graph was a straight line. In fact, we may say in general that

> The graph of an equation of the form
> $$ax + by = c$$
> is a straight line.

Let us try one more example in which we graph

$$y = 5x$$

Since y is isolated already, we shall use a table to organize our work. Remember, we must "choose an x and find a y" to obtain points.

x	$y = 5x$	(x, y)
-1	$5(-1) = -5$	$(-1, -5)$
0	$5(0) = 0$	$(0, 0)$
$\dfrac{1}{2}$	$5\left(\dfrac{1}{2}\right) = \dfrac{5}{2}$	$\left(\dfrac{1}{2}, \dfrac{5}{2}\right)$
1	$5(1) = 5$	$(1, 5)$

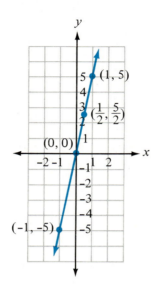

The equation $y = 5x$ may not have appeared to be of the general form

$$ax + by = c$$

but in fact it certainly is. By subtracting $5x$ from both sides of the equation, we may rewrite $y = 5x$ as

$$-5x + y = 0$$

which is general first-degree form (with c being 0).

Because of their graphs, equations of the form

$$ax + by = c$$

are sometimes called **linear equations**. Can you see why? The method we have developed for graphing linear equations is sometimes called the **arbitrary point** method because the x-values used are chosen arbitrarily by the grapher. We have also called this method "choose an x and find a y."

To graph equations of the form

$$ax + by = c$$

we have followed an orderly procedure.

1. Isolate y in the equation.
2. Choose at least three x-values and find the corresponding y-values by substituting into the equation with y isolated.
3. Write each solution as an ordered pair (x, y).
4. Plot the point in the Cartesian coordinate plane corresponding to each ordered pair.
5. Join the points with a straight line.

By the way, you may recall from previous work with geometry that two points are all that is needed to construct a straight line. This is certainly true, but in the "choose an x, find a y" method, we shall continue to use three points just to check our arithmetic. If your three points do not lie in a straight line, that is a signal that you have made an error in substitution and you should check each of your points again.

Example 5. Graph the linear equation $3x + 6y = 4$.

Solution. Isolating y, we have

$$3x + 6y = 4$$
$$3x + 6y - 3x = 4 - 3x$$
$$6y = 4 - 3x$$
$$\frac{6y}{6} = \frac{4 - 3x}{6}$$
$$y = \frac{4 - 3x}{6}$$

x	$y = \dfrac{4 - 3x}{6}$	(x, y)
-1	$\dfrac{4 - 3(-1)}{6} = \dfrac{4 + 3}{6} = \dfrac{7}{6}$	$\left(-1, \dfrac{7}{6}\right)$
0	$\dfrac{4 - 3(0)}{6} = \dfrac{4 - 0}{6} = \dfrac{4}{6} = \dfrac{2}{3}$	$\left(0, \dfrac{2}{3}\right)$
2	$\dfrac{4 - 3(2)}{6} = \dfrac{4 - 6}{6} = \dfrac{-2}{6} = \dfrac{-1}{3}$	$\left(2, -\dfrac{1}{3}\right)$

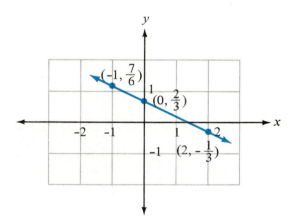

Notice from this example that the lengths of units marked off on the x-axis need not match the lengths of units marked off on the y-axis. Sometimes it's convenient to spread out or squeeze together the units on one axis or the other. Just be sure to label your points for the benefit of an observer.

Example 6. Graph the linear equation $y = 50x$.

Solution. Since y is isolated, we make a table.

x	$y = 50x$	(x, y)
-1	$50(-1) = -50$	$(-1, -50)$
0	$50(0) = 0$	$(0, 0)$
2	$50(2) = 100$	$(2, 100)$

Here we would be wise to mark off the y-axis in large units, say 25.

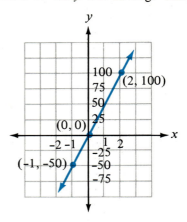

Trial Run   *Graph each equation using "choose an x and find a y" method.*

1. $x + y = -3$

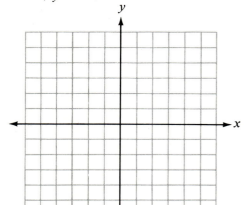

2. $y = 2x - 1$

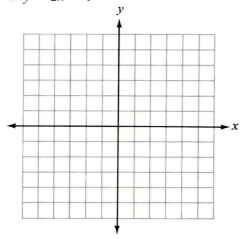

3. $2x - y = 3$

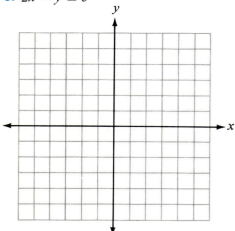

4. $y = -16x$

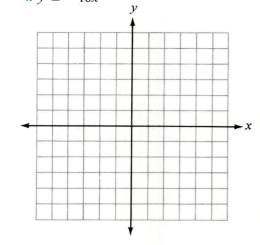

1.

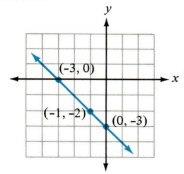

2.

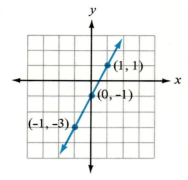

3.

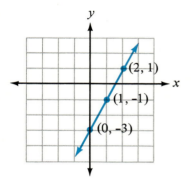

4.

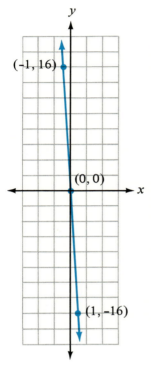

Remember, if your three or four points do not lie in a straight line, you should check each of your ordered pairs to discover your mistake. For every equation in this chapter the graph should be a straight line.

EXERCISE SET 10.2

Graph the following points.

1. A: $(2, 0)$

2. B: $(-3, 1)$

3. C: $(0, -5)$

4. D: $(-3, -4)$

5. E: $\left(\dfrac{1}{2}, 4\right)$

6. F: $(0, 0)$

7. G: $(6, 5)$

8. H: $(3, -2)$

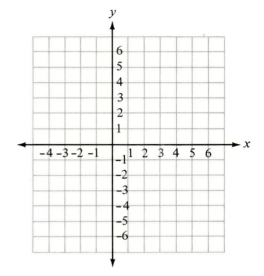

Give the coordinates of each point.

_____ 9. A

_____ 10. B

_____ 11. C

_____ 12. D

_____ 13. E

_____ 14. F

_____ 15. G

_____ 16. H

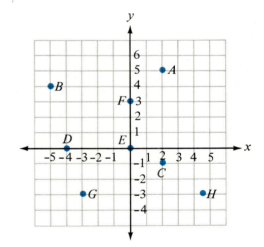

Graph each equation.

17. $x + y = 2$

18. $x + y = 5$

19. $y = 25x$

20. $3y = 15x$

21. $y = 4x - 2$

22. $y = 3x - 1$

23. $2x + y = 1$

24. $3x + y = 2$

25. $x + 6y = 8$

26. $x + 4y = 5$

27. $2x - y = -5$

28. $3x - y = -6$

Name _____ **Date** _____

CHECKUP 10.2

Graph each of the following points.

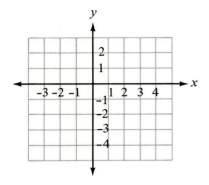

1. *A*: (3, 0)

2. *B*: (−2, 1)

3. *C*: (2, −3)

Give the coordinates of each point.

_____ **4.** *A*

_____ **5.** *B*

_____ **6.** *C*

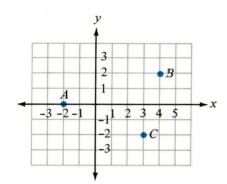

Graph each equation.

7. *y* = 2*x*

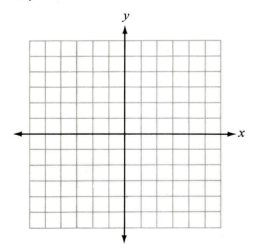

8. *x* + *y* = 1

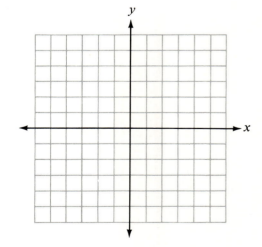

9. $x - 2y = 4$

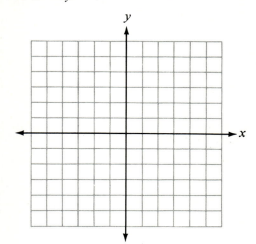

10. $3x - y = 0$

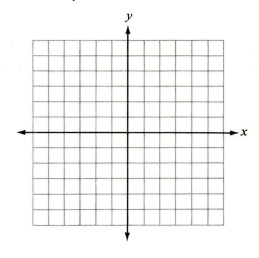

10.3 Graphing More Equations in Two Variables

We have noticed that when an equation appears in the form

$$ax + by = c$$

and we isolate y, substituting into the final form of the equation sometimes becomes awkward. Perhaps we can arrive at a less complicated method for graphing linear equations in which y is not already isolated.

10.3-1 Graphing Linear Equations by the Intercepts Method

If you glance back at the linear equations we have already graphed, you will notice that each straight line crosses the x-axis at some point and also crosses the y-axis at some point. These points are especially interesting to us and are called the **intercepts** for a graph.

> x-intercept: point where a graph crosses the x-axis.
>
> y-intercept: point where a graph crosses the y-axis.

We observed earlier that all the points on the x-axis had something in common: namely, we saw that each point on the x-axis had a y-coordinate of 0. Similarly, we observed that for every point on the y-axis, the x-coordinate was 0.

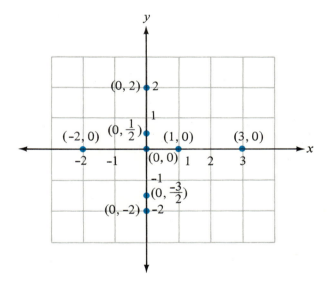

Putting this observation together with the idea of intercepts, we may say that

> The y-intercept for a graph is the point that has an x-coordinate of 0.
>
> The x-intercept for a graph is the point that has a y-coordinate of 0.

To find the y-intercept for a graph, we must let $x = 0$ in the given equation. To find the x-intercept for a graph, we must let $y = 0$ in the given equation.

Let's find the intercepts for the graph of

$$3x + 2y = 6$$

To find the y-intercept, we substitute 0 for x.

$$3x + 2y = 6$$
$$3(0) + 2y = 6$$
$$2y = 6$$
$$\frac{2y}{2} = \frac{6}{2}$$
$$y = 3$$

So the y-intercept is the point (0, 3).

To find the x-intercept, we must substitute 0 for y.

$$3x + 2y = 6$$
$$3x + 2(0) = 6$$
$$3x = 6$$
$$\frac{3x}{3} = \frac{6}{3}$$
$$x = 2$$

So the x-intercept is the point (2, 0).

Because there is less chance of error when we substitute 0 for a variable, we trust these two points to be accurate and use them to graph our line. (Remember that two points are all that are needed to construct a straight line.)

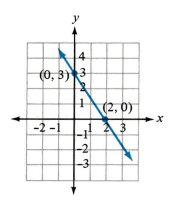

Example 1. Graph $5x - 2y = 15$ by intercepts.

Solution

$$5x - 2y = 15$$
$$y\text{-intercept: let } x = 0$$
$$5(0) - 2y = 15$$
$$-2y = 15$$
$$\frac{-2y}{-2} = \frac{15}{-2}$$
$$y = \frac{-15}{2} \qquad \left(0, \frac{-15}{2}\right)$$
$$x\text{-intercept: let } y = 0$$
$$5x - 2(0) = 15$$
$$5x = 15$$
$$\frac{5x}{5} = \frac{15}{5}$$
$$x = 3 \qquad (3, 0)$$

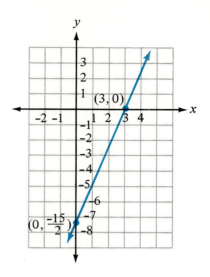

Example 2. Graph $-x - 4y = 6$ by intercepts.

Solution

$$-x - 4y = 6$$

y-intercept: let $x = 0$

$$-0 - 4y = 6$$
$$-4y = 6$$
$$\frac{-4y}{-4} = \frac{6}{-4}$$
$$y = \frac{-6}{4} = \frac{-3}{2}$$

The y-intercept is $\left(0, \frac{-3}{2}\right)$.

x-intercept: let $y = 0$

$$-x - 4(0) = 6$$
$$-x = 6$$
$$\frac{-x}{-1} = \frac{6}{-1}$$
$$x = -6$$

The x-intercept is $(-6, 0)$.

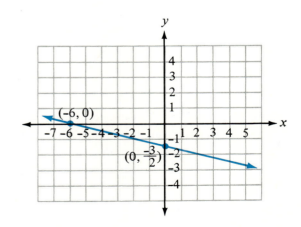

Example 3. Graph $2x + 4y = 0$.

Solution

$$2x + 4y = 0$$

y-intercept: let $x = 0$

$$2(0) + 4y = 0$$
$$0 + 4y = 0$$
$$4y = 0$$
$$y = \frac{0}{4}$$
$$y = 0$$

The y-intercept is $(0, 0)$.

x-intercept: let $y = 0$

$$2x + 4(0) = 0$$
$$2x + 0 = 0$$
$$2x = 0$$
$$x = \frac{0}{2}$$
$$x = 0$$

The x-intercept is $(0, 0)$.

In this example, the x-intercept and y-intercept are both $(0, 0)$. Since one point will not give us a straight line, we must choose another x-value and find the corresponding y-value. Let

$$x = 2$$
$$2x + 4y = 0$$
$$2(2) + 4y = 0$$
$$4 + 4y = 0$$
$$4y = -4$$
$$y = -1$$

Our second point is $(2, -1)$. We can now use our intercept $(0, 0)$ and the point $(2, -1)$ to graph our line.

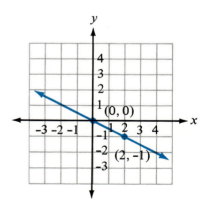

Although the intercept method can be used to graph a linear equation in almost any form, it is most useful for equations in which y is *not* isolated. Whenever we wish to graph an equation in which y is isolated already, we shall continue to use the "choose an x and find a y" method.

Trial Run   *Graph each equation by the intercept method.*

1. $5x - 3y = -15$

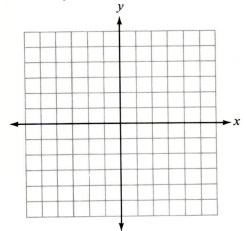

2. $-2x - 3y = 6$

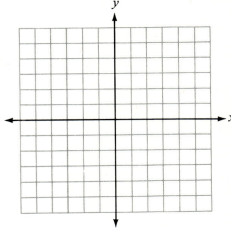

3. $-3x + 8y = -12$

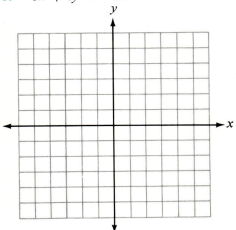

4. $4x + y = 4$

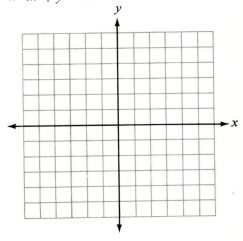

ANSWERS

1.

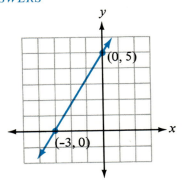

2.

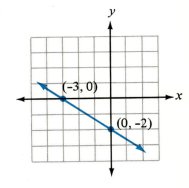

3.

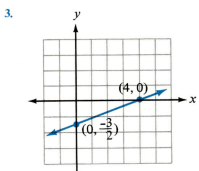

4.

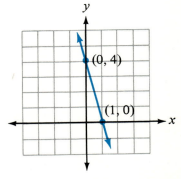

GRAPHING MORE EQUATIONS IN TWO VARIABLES [Sec. 10.3] **449**

10.3-2 Graphing Constant Equations

In looking at the general first-degree equation for a line

$$ax + by = c$$

we agreed that a, b, and c represent constant numbers. We have graphed several equations in which c was zero. For instance, we looked at the lines described by

$$y = 5x \qquad \text{or} \qquad -5x + y = 0$$

$$y = \frac{3}{2}x \qquad \text{or} \qquad -3x + 2y = 0$$

Until now, however, we have not considered any equations in which a or b is zero. In other words, we have not discussed equations in which the entire x-term or the entire y-term is missing. We shall now investigate what happens when we use the Cartesian coordinate plane to graph equations such as

$$y = 4$$
$$y = -2$$
$$x = 1$$
$$x = -3$$

The equation

$$y = 4$$

states that y is always 4. What about x? This equation says that "no matter what x is, y is *always* 4."

$$\text{If } x = 3, \qquad y = 4$$
$$\text{If } x = 0, \qquad y = 4$$
$$\text{If } x = -2, \quad y = 4$$
$$\text{If } x = \frac{7}{8}, \qquad y = 4$$

and so on. Some ordered pairs for our graph are

$$(3, 4) \qquad (0, 4) \qquad (-2, 4) \qquad \left(\frac{7}{8}, 4\right)$$

and plotting these points, we have

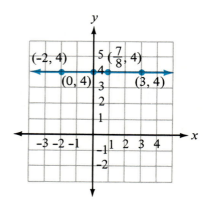

The graph of $y = 4$ in the Cartesian plane is a horizontal line (parallel to the x-axis) crossing the y-axis at $(0, 4)$.

To graph the equation

$$y = -2$$

we again realize that "no matter what x is, y is always -2." Some ordered pairs might be

$$(0, -2) \qquad (5, -2) \qquad (-2, -2) \qquad (1, -2)$$

Plotting these points gives

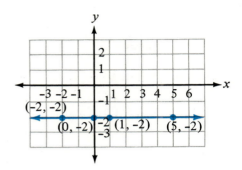

Once again, the graph of $y = -2$ is a horizontal line (parallel to the x-axis) crossing the y-axis at $(0, -2)$.

You should agree that the following equations can be graphed on sight without many steps.

Example 3. Graph $y = 1$.

Solution. The graph of $y = 1$ is a horizontal line crossing the y-axis at $(0, 1)$.

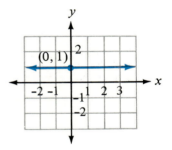

Example 4. Graph $y + 3 = 0$.

Solution. First we isolate y.

$$
\begin{aligned}
y + 3 &= 0 \\
y + 3 - 3 &= 0 - 3 \\
y &= -3
\end{aligned}
$$

and the graph is a horizontal line crossing the y-axis at $(0, -3)$.

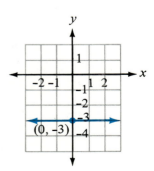

Example 5. Graph $2y - 1 = 0$.

Solution. Isolating y, we have

$$2y - 1 = 0$$
$$2y - 1 + 1 = 0 + 1$$
$$2y = 1$$
$$\frac{2y}{2} = \frac{1}{2}$$
$$y = \frac{1}{2}$$

and the graph is a horizontal line that crosses the y-axis at $\left(0, \frac{1}{2}\right)$.

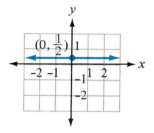

In general, we state that

> The graph of an equation of the form
>
> $$y = \text{a constant}$$
>
> is a horizontal line (parallel to the x-axis) crossing the y-axis at the constant.

Suppose that we consider the equation

$$x = 1$$

which states that x is always 1. No matter what the value of y, x will always have value 1. Some points on our graph will be

$$(1, 0) \qquad (1, 1) \qquad (1, 3) \qquad (1, -3)$$

and so on. Plotting these points, we have

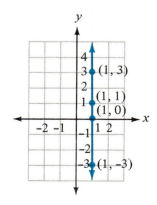

Our graph for $x = 1$ is a vertical line (parallel to the y-axis) crossing the x-axis at $(1, 0)$. Now consider the graph of

$$x = -3$$

Here we see that "no matter what y is, x is always -3." Some points on our graph are

$$(-3, 0) \qquad (-3, 2) \qquad \left(-3, \frac{1}{2}\right) \qquad (-3, -1)$$

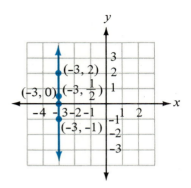

Again the graph of $x = -3$ is a vertical line (parallel to the y-axis) crossing the x-axis at $(-3, 0)$.

In general, we agree that

> The graph of an equation of the form
>
> $$x = \text{a constant}$$
>
> is a vertical line (parallel to the y-axis) crossing the x-axis at the constant.

Example 6. Graph $x - 6 = 0$.

Solution. Isolate x first.

$$x - 6 = 0$$
$$x - 6 + 6 = 0 + 6$$
$$x = 6$$

and the graph is a vertical line crossing the x-axis at $(6, 0)$.

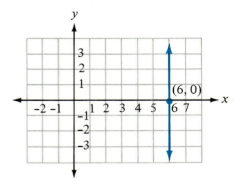

Example 7. Graph $3x + 4 = 0$.

Solution. Isolating x, we have

$$3x + 4 = 0$$
$$3x + 4 - 4 = 0 - 4$$
$$3x = -4$$
$$\frac{3x}{3} = \frac{-4}{3}$$
$$x = \frac{-4}{3}$$

so the graph is a vertical line crossing the x-axis at $\left(\frac{-4}{3}, 0\right)$.

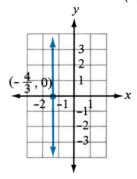

Trial Run *Graph each of the equations.*

1. $y = 3$

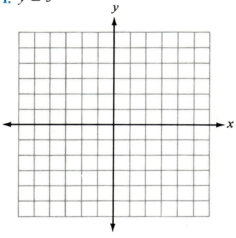

2. $3y + 8 = 10$

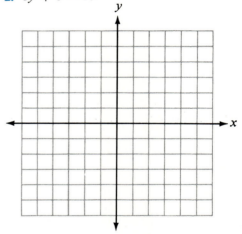

3. $x + 5 = 0$

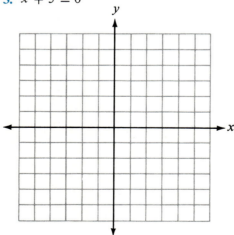

4. $4x - 8 = 0$

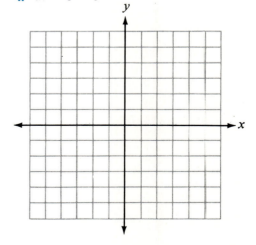

1.

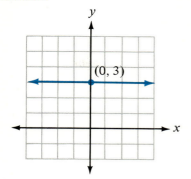

(0, 3)

2.

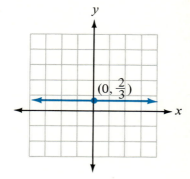

$(0, \frac{2}{3})$

3.

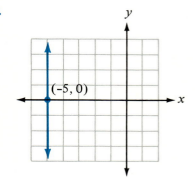

(−5, 0)

4.

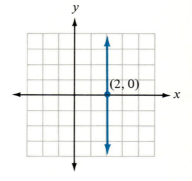

(2, 0)

EXERCISE SET 10.3

Graph each equation by the intercept method.

1. $x + y = 3$

2. $x + y = -2$

3. $x - y = -5$

4. $x - y = 4$

5. $3x + y = 6$

6. $5x + y = 5$

7. $x - 2y = 8$

8. $x - 3y = -6$

9. $5x + 3y = -15$

10. $3x + 4y = 12$

11. $x + 2y = 5$

12. $x + 4y = 6$

13. $3x - y = 7$

14. $2x - y = -3$

15. $3x - 4y = -14$

16. $2x - 3y = 10$

17. $-5x + 2y = 15$

18. $-7x + 3y = 14$

19. $-x - y = 3$

20. $-x - y = 1$

Graph the lines represented by the following equations.

21. $y = 4$

22. $y = -3$

23. $x - 2 = 0$

24. $x - 5 = 0$

25. $y + 4 = 0$

26. $y + 1 = 0$

27. $3y - 9 = 0$

28. $5y - 10 = 0$

29. $2x = 7$

30. $3x = 5$

31. $2x + 5 = 0$

32. $4x + 18 = 0$

33. $6y - 16 = 0$

34. $8y - 28 = 0$

35. $9x - 27 = 0$

36. $7x + 35 = 0$

CHECKUP 10.3

Graph each of the following equations.

1. $2x - y = 4$

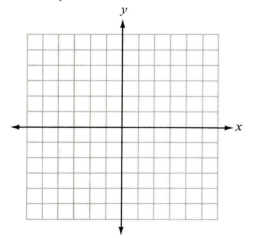

2. $x + 2y = 5$

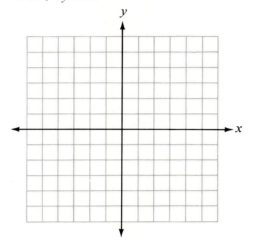

3. $-2x + 3y = -12$

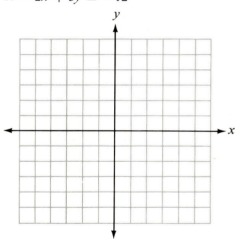

4. $y - 4 = 0$

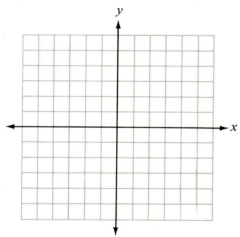

5. $3x = 8$

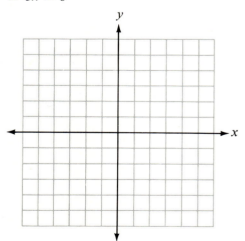

6. $3y + 6 = 0$

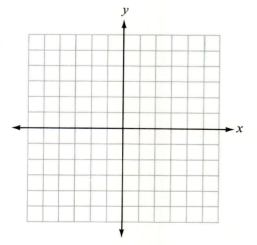

SUMMARY

In this chapter we have considered equations containing two variables. In particular, we investigated first-degree equations of the form

$$ax + by = c$$

where a, b, and c are constants and a and b are not both zero.

We discovered that solutions to such equations are ordered pairs of numbers (x, y) that can be graphed as points in the Cartesian coordinate plane. Every first-degree equation will yield a graph that is a straight line, and we learned to graph straight lines by two different methods.

1. If y can be easily isolated in the given equation, we find three points by the method of "choose an x and find a y."
2. If y cannot be easily isolated in the given equation, we find the x-intercept (by letting $y = 0$) and the y-intercept (by letting $x = 0$).

In graphing constant equations, we noticed the pattern that occurred.

1. The graph of an equation of the form y equals a constant is a horizontal line (parallel to the x-axis) crossing the y-axis at the constant.
2. The graph of an equation of the form x equals a constant is a vertical line (parallel to the y-axis) crossing the x-axis at the constant.

Name _____ **Date** _____

REVIEW EXERCISES

<small>SECTION</small> 10.1

Decide if the given point satisfies the equation.

_____ **1.** $y = \frac{1}{4}x - 3$, $(-8, -5)$ _____ **2.** $4x - y = 5$, $(0, -5)$

Find three ordered pairs that satisfy each equation.

_____ **3.** $2x - y = 6$ _____ **4.** $x - 3y = 0$

_____ **5.** $y = -2x + 5$ _____ **6.** $x - y = -2$

Write each of the following word statements as an equation containing two variables, then list three ordered pairs that will satisfy the equation.

_____ **7.** The length of a rectangle is 5 less than 3 times the width.

_____ **8.** Regina sold some of her blouses at a yard sale. For some she got $5 and for others $3. The total amount she made from the sale of her blouses was $45.

<small>SECTION</small> 10.2

9. Graph the following points.

$$A: (-4, 2) \qquad B: (0, -5) \qquad C: (-2, -4)$$

10. Give the coordinates of each point.

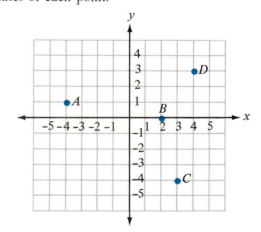

Graph each equation.

11. $x + y = 5$ **12.** $y = -3x$

13. $4x - y = 0$ **14.** $3x - y = 6$

<small>SECTION</small> 10.3

Graph each equation by the intercept method.

15. $x + y = 6$ **16.** $x - 3y = -6$

17. $2x - y = 7$ **18.** $5x - 3y = 15$

19. $4x + y = 8$ **20.** $3x + 2y = -12$

11 Working with Systems of Linear Equations

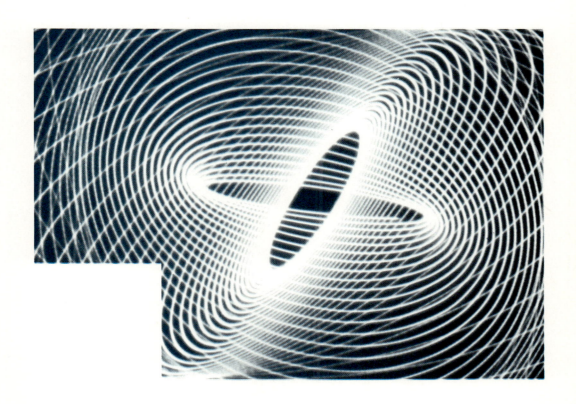

Sometimes we would like to find out what ordered pairs (if any) will satisfy *two* first-degree equations at the same time. This is called solving a **system of linear equations**; in this chapter we learn how to

1. Solve a system of linear equations by graphing.
2. Solve a system of linear equations by substitution.
3. Solve a system of linear equations by elimination.
4. Switch from words to a system of linear equations.

11.1 Solving a System of Equations by Graphing

One way in which we can try to locate the common solutions for a system of two equations is by looking at the *graphs* of the two equations. We shall graph two equations in the same Cartesian plane and see what conclusions we can draw.

Consider the system of two linear equations.

$$(1) \quad x + y = 7$$
$$(2) \quad 2x - y = 5$$

First we shall graph each equation by the intercept method. For equation (1):

$$\text{if } x = 0, \quad y = 7 \quad (0, 7)$$
$$\text{if } y = 0, \quad x = 7 \quad (7, 0)$$

For equation (2):

$$\text{if } x = 0, \quad y = -5 \quad (0, -5)$$
$$\text{if } y = 0, \quad x = \frac{5}{2} \quad \left(\frac{5}{2}, 0\right)$$

Now graphing both equations in the same plane and labeling the lines (1) and (2), we see

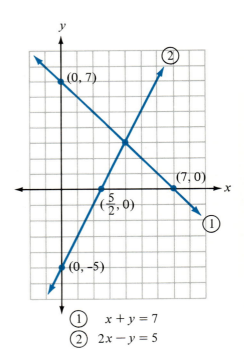

$$\textcircled{1} \quad x + y = 7$$
$$\textcircled{2} \quad 2x - y = 5$$

It appears that these lines have just one point in common—the point where the lines cross. That point appears to be (4, 3). We can see if that point is shared by both equations by checking the ordered pair (4, 3) in each one.

$$(1) \quad x + y = 7 \qquad (2) \quad 2x - y = 5$$
$$4 + 3 \stackrel{?}{=} 7 \qquad 2(4) - 3 \stackrel{?}{=} 5$$
$$7 = 7 \qquad 8 - 3 \stackrel{?}{=} 5$$
$$5 = 5$$

Indeed, the ordered pair (4, 3) does satisfy both equations. Are there any other points common to both lines? Certainly not. Therefore, the ordered pair (4, 3) is the only solution for the system of equations.

Let's use graphing to solve the system

$$(1) \quad 2x + 3y = 12$$
$$(2) \qquad y = x - 1$$

To obtain our graph, we can use the intercept method for equation (1).

$$\text{If } x = 0, \quad y = 4 \qquad (0, 4)$$
$$\text{If } y = 0, \quad x = 6 \qquad (6, 0)$$

For equation (2) we can use the "choose an x, find a y" method.

$$\text{If } x = 0, \qquad y = 0 - 1 = -1 \qquad (0, -1)$$
$$\text{If } x = -2, \quad y = -2 - 1 = -3 \qquad (-2, -3)$$
$$\text{If } x = 2, \qquad y = 2 - 1 = 1 \qquad (2, 1)$$

Graphing both lines in the same plane, we have

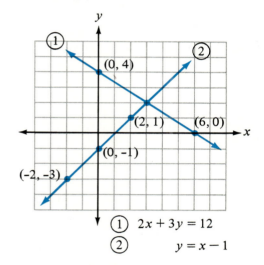

$$\textcircled{1} \quad 2x + 3y = 12$$
$$\textcircled{2} \qquad y = x - 1$$

The point (3, 2) seems to be common to both lines, but we should check both equations to see if the ordered pair (3, 2) satisfies each one.

$$(1) \quad 2x + 3y = 12 \qquad (2) \quad y = x - 1$$
$$2(3) + 3(2) \stackrel{?}{=} 12 \qquad 2 \stackrel{?}{=} 3 - 1$$
$$6 + 6 \stackrel{?}{=} 12 \qquad 2 = 2$$
$$12 = 12$$

So the ordered pair (3, 2) is the solution for this system.

Let's find all solutions for the system

$$(1) \quad x + y = 3$$
$$(2) \quad x + y = -1$$

We can graph both lines by the intercept method. For equation (1),

$$\text{if } x = 0, \quad y = 3 \qquad (0, 3)$$
$$\text{if } y = 0, \quad x = 3 \qquad (3, 0)$$

For equation (2),

$$\text{if } x = 0, \quad y = -1 \qquad (0, -1)$$
$$\text{if } y = 0, \quad x = -1 \qquad (-1, 0)$$

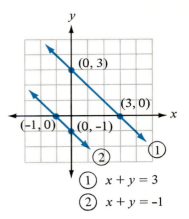

① $x + y = 3$
② $x + y = -1$

Where do these graphs cross? Nowhere! The lines are parallel. A system such as this has *no solution* and is called an **inconsistent system**.

Now let's solve the system

$$(1)\ y = 5x - 2$$
$$(2)\ y = 3$$

We can graph equation (1) by the "choose an x, find a y" method.

$$\text{If } x = 0, \quad y = 5(0) - 2 = 0 - 2 = -2 \qquad (0, -2)$$
$$\text{If } x = -1, \quad y = 5(-1) - 2 = -5 - 2 = -7 \qquad (-1, -7)$$
$$\text{If } x = 2, \quad y = 5(2) - 2 = 10 - 2 = 8 \qquad (2, 8)$$

We recognize the graph of equation (2) as a horizontal line (parallel to the x-axis) crossing the y-axis at $(0, 3)$.

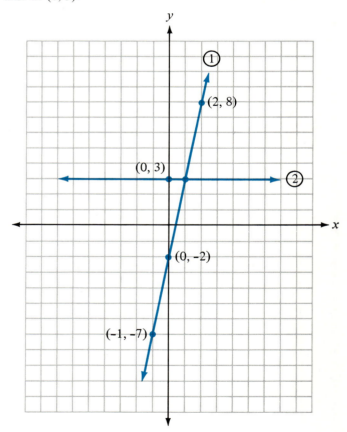

① $y = 5x - 2$
② $y = 3$

The point (1, 3) appears to be a common solution. We check by substituting in each equation.

$$\begin{array}{ll}(1)\ y = 5x - 2 & (2)\ y = 3 \\ \quad\ 3 \overset{?}{=} 5(1) - 2 & \quad\ 3 \overset{?}{=} 3 \\ \quad\ 3 \overset{?}{=} 5 - 2 & \quad\ 3 = 3 \\ \quad\ 3 = 3 \end{array}$$

The ordered pair (1, 3) satisfies both equations, so it is the solution for the system. Let's try another problem:

$$\begin{array}{ll}(1) & y = 3 - x \\ (2) & 2x + 2y = 6\end{array}$$

Finding points for equation (1), we see that

$$\begin{array}{lll} \text{if } x = -1, & y = 3 - (-1) = 4 & (-1, 4) \\ \text{if } x = 0, & y = 3 - 0 = 3 & (0, 3) \\ \text{if } x = 1, & y = 3 - 1 = 2 & (1, 2) \end{array}$$

Finding intercepts for equation (2), we see that

$$\begin{array}{ll} \text{if } x = 0, & 2(0) + 2y = 6 \\ & \qquad\quad 2y = 6 \\ & \qquad\quad\ y = 3 \quad (0, 3) \\ \text{if } y = 0, & 2x + 2(0) = 6 \\ & \qquad\quad 2x = 6 \\ & \qquad\quad\ x = 3 \quad (3, 0) \end{array}$$

Graphing both lines in the same plane, we have

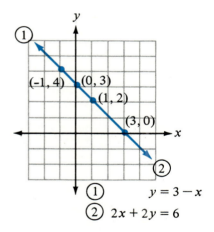

$$\textcircled{1}\ y = 3 - x$$
$$\textcircled{2}\ 2x + 2y = 6$$

The graph for both equations is the same line! Any solution for one of the equations will be a solution for the other equation. Such a system is said to be a **dependent system**.

EXERCISE SET 11.1

Solve each system by graphing.

1. $x - y = -2$
   $x + y = 4$

2. $x + y = 6$
   $x - y = -4$

3. $3x + 2y = 6$
   $2x + y = 2$

4. $5x - 4y = -60$
   $x + y = 6$

5. $3x - y = 6$
   $y = 6x$

6. $-2x + y = 8$
   $y = 4x$

7. $x + y = -5$
   $-2x - 2y = 8$

8. $x - y = 3$
   $3x - 3y = -6$

9. $4x - y = 8$
   $2x + y = 4$

10. $-3x + y = 9$
    $-2x - y = 6$

11. $2x - y = 2$
    $-6x + 3y = -6$

12. $-3x + 3y = 6$
    $2x - 2y = -4$

13. $3x - 2y = 0$
    $x - y = -1$

14. $4x + 3y = 0$
    $x + y = -1$

15. $2x - y = 10$
    $y = 4$

16. $3x + y = 9$
    $y = -3$

17. $2x + y = 9$
    $x - y = 3$

18. $6x + y = -3$
    $x + y = 7$

19. $4x - 2y = 16$
    $-2x + y = -4$

20. $-3x + 2y = -6$
    $6x - 4y = 8$

CHECKUP 11.1

Solve each system by graphing.

1. $x - y = 2$
  $x + y = 0$

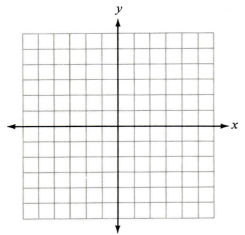

2. $-4x + 2y = 8$
  $y = 2x - 2$

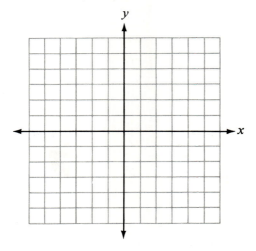

3. $2x + 3y = 6$
  $x = -3$

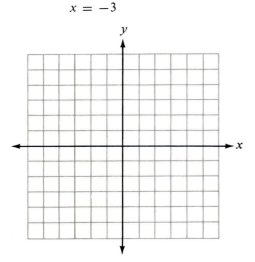

4. $-x + 2y = 4$
  $5x - 10y = -20$

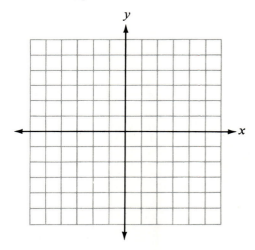

11.2 Solving a System of Equations by Substitution

We have seen that the solution for a system of equations can often be found by graphing each equation and finding the point common to both graphs. Sometimes, however, it is difficult (if not impossible) to find the exact coordinates of the point where two lines cross.

For instance, suppose that we try to use a graph to solve the system

$$(1) \; y = 3x + 1$$
$$(2) \; y = 2 - x$$

We shall graph both equations using several points for each. For equation (1), $y = 3x + 1$.

$$\text{if } x = -1, \quad y = 3(-1) + 1 = -2 \qquad (-1, -2)$$
$$\text{if } x = 0, \qquad y = 3(0) + 1 = 1 \qquad (0, 1)$$
$$\text{if } x = 1, \qquad y = 3(1) + 1 = 4 \qquad (1, 4)$$

For equation (2), $y = 2 - x$.

$$\text{if } x = -1, \quad y = 2 - (-1) = 3 \qquad (-1, 3)$$
$$\text{if } x = 0, \qquad y = 2 - 0 = 2 \qquad (0, 2)$$
$$\text{if } x = 1, \qquad y = 2 - 1 = 1 \qquad (1, 1)$$

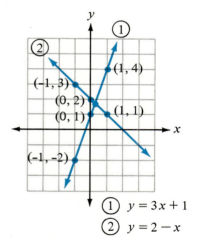

① $y = 3x + 1$
② $y = 2 - x$

It is difficult to decide the exact coordinates of the point where the lines cross. Using a graph is not the best way to solve such a system. We need an algebraic method to find a solution without graphing.

In finding the solution for a system of two equations, remember that we are looking for an ordered pair that is a common solution for both equations. We must find an x-value and a y-value that satisfy *both* equations.

Let's see how this idea can help us find the solution to the system

$$(1) \; y = 3x + 1$$
$$(2) \; y = 2 - x$$

without graphing.

In equation (1), y is equal to $3x + 1$, and in equation (2), y is equal to $2 - x$. Remember, we are looking for the point where the y-values for the equations are the same. We want the y-value in equation (1) to equal the y-value in equation (2). Equating those y-values, we have

$$3x + 1 = 2 - x$$

Now we must solve this equation for x by the methods learned earlier.

$$3x + 1 = 2 - x$$
$$3x + 1 + x = 2 - x + x$$
$$4x + 1 = 2$$
$$4x + 1 - 1 = 2 - 1$$
$$4x = 1$$
$$x = \frac{1}{4}$$

We now know that the common x-value for the equations is

$$x = \frac{1}{4}$$

To find the y-value, we can substitute this x-value into equation (1),

$$y = 3x + 1$$
$$y = 3\left(\frac{1}{4}\right) + 1$$
$$y = \frac{3}{4} + 1$$
$$y = \frac{3}{4} + \frac{4}{4}$$
$$y = \frac{7}{4}$$

or into equation (2),

$$y = 2 - x$$
$$y = 2 - \frac{1}{4}$$
$$y = \frac{8}{4} - \frac{1}{4}$$
$$y = \frac{7}{4}$$

The y-value of $\frac{7}{4}$ corresponds to the x-value of $\frac{1}{4}$ in each equation. The common solution for both equations is therefore

$$\left(\frac{1}{4}, \frac{7}{4}\right)$$

Looking at the graph again, doesn't this ordered pair seem reasonable?

The method just discussed is called **solving a system by substitution**. In this method, we find an expression for one of the variables (either x or y) in one of the equations, and then we substitute that expression for that variable in the other equation. Such a substitution will leave us with one equation containing just one variable.

The method of substitution works especially well when one variable is already isolated in one of the original equations or when one of the variables has a coefficient of 1.

For instance, substitution works nicely in the system

$$\begin{aligned} (1) \qquad y &= 2x - 1 \\ 2x + 3y &= 5 \end{aligned}$$

In equation (1), y is isolated. We can now substitute the expression $2x - 1$ for y in equation (2) and then

$$(2) \quad 2x + 3y = 5$$

becomes

$$2x + 3(2x - 1) = 5$$

Equation (2) now contains the single variable x, so we can solve for x.

$$
\begin{aligned}
2x + 3(2x - 1) &= 5 \\
2x + 6x - 3 &= 5 \\
8x - 3 &= 5 \\
8x &= 8 \\
x &= 1
\end{aligned}
$$

This is the x-value for the common solution of equations (1) and (2). We can substitute 1 for x in either of the original equations to find the corresponding y-value. Let's use equation (1).

$$(1) \quad y = 2x - 1$$

We know $x = 1$, so that

$$
\begin{aligned}
y &= 2(1) - 1 \\
y &= 2 - 1 \\
y &= 1
\end{aligned}
$$

Our common solution for the system is the ordered pair $(x, y) = (1, 1)$. Graphically, this tells us that our two lines intersect at the point $(1, 1)$. You should check to see that the ordered pair $(1, 1)$ also satisfies equation (2).

In the system

$$
\begin{aligned}
(1) \quad & 5x + 2y = 9 \\
(2) \quad & x + 3y = 4
\end{aligned}
$$

neither variable is isolated in either equation. However, in equation (2) the variable x appears with a coefficient of 1. Let's isolate x in equation (2).

$$
\begin{aligned}
(2) \quad x + 3y &= 4 \\
x + 3y - 3y &= 4 - 3y \\
x &= 4 - 3y
\end{aligned}
$$

Now we can substitute this value for x in equation (1), and

$$(1) \quad 5x + 2y = 9$$

becomes

$$5(4 - 3y) + 2y = 9$$

We now solve for y.

$$
\begin{aligned}
5(4 - 3y) + 2y &= 9 \\
20 - 15y + 2y &= 9 \\
20 - 13y &= 9 \\
-13y &= 9 - 20 \\
-13y &= -11 \\
y &= \frac{-11}{-13} \\
y &= \frac{11}{13}
\end{aligned}
$$

To find the corresponding x-value, we return to equation (2) with x isolated and substitute $\frac{11}{13}$ for y.

$$(2) \quad x = 4 - 3y$$

$$x = 4 - 3\left(\frac{11}{13}\right)$$

$$x = \frac{4}{1} - \frac{33}{13}$$

$$x = \frac{52}{13} - \frac{33}{13} = \frac{52 - 33}{13}$$

$$x = \frac{19}{13}$$

Our common solution is $(x, y) = \left(\frac{19}{13}, \frac{11}{13}\right)$, which is the point where the graphs of our equations intersect. As you can see, it would have been very difficult to find this solution using a graph, but it was not difficult to find it algebraically.

Example 1. Solve the system

$$(1) \quad 2x + 3y = 4$$
$$(2) \quad 3x + y = 7$$

Solution. First, let's isolate y in equation (2).

$$3x + y = 7$$
$$y = 7 - 3x$$

Now we substitute $7 - 3x$ for y in equation (1).

$$(1) \qquad 2x + 3y = 4$$

becomes

$$2x + 3(7 - 3x) = 4$$

Now solve this equation for x.

$$2x + 21 - 9x = 4$$
$$21 - 7x = 4$$
$$-7x = 4 - 21$$
$$-7x = -17$$
$$x = \frac{17}{7}$$

Substituting $x = \frac{17}{7}$ into $y = 7 - 3x$, we find

$$y = 7 - 3\left(\frac{17}{7}\right)$$

$$y = 7 - \frac{51}{7}$$

$$y = \frac{49}{7} - \frac{51}{7} = \frac{49 - 51}{7}$$

$$y = \frac{-2}{7}$$

The solution is $\left(\frac{17}{7}, \frac{-2}{7}\right)$.

Now let's consider the system

$$(1) \quad y = 3x - 1$$
$$(2) \quad y = 3x + 2$$

Substitution will work very nicely for this system, since y is isolated in both equations. Equating the y-values, we have

$$3x - 1 = 3x + 2$$

Solving this equation for x by our usual methods, we find

$$3x - 1 - 3x = 3x + 2 - 3x$$
$$-1 = 2$$

This result is very disturbing, since we know that

$$-1 \neq 2$$

To determine what this result means for our system, let's graph our original equations and see what happens. The method of "choose an x, find a y" will work well for both equations.

For equations (1), $y = 3x - 1$,

if $x = -1$, $y = 3(-1) - 1 = -3 - 1 = -4$ $(-1, -4)$

if $x = 0$, $y = 3(0) - 1 = 0 - 1 = -1$ $(0, -1)$

if $x = 1$, $y = 3(1) - 1 = 3 - 1 = 2$ $(1, 2)$

For equation (2), $y = 3x + 2$,

if $x = -1$, $y = 3(-1) + 2 = -3 + 2 = -1$ $(-1, -1)$

if $x = 0$, $y = 3(0) + 2 = 0 + 2 = 2$ $(0, 2)$

if $x = 1$, $y = 3(1) + 2 = 3 + 2 = 5$ $(1, 5)$

Graphing both lines in the same plane, we have

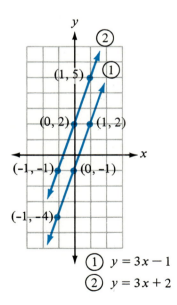

① $y = 3x - 1$
② $y = 3x + 2$

Our two lines are *parallel;* the equations have no ordered pair in common. There is *no solution* for the system. In general, if the method of substitution yields the result that one constant equals a different constant, we must conclude that the original system has no solution. The graphs of the equations are parallel lines and the system is *inconsistent.*

Let's consider solving the system

(1) $y = 5 - 2x$
(2) $4x + 2y = 10$

Since y is isolated in equation (1), we can substitute $5 - 2x$ for y in equation (2), and

$$(2) \qquad 4x + 2y = 10$$

becomes

$$4x + 2(5 - 2x) = 10$$
$$4x + 10 - 4x = 10$$

Combining like terms, we have

$$10 = 10$$

which is a true statement. Unfortunately, this equation contains no variable, so we have not found either coordinate of a common point. What does this result tell us? It is not like the last example, in which we found

$$-1 \neq 2$$

Let's graph our original equations and see what happens.

To graph equation (1), $y = 5 - 2x$, we can use the method of "choose an x, find a y."

$$\text{If } x = -1, \quad y = 5 - 2(-1) = 5 + 2 = 7 \qquad (-1, 7)$$
$$\text{If } x = 0, \quad y = 5 - 2(0) = 5 - 0 = 5 \qquad (0, 5)$$
$$\text{If } x = 1, \quad y = 5 - 2(1) = 5 - 2 = 3 \qquad (1, 3)$$

For equation (2), $4x + 2y = 10$, we can use the intercept method.

$$\text{If } x = 0, \quad y = 5 \qquad (0, 5)$$
$$\text{If } y = 0, \quad x = \frac{10}{4} = \frac{5}{2} \qquad \left(\frac{5}{2}, 0\right)$$

Graphing both lines in the same plane, we have

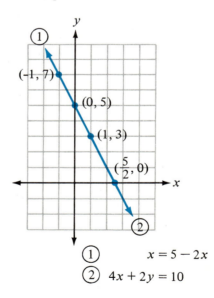

① $\quad x = 5 - 2x$

② $\quad 4x + 2y = 10$

Both equations graph into the *same line!* Our conclusion is that our equations are satisfied by the same ordered pairs; any solution for one equation is a solution for the other equation. In general, whenever the method of substitution yields the result that a constant equals itself, we must conclude that the equations have all ordered pairs in common, and the system is *dependent*.

Example 2. Solve the system

$$(1) \quad x + 2y = 1$$
$$(2) \qquad 10y = 3 - 5x$$

Solution. We can isolate x in equation (1).

$$x + 2y = 1$$
$$x = 1 - 2y$$

Now we substitute $1 - 2y$ for x in equation (2).

$$10y = 3 - 5x$$

becomes

$$10y = 3 - 5(1 - 2y)$$
$$10y = 3 - 5 + 10y$$

Combining like terms, we have

$$10y = -2 + 10y$$
$$10y - 10y = -2 + 10y - 10y$$
$$0 = -2$$

Since $0 \neq -2$, the system has no solution; its graph is a pair of parallel lines.

Example 3. Solve the system

$$(1) \qquad 3x = 3y - 15$$
$$(2) \quad y - x = 5$$

Solution. First we isolate y in equation (2).

$$y - x = 5$$
$$y = 5 + x$$

Now we substitute $5 + x$ for y in equation (1).

$$3x = 3y - 15$$

becomes

$$3x = 3(5 + x) - 15$$
$$3x = 15 + 3x - 15$$
$$3x = 3x$$
$$3x - 3x = 3x - 3x$$
$$0 = 0$$

Since $0 = 0$, we know the graphs of our original equations are the same line. Any solution for one equation is a solution for the other equation.

In general, we note the two unusual results that can occur in solving a system by substitution.

If substitution yields the result "constant $=$ different constant," the system has *no solution*. The graphs are *parallel lines* with no common points, and the system is *inconsistent*.

If substitution yields the result "constant $=$ itself," the system has an *infinite number* of solutions. Any solution for one equation is a solution for both equations. The graphs are the *same line*, with all points in common, and the system is *dependent*.

If, however, solving a system yields *one* ordered pair, the system is said to be **independent and consistent**.

Trial Run *Solve each system by the substitution method.*

_____ **1.** $y = 7x - 3$
$y = 6 - 2x$

_____ **2.** $y = 3x$
$3x - 2y = 6$

_____ **3.** $x + 4y = 5$
$2y - x = 7$

_____ **4.** $y + 2 = 4x$
$3y + 2x = 1$

_____ **5.** $y = 3x + 7$
$2y - 6x = 5$

_____ **6.** $x = 4y + 1$
$8y = 2x - 2$

_____ **7.** $2y = 1 - 2x$
$x + y = 3$

ANSWERS

1. $(1, 4)$.   **2.** $(-2, -6)$.   **3.** $(-3, 2)$.   **4.** $\left(\frac{1}{2}, 0\right)$.   **5.** No solution.   **6.** All ordered pairs that satisfy $x = 4y + 1$.   **7.** No solution.

To solve a system of equations by the method of *substitution,* we should keep the following steps in mind:

> **1.** Look for a variable appearing in an equation with a coefficient of 1, and isolate that variable in its equation.
> **2.** Substitute the obtained expression for that isolated variable into the other equation.
> **3.** Solve for the remaining variable.
> **4.** Substitute that solution into either of the original equations and solve for the other variable.
> **5.** Write the solution as an ordered pair.

EXERCISE SET 11.2

Solve each system by the substitution method.

_____ **1.** $8x - 3y = 7$
$\qquad x = 2$

_____ **2.** $7x - 11y = -1$
$\qquad x = 3$

_____ **3.** $3x - 8y = 13$
$\qquad y = 2x$

_____ **4.** $5x - 3y = 28$
$\qquad y = -3x$

_____ **5.** $\qquad y = 4x - 5$
$5x - y = 11$

_____ **6.** $\qquad y = 3x + 8$
$7x - y = -4$

_____ **7.** $2x - 5y = 9$
$\qquad x - 3y = 2$

_____ **8.** $3x - 2y = 15$
$\qquad x - 2y = 1$

_____ **9.** $10x - 5y = 20$
$\qquad y = 2x - 4$

_____ **10.** $3x - 15y = 15$
$\qquad x = 5y + 5$

_____ **11.** $4x - 3y = 7$
$2x - y = -1$

_____ **12.** $5x - 2y = 12$
$4x - y = 3$

_____ **13.** $x + 2y = -1$
$4x + 6y = 5$

_____ **14.** $x - 3y = 5$
$3x - 7y = 4$

_____ **15.** $3x - 2y = 12$
$x + y = 6$

_____ **16.** $2x - 5y = 11$
$x + y = 3$

_____ **17.** $x + 3y = 10$
$x - 3 = 0$

_____ **18.** $x + 9y = 1$
$x + 5 = 0$

_____ **19.** $14x - 2y = 10$
$7x - y = 2$

_____ **20.** $10x + 2y = 3$
$5x + y = -1$

_____ **21.** $x - y = -2$
$11x - 3y = 2$

_____ **22.** $x - y = -4$
$7x + 5y = -4$

_____ **23.** $x - 2y = -3$
$3x - 7y = 1$

_____ **24.** $x + 5y = 1$
$2x + 8y = -6$

_____ **25.** $3x + y = 2$
$4x - 2y = -3$

_____ **26.** $5x + y = -1$
$13x + 3y = 2$

Name _____ Date _____

CHECKUP 11.2

Solve each system by the substitution method.

_____ **1.** $5x - 2y = 3$
$\quad\quad\quad\quad y = 2x$

_____ **2.** $\quad x - y = -3$
$\quad\quad\quad -4x + 4y = \quad 1$

_____ **3.** $\quad x - 6 \;= 0$
$\quad\quad\quad 3x - 4y = 8$

_____ **4.** $9x - 4y = 10$
$\quad\quad\quad 2x + \;y = \;6$

_____ **5.** $4x + 7y = 16$
$\quad\quad\quad\; x + 4y = \;7$

11.3 Solving a System of Equations by Elimination

Besides graphing and substitution, there is another method for solving a system of first-degree equations. This method uses two properties of equations with which we are already familiar.

> 1. We may multiply both sides of an equation by any constant (except zero).
> 2. We may add equal quantities to both sides of an equation without changing the solutions of the original equation.

Suppose that we wish to solve the system

$$(1) \quad 3x + 2y = 9$$
$$(2) \quad x - 2y = 7$$

Notice what happens if we add the left-hand side of equation (1) to the left-hand side of equation (2) *and* add the right-hand side of equation (1) to the right-hand side of equation (2). (We know that this is permissible because of the second property stated above.)

$$(1) \quad 3x + 2y = 9$$
$$(2) \quad x - 2y = 7$$

becomes

$$(3x + x) + (2y - 2y) = (9 + 7)$$
$$4x \quad + \quad 0 \quad = 16$$

Notice that we have completely eliminated the variable y; our new equation contains the single variable x. Let's solve for x.

$$4x + 0 = 16$$
$$4x = 16$$
$$\frac{4x}{4} = \frac{16}{4}$$
$$x = 4$$

We may find the corresponding y-value by substituting our x value of 4 into either equation (1) or equation (2). Let's use equation (2).

$$(2) \quad x - 2y = 7$$
$$4 - 2y = 7$$
$$-2y = 3$$
$$y = \frac{3}{-2}$$
$$y = \frac{-3}{2}$$

It seems that the solution for our system is the ordered pair

$$(x, y) = \left(4, \frac{-3}{2}\right)$$

Just to be sure, let's check this ordered pair in equation (1).

$$(1) \qquad 3x + 2y = 9$$
$$3(4) + 2\left(\frac{-3}{2}\right) \stackrel{?}{=} 9$$
$$12 - 3 \stackrel{?}{=} 9$$
$$9 = 9$$

Our solution checks.

This method, called the **method of elimination**, works nicely if one of the variables appears in each equation with coefficients that are *opposites*. This is an important observation.

Example 1. Solve by elimination.

$$(1) \quad x + 3y = 5$$
$$(2) \quad -x - 7y = 3$$

Solution. Noting that the coefficients of x are opposites, we shall add corresponding sides of the equations together and eliminate the variable x. Let's do our addition in columns.

$$
\begin{array}{ll}
(1) & x + 3y = 5 \\
(2) & \underline{-x - 7y = 3} \\
& 0 - 4y = 8 \\
& \quad -4y = 8 \\
& \quad \dfrac{-4y}{-4} = \dfrac{8}{-4} \\
& \quad y = -2
\end{array}
$$

Substituting this y value into equation (1), we have

$$
\begin{array}{ll}
(1) & x + 3y = 5 \\
& x + 3(-2) = 5 \\
& x - 6 = 5 \\
& x = 11
\end{array}
$$

And our solution is the ordered pair

$$(x, y) = (11, -2)$$

which we check using equation (2).

$$
\begin{array}{ll}
(2) & -x - 7y = 3 \\
& -11 - 7(-2) \overset{?}{=} 3 \\
& -11 + 14 \overset{?}{=} 3 \\
& 3 = 3
\end{array}
$$

Example 2. Solve by elimination.

$$(1) \quad 5x - 2y = 10$$
$$(2) \quad \quad\;\; 2y = 15$$

Solution. Since the coefficients of y are opposites, we shall eliminate the variable y.

$$
\begin{array}{ll}
(1) & 5x - 2y = 10 \\
(2) & \underline{\quad\;\; 2y = 15} \\
& 5x + 0 = 25 \\
& \quad 5x = 25 \\
& \quad\; x = 5
\end{array}
$$

Now we shall substitute 5 for x in equation (1) to find y.

$$
\begin{array}{ll}
(1) & 5x - 2y = 10 \\
& 5(5) - 2y = 10 \\
& 25 - 2y = 10 \\
& -2y = 10 - 25 \\
& -2y = -15 \\
& y = \dfrac{15}{2}
\end{array}
$$

Our solution is the ordered pair

$$(x, y) = \left(5, \frac{15}{2}\right)$$

which we check in equation (2).

$$(2) \qquad 2y = 15$$

$$2\left(\frac{15}{2}\right) \overset{?}{=} 15$$

$$15 = 15$$

Notice that equation (2) contained no x-term. Of course, you recall that the graph of such an equation is a horizontal line; here, $y = \dfrac{15}{2}$ no matter what x is.

Elimination seems to work well when one of the variables has opposite coefficients in the two equations. But can the elimination method be used when the original coefficients are not opposites? Yes, provided that you *make* the coefficients opposites. Consider the following system.

$$(1) \ 2x + 3y = 7$$
$$(2) \ \ x + 5y = 4$$

Suppose that we decide to use elimination here. The only time we can eliminate one of the variables is when its coefficients are opposites. Look at the coefficients of the variable x. What would the coefficient of x in equation (2) have to be to allow us to use elimination? We would like the x-term in equation (2) to be $-2x$. If we multiply both sides of equation (2) by -2, we will have what we want. (Remember that it is permissible to multiply both sides of an equation by any constant except zero.)

$$(2) \qquad x + 5y = 4$$

becomes

$$-2[x + 5y = 4]$$
$$-2x - 10y = -8$$

and our system becomes

$$(1) \qquad 2x + 3y = \ \ \ 7$$
$$(2) \ -2x - 10y = -8$$

We may now eliminate the variable x and solve for y.

$$(1) \qquad 2x + 3y = \ \ \ 7$$
$$(2) \ \underline{-2x - 10y = -8}$$
$$0 - 7y = -1$$
$$y = \frac{-1}{-7}$$
$$y = \frac{1}{7}$$

Substituting $\dfrac{1}{7}$ for y in the original equation (2), we have

$$x + 5y = 4$$
$$x + 5\left(\frac{1}{7}\right) = 4$$
$$x + \frac{5}{7} = 4$$

$$x = 4 - \frac{5}{7}$$

$$x = \frac{28}{7} - \frac{5}{7}$$

$$x = \frac{23}{7}$$

and our solution is the ordered pair

$$(x, y) = \left(\frac{23}{7}, \frac{1}{7}\right)$$

Let's check this solution in equation (1).

$$2x + 3y = 7$$

$$2\left(\frac{23}{7}\right) + 3\left(\frac{1}{7}\right) \overset{?}{=} 7$$

$$\frac{46}{7} + \frac{3}{7} \overset{?}{=} 7$$

$$\frac{49}{7} \overset{?}{=} 7$$

$$7 = 7$$

Suppose that we had decided to try to eliminate the variable y in the last system.

$$(1) \quad 2x + 3y = 7$$
$$(2) \quad x + 5y = 4$$

We would like the coefficients of y to be opposites. Can you see that multiplying equation (1) by 5 and multiplying equation (2) by -3 will make the coefficients of y become 15 and -15? Let's proceed with that idea.

$$5[2x + 3y = 7]$$
$$-3[x + 5y = 4]$$

Our system becomes

$$10x + 15y = \quad 35$$
$$-3x - 15y = -12$$

and we may now eliminate the variable y.

$$\begin{aligned} 10x + 15y &= \quad 35 \\ -3x - 15y &= -12 \\ \hline 7x + 0 &= 23 \\ 7x &= 23 \\ x &= \frac{23}{7} \end{aligned}$$

Notice that this is the same x-value we obtained when we solved the system before. Again we find the corresponding y-value by substituting $\frac{23}{7}$ for x in either of the original equations. Let's use equation (1).

$$2x + 3y = 7$$

$$2\left(\frac{23}{7}\right) + 3y = 7$$

$$\frac{46}{7} + 3y = 7$$

$$3y = 7 - \frac{46}{7}$$

$$3y = \frac{49}{7} - \frac{46}{7}$$

$$3y = \frac{3}{7}$$

$$y = \frac{3}{7} \div 3$$

$$y = \frac{3}{7} \cdot \frac{1}{3}$$

$$y = \frac{1}{7}$$

The solution is the same ordered pair as before.

$$(x, y) = \left(\frac{23}{7}, \frac{1}{7} \right)$$

Let's solve another system using the method of elimination.

$$(1) \ 4x - 7y = 20$$
$$(2) \ 5x + 2y = 3$$

In order to eliminate either variable in these equations, we must multiply both equations by a constant. Since the y coefficients are already opposite in sign, let's work to eliminate the variable y. We may do so by multiplying equation (1) by the coefficient of y from equation (2): multiply equation (1) by 2. Then multiply equation (2) by the positive version of the coefficient of y from equation (1): multiply equation (2) by 7.

$$2[4x - 7y = 20]$$
$$7[5x + 2y = 3]$$

Our system becomes

$$8x - 14y = 40$$
$$35x + 14y = 21$$

Adding the equations and solving for x, we have

$$43x + 0 = 61$$
$$43x = 61$$
$$x = \frac{61}{43}$$

The thought of substituting this value for x into either equation is not very appealing. It might be easier to start all over again and this time try to eliminate the variable x (so that we are left only with the variable y).

$$(1) \ 4x - 7y = 20$$
$$(2) \ 5x + 2y = 3$$

To accomplish the elimination of the variable x, you should agree that we can multiply equation (1) by 5 and multiply equation (2) by -4.

$$5[4x - 7y = 20]$$
$$-4[5x + 2y = 3]$$

Our system becomes

$$20x - 35y = 100$$
$$-20x - 8y = -12$$

Adding the equations and solving for y, we find

$$0 - 43y = 88$$
$$-43y = 88$$
$$y = \frac{88}{-43}$$
$$y = \frac{-88}{43}$$

Our solution is the ordered pair

$$(x, y) = \left(\frac{61}{43}, \frac{-88}{43}\right)$$

In general, to solve a system of equations by the method of elimination, we should keep these steps in mind:

---

1. Write both equations in the form $ax + by = c$.
2. Look to see if the coefficient of one of the variables in one equation is the opposite of the coefficient of the same variable in the other equation.
3. If the coefficients are not opposites, multiply one or both of the equations by constants that will make the coefficients opposites.
4. Add the equations together to eliminate one of the variables.
5. Solve for the remaining variable.
6. Substitute that value for the variable into either of the original equations and solve for the other variable; *or*
   Repeat the elimination process, eliminating the other variable.
7. Write the solution as an ordered pair.

Trial Run *Solve each system by the elimination method.*

_____ 1. $3x - y = 2$
$\ 2x + y = 8$

_____ 2. $4x - 3y = 16$
$\ 3y = 8$

_____ 3. $9x - 2y = 4$
$\ -9x + 4y = -3$

_____ 4. $11x - 8y = 4$
$\ -5x + 8y = -4$

_____ 5. $-3x + 10y = -4$
$\ x - 2y = 8$

_____ 6. $5x - y = 7$
$\ 3x + 4y = -5$

_____ 7. $-2x + y = 1$
$\ 4x + 3y = 8$

_____ 8. $7x - 3y = 9$
$\ 2x + 5y = -4$

Consider the following system of equations:

$$(1)\quad x + 3y = 9$$
$$(2)\quad 2x + 6y = 6$$

In order to eliminate the variable x, we can multiply equation (1) by -2.

$$-2[x + 3y = 9]$$
$$-2x - 6y = -18$$

Our system becomes

$$-2x - 6y = -18$$
$$2x + 6y = 6$$

and adding the equations, we have

$$0 + 0 = -12$$
$$0 = -12$$

This result is very disturbing, since we know that

$$0 \neq -12$$

As before, our conclusion is that there is *no solution* common to both equations. Suppose that we look at the graphs of these two equations for a better understanding of this situation. We shall use the intercepts method to graph each line. For equation (1), $x + 3y = 9$:

$$\text{if } x = 0, \quad y = 3 \quad (0, 3)$$
$$\text{if } y = 0, \quad x = 9 \quad (9, 0)$$

For equation (2), $2x + 6y = 6$:

$$\text{if } x = 0, \quad y = 1 \quad (0, 1)$$
$$\text{if } y = 0, \quad x = 3 \quad (3, 0)$$

Graphing both equations in the same plane, we have

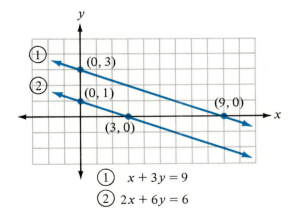

Indeed, our two lines are *parallel;* the lines have no points in common. There is *no solution* for the system.

Example 3. Solve the system by elimination.

$$(1) \quad 2x + 2y = 1$$
$$(2) \quad 1 - 5x = 5y$$

Solution. First we must rearrange the equations to line up like variables beneath like variables.

$$(1) \quad\ \ 2x + 2y =\ \ \ 1$$
$$(2) \quad -5x - 5y = -1$$

To eliminate the variable y, we can multiply equation (1) by 5 and multiply equation (2) by 2.

$$5[2x + 2y =\ \ \ 1]$$
$$2[-5x - 5y = -1]$$

Our system becomes

$$\begin{array}{r} 10x + 10y =\ \ \ 5 \\ -10x - 10y = -2 \\ \hline \end{array}$$

$$0 + 0 = 3$$
$$0 = 3$$

But we know that $0 \neq 3$, so we agree that this system has *no solution*. If we were to graph this system, we would find that the two lines are *parallel*.

Let's look at another system of equations to be solved by elimination.

$$(1) \qquad\qquad y = 3x + 1$$
$$(2) \quad 6x - 2y + 2 = 0$$

To solve by elimination, we must again line up like variables beneath like variables.

$$(1) \quad -3x +\ \ y =\ \ \ 1$$
$$(2) \qquad 6x - 2y = -2$$

Let's eliminate x by multiplying equation (1) by 2.

$$2[-3x + y = 1]$$
$$-6x + 2y = 2$$

Our system becomes

$$(1) \quad -6x + 2y =\ \ \ 2$$
$$(2) \qquad\underline{6x - 2y = -2}$$

$$0 + 0 = 0$$
$$0 = 0$$

We seem to have eliminated both variables by accident! What does this mean? This is not the same as the situations just encountered when we arrived at the false statements:

$$0 = -12$$
$$0 = 3$$

The statement $0 = 0$ is certainly a true statement, but what is the solution to our system? Perhaps graphing the original equations would help. We graph equation (1), $y = 3x + 1$, by the "choose an x, find a y" method.

If $x = -1$, $\quad y = 3(1) + 1 = -3 + 1 = -2 \qquad (-1, -2)$

If $x = 0$, $\qquad y = 3(0) + 1 = 0 + 1 = 1 \qquad\ \ (0, 1)$

If $x = 1$, $\qquad y = 3(1) + 1 = 3 + 1 = 4 \qquad\ \ (1, 4)$

We graph equation (2), $6x - 2y = -2$, using the intercept method.

$$\text{If } x = 0, \quad y = 1 \qquad (0, 1)$$

$$\text{If } y = 0, \quad x = -\frac{1}{3} \qquad \left(-\frac{1}{3}, 0\right)$$

Graphing both lines in the same plane, we have

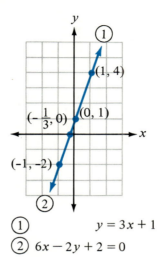

$$\text{(1)} \qquad\qquad y = 3x + 1$$
$$\text{(2)} \quad 6x - 2y + 2 = 0$$

The graph of each equation is the same line! Our conclusion is that these equations have *all* ordered pairs in common; any solution for one equation is a solution for the other equation.

In general, we should note the two unusual results that can occur when solving a system of equations by the elimination method.

---

1. If the elimination method yields the result "zero = nonzero constant," the system has *no solution*. The lines are *parallel* with no common points. The system is *inconsistent*.
2. If the elimination method yields the result "zero = zero," the system has an *infinite number* of solutions. Any solution for one equation is a solution for both equations. The lines have all points in common. The system is *dependent*.

Example 4. Solve by elimination.

$$\text{(1)} \qquad 5y = x$$
$$\text{(2)} \quad 3x - 15y = 4$$

Solution. First we must line up the equations.

$$\text{(1)} \quad -x + 5y = 0$$
$$\text{(2)} \quad 3x - 15y = 4$$

Now we multiply equation (1) by 3 to eliminate the variable x.

$$3[-x + 5y = 0]$$
$$\text{(1)} \quad -3x + 15y = 0$$
$$\text{(2)} \quad \underline{3x - 15y = 4}$$

$$0 + 0 = 4$$
$$0 = 4$$

Our system has *no solution*.

Example 5. Solve by elimination.

$$(1) \quad 8y = 4x$$
$$(2) \quad y = \frac{1}{2}x$$

Solution. First we line up the equations.

$$(1) \quad -4x + 8y = 0$$
$$(2) \quad \frac{-1}{2}x + y = 0$$

To eliminate the variable y, we must multiply equation (2) by -8.

$$-8\left[-\frac{1}{2}x + y = 0\right]$$
$$4x - 8y = 0$$

Our system becomes

$$(1) \quad -4x + 8y = 0$$
$$(2) \quad \underline{4x - 8y = 0}$$
$$0 = 0$$

Any solution for one of the equations is a solution for both equations. All ordered pairs that satisfy $8y = 4x$ are solutions for the system.

Trial Run   *Solve each system by the elimination method.*

_____ **1.** $3x - 12y = 15$
                $x - 4y = -7$

_____ **2.** $5x + 6y = 4$
                $4x - 3y = -2$

_____ **3.** $5x - 10y = -15$
                $-4x + 8y = 12$

_____ **4.** $-9x + 12y = -6$
                $15x - 20y = 1$

ANSWERS

1. No solution.   **2.** $\left(0, \frac{2}{3}\right)$.   **3.** All ordered pairs that satisfy $5x - 10y = -15$.

4. No solution.

EXERCISE SET 11.3

Solve each system by the elimination method.

_____ **1.** $3x - y = 4$
$2x + y = 1$

_____ **2.** $4x - y = 5$
$3x + y = 9$

_____ **3.** $-2x - y = 3$
$2x + 7y = -9$

_____ **4.** $4x - 5y = 6$
$-4x - y = 6$

_____ **5.** $2x - 4y = 1$
$4y = 7$

_____ **6.** $4x - 9y = -9$
$9y = 1$

_____ **7.** $3x + 4y = -2$
$2x - 4y = 12$

_____ **8.** $5x + 3y = -1$
$x - 3y = 7$

_____ **9.** $6x - 4y = 1$
$-6x - 4y = 7$

_____ **10.** $-5x - 3y = 2$
$5x - 3y = 10$

_____ **11.** $3x - 2y = 0$
$5x + 2y = 4$

_____ **12.** $-4x + 5y = 0$
$4x + y = 2$

_____ **13.** $7x - 10y = 4$
$x + 5y = 7$

_____ **14.** $8x - 4y = 6$
$3x + 2y = 4$

_____ **15.** $3x + 4y = 5$
$-9x - 7y = 5$

_____ **16.** $5x + 3y = 7$
$-10x - 13y = 7$

_____ **17.** $2x - 8y = 9$
$-x + 4y = -6$

_____ **18.** $-3x + 3y = -1$
$x - y = 8$

_____ **19.** $5x - 2y = 5$
$4x - 6y = 15$

_____ **20.** $10x - 9y = 12$
$2x - 3y = 4$

_____ **21.** $3x + 4y = 9$
$9x + 4y = 5$

_____ **22.** $7x + 6y = 10$
$7x + 12y = 9$

_____ **23.** $5x - 4y = 1$
$2x + 3y = 5$

_____ **24.** $7x - 6y = 20$
$3x + 5y = 1$

_____ **25.** $14x + 7y = 7$
$-6x - 3y = -3$

_____ **26.** $5x - 15y = 10$
$-6x + 18y = -12$

_____ **27.** $11x - 9y = 14$
$7x + 3y = 6$

_____ **28.** $3x - 5y = 15$
$11x + 15y = -5$

_____ **29.** $7x - 5y = 2$
$2x - 3y = -1$

_____ **30.** $2x - 3y = 5$
$5x - 2y = -4$

CHECKUP 11.3

Solve each system by the elimination method.

_____ **1.** $3x + y = 6$
$5x - y = 10$

_____ **2.** $-3x + 2y = -2$
$3x - 8y = 14$

_____ **3.** $2x - 3y = 5$
$-4x + 6y = -7$

_____ **4.** $7x + y = 6$
$7x = 8$

_____ **5.** $4x - 10y = 1$
$3x - 5y = 3$

11.4 Switching from Words to Systems of Equations

Throughout our study of algebra, we have practiced switching word statements into variable equations. Sometimes we may be asked to solve a problem containing more than one unknown quantity; such a situation requires that we use two variables. In dealing with two variables, we must translate the given information into *two* equations and solve the resulting system. Let's try our hand at working such problems.

Problem 1

For a recent chili supper a total of 191 tickets were sold. Adult tickets sold for $2 each and children's tickets sold for $1 each. If $304 worth of tickets were sold, how many adult tickets and how many children's tickets were sold?

In this problem we are asked to find two unknown quantities, so we can let

$$x = \text{number of adult tickets sold}$$
$$y = \text{number of children's tickets sold}$$
$$x + y = \text{total number of tickets sold}$$

From the first sentence in our problem, we know that

$$x + y = 191$$

The second sentence deals with money. Since each adult ticket sold for $2 and each child's ticket sold for $1, we know that

$$2 \cdot x = \text{dollars from adult tickets}$$
$$1 \cdot y = \text{dollars from children's tickets}$$
$$2x + 1y = \text{total dollars received}$$

Since a total of $304 was received from ticket sales, we know that

$$2x + y = 304$$

We now have two equations that contain the variables x and y, so we may set up our system:

$$(1) \quad x + y = 191$$
$$(2) \quad 2x + y = 304$$

Let's use elimination to solve this system.

$$-1[x + y = 191]$$
$$(1) \quad -x - y = -191$$
$$(2) \quad 2x + y = 304$$
$$\overline{}$$
$$x = 113$$

We conclude that 113 adult tickets were sold, and we find y by substituting 113 for x in equation (1).

$$x + y = 191$$
$$113 + y = 191$$
$$y = 191 - 113$$
$$y = 78$$

We know now that

$$x = 113 \text{ adult tickets}$$
$$y = 78 \text{ children's tickets}$$

As before, we discover that switching from words to variable equations requires a systematic approach in which we write down as much as possible about the variables before attempting a solution. Let's try another problem.

Problem 2

In Mario's psychology class, the male students outnumber the female students by 10. If there are 46 students altogether in the class, how many are male and how many are female?

Once again, there are two unknown quantities, so we can let

$$x = \text{number of females}$$
$$y = \text{number of males}$$

From the first sentence, we see that the number of male students is 10 more than the number of female students. We may therefore write

$$\text{males} = \text{females} + \text{ten}$$
$$y = x + 10$$

The second sentence says that

$$\text{males} + \text{females} = 46 \text{ students}$$
$$y + x = 46$$

Our system of equations becomes

$$(1) \qquad y = x + 10$$
$$(2) \ y + x = 46$$

Since the variable y is isolated in equation (1), it seems sensible to solve our system by substitution. In equation (2) we can replace y by the expression $x + 10$ and

$$(2) \ y + x = 46$$

becomes

$$(x + 10) + x = 46$$
$$2x + 10 = 46$$
$$2x = 46 - 10$$
$$2x = 36$$
$$\frac{2x}{2} = \frac{36}{2}$$
$$x = 18$$

So we know there are 18 females in the class. Since the number of males is expressed as

$$y = x + 10$$

we know that

$$y = 18 + 10$$
$$y = 28$$

There are 18 females and 28 males in Mario's class.

After one more example, you should be ready to try some problems on your own.

Problem 3

Suppose that the sum of two numbers is 18 and twice the difference of the two numbers is 100. Find the numbers.

We can let

$$x = \text{larger number}$$
$$y = \text{smaller number}$$

Then

$$x + y = \text{sum of the numbers}$$
$$x - y = \text{difference of the numbers}$$
$$2(x - y) = \text{twice the difference of the numbers}$$

From the first part of our sentence we know that

$$x + y = 18$$

From the second part of our sentence we know that

$$2(x - y) = 100$$

Our system becomes

(1) $x + y = 18$
(2) $2(x - y) = 100$

or

(1) $x + y = 18$
(2) $2x - 2y = 100$

To solve this system by elimination, we can multiply equation (1) by 2 in order to eliminate the variable y.

$$2[x + y = 18]$$
$$2x + 2y = 36$$

So we have

$$2x + 2y = 36$$
$$\underline{2x - 2y = 100}$$
$$4x + 0 = 136$$
$$4x = 136$$
$$x = \frac{136}{4}$$
$$x = 34$$

So the larger number is 34. To find the smaller number we can substitute 34 for x in equation (1).

$$x + y = 18$$
$$34 + y = 18$$
$$y = 18 - 34$$
$$y = -16$$

Our two numbers are 34 and -16.

Let's check our solution with the words of the original problem. We think our two numbers are 34 and -16. Is the sum of these numbers 18?

$$34 + (-16) \overset{?}{=} 18$$
$$18 = 18$$

Yes. Now we must be sure that twice the difference of the two numbers is 100.

$$2[34 - (-16)] \overset{?}{=} 100$$
$$2[34 + 16] \overset{?}{=} 100$$
$$2[50] \overset{?}{=} 100$$
$$100 = 100$$

So our numbers do satisfy the words of the original problem.

Name _____ Date _____

EXERCISE SET 11.4

Change each word problem to a system of equations using two variables. Solve each system by the better method (substitution or elimination).

_____ **1.** One number is 9 more than twice the other. The sum of the two numbers is 6. Find the numbers.

_____ **2.** The sum of two numbers is 10. One number is 14 less than 3 times the other. Find the numbers.

_____ **3.** The length of a rectangle is 5 feet more than twice the width. If the perimeter of the rectangle is 58 feet, find the dimensions.

_____ **4.** For her flower bed, Nancy needed 3 times as many boxes of petunias as she did marigolds. If she bought 12 boxes of flowers, how many boxes of each kind did she buy?

_____ **5.** The cashier at Mary's Market has a stack of 25 bills. Some are five-dollar bills and some are one-dollar bills. The total value of the stack of bills is $77. How many bills of each denomination does the cashier have?

_____ **6.** The length of a carpet is 20 feet less than 3 times the width. The difference between the length and the width is 4 feet. Find the dimensions of the carpet.

_____ **7.** Jay sold a box of books at a yard sale for $51. He sold some of the books for $2 each and some for $3 each. If there were 20 books in the box, how many did he sell at each price?

_____ **8.** When Hector went to see his father he made part of the trip by bus and the other part by taxi. The time he traveled by taxi was twice the time he rode in the bus. He averaged 40 miles per hour on the bus and 30 miles per hour in the taxi. If the total trip was 25 miles, how much time did he spend traveling by bus and how much time by taxi?

_____ **9.** A. J. has money invested in two different companies. The sum of twice the amount invested in Abell's Alarm Service and the amount invested in Betti's Boutique is $11,000. The amount invested in Betti's Boutique is $2000 more than the amount invested in Abell's Alarm Service. Find the amount of each investment.

_____ **10.** The sum of 3 times one number and 4 times a second number is −17. The difference between twice the first number and the second number is 18. Find the two numbers.

_____ 11. At a sidewalk sale Chris bought 3 pairs of jeans and 2 tops at a cost of $72. Kelly bought 5 pairs of jeans and 3 tops from the same racks for $116. What was the price of each pair of jeans and each top?

_____ 12. For a concert, the sum of the 3 times the number of tickets sold in advance and the number of tickets sold at the door was 2825. The advance tickets sold for $8 and tickets at the door cost $10. If the total ticket receipts were $16,700, how many tickets of each kind were sold?

_____ 13. The sum of the measures of the three angles of any triangle is 180°. One angle measures 90°, one of the remaining angles measures 6° less than 3 times the other. Find the measure of each angle.

_____ 14. Jose works part-time mowing lawns and lifeguarding at the swimming pool. When he mows for 5 hours and guards for 6 hours, he earns $42. When he mows for 11 hours and guards for 4 hours, he earns $51. What is his hourly wage for each job?

Summary

In this chapter we have discussed three methods for solving systems of two equations involving two variables.

We discovered that finding a common solution by *graphing* works well only when the point where the lines cross is easy to identify. In most cases we agreed that algebraic methods would give a more accurate solution.

The method of *substitution* works best when one of the variables is easily isolated in one of the equations. Finally, the method of *elimination* requires that we make the coefficient of one of the variables in one of the equations the opposite of the coefficient of the same variable in the other equation.

No matter which method we used, we discovered three possible situations that could occur for systems of two first-degree equations.

1. The solution is an ordered pair (x, y), and the graphs are lines crossing at exactly one point. (Independent and consistent system)
2. There is no solution, and the graphs are parallel lines. (Inconsistent system)
3. Any solution for one equation is a solution for the other, and both graphs are the same line. (Dependent system)

Once we learned how to solve systems of equations algebraically, we put our methods to practical use in solving problems stated in words.

REVIEW EXERCISES

Solve each system by graphing.

SECTION 11.1

_____ **1.** $2x - y = 1$
$\quad\quad x + 2y = 8$

_____ **2.** $5x - 6y = 4$
$\quad\quad\quad\quad x = 2$

_____ **3.** $x - 3y = 7$
$\quad\quad x - 2y = 0$

_____ **4.** $3x + y = -1$
$\quad\quad 2x - y = -4$

Solve each system by the substitution method.

SECTION 11.2

_____ **5.** $6x + y = -3$
$\quad\quad\quad\quad y = \quad 3$

_____ **6.** $9x - 5y = \quad 12$
$\quad\quad\quad\quad\quad y = -3x$

_____ **7.** $4x - y = \quad 0$
$\quad\quad 3x - 2y = 10$

_____ **8.** $\quad x - 3y = \quad 8$
$\quad\quad -2x + 6y = 20$

_____ **9.** $\quad\quad x = 2y + 7$
$\quad\quad 2x - y = 23$

_____ **10.** $x - \quad y = -\ 5$
$\quad\quad x - 9y = -13$

Solve each system by the elimination method.

SECTION 11.3

_____ **11.** $2x + y = 12$
$\quad\quad 3x - y = 13$

_____ **12.** $\quad x + 3y = 9$
$\quad\quad 4x - 3y = 6$

_____ **13.** $\quad 2x - \quad y = -6$
$\quad\quad -4x + 2y = -5$

_____ **14.** $\quad\quad x + 11y = \quad\ 0$
$\quad\quad -2x + \quad y = -23$

_____ **15.** $-6x + \quad 5y = \quad\ 9$
$\quad\quad 24x - 20y = -36$

_____ **16.** $5x + 6y = \quad 17$
$\quad\quad 8x - 9y = -10$

Solve by any method you choose.

_____ **17.** $\quad 3x - \quad y = -8$
$\quad\quad -5x + 4y = \quad 32$

_____ **18.** $\quad x - \quad 2y = 0$
$\quad\quad 6x - 13y = 5$

_____ **19.** $4x + 15y = -18$
$\quad\quad 6x - \quad 5y = \quad 28$

_____ **20.** $2x - 3y = \quad 4$
$\quad\quad 5x + 4y = 33$

Change each word problem to a system of equations using two variables. Solve each system by the better method (substitution or elimination).

SECTION 4.4

_____ **21.** The sum of two numbers is 22. The difference between twice the first number and the second number is 8. Find both numbers.

_____ **22.** The length of a rectangular field is 10 meters more than the width. The perimeter of the field is 116 meters. Find the length and width of the field.

_____ **23.** Chico has savings accounts at two different banks. The amount at the Union Bank is $100 less than 3 times the amount at the National Bank. The sum of the two accounts is $1700. Find the amount in each savings account.

12 Working with Exponents and Square Roots

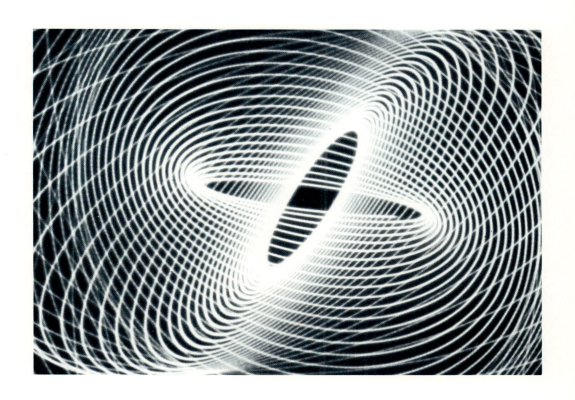

In earlier chapters we discussed exponents and learned several basic definitions and laws for working with exponents which were whole numbers. In this chapter we shall

1. Review the laws of exponents.
2. Discuss the use of negative exponents.
3. Define and operate with square roots.
4. Solve more quadratic equations.

For handy reference, we shall again state the definitions and laws that we used in working with exponents. Remember that a stands for any base, and m and n represent whole numbers.

1. Definition:

$$a^n = \underbrace{a \cdot a \cdot \,\cdots\, \cdot a}_{n \text{ factors}}$$

examples: $2^5 = 2 \cdot 2 \cdot 2 \cdot 2 \cdot 2 = 32$
$x^3 = x \cdot x \cdot x$

2. Definition:

$a^0 = 1$ (provided that $a \neq 0$)

examples: $17^0 = 1$
$x^0 = 1$

3. First Law of Exponents:

$$a^m \cdot a^n = a^{m+n}$$

examples: $2^5 \cdot 2^3 = 2^{5+3} = 2^8$
$x^2 \cdot x = x^{2+1} = x^3$

4. Second Law of Exponents:

$$(a^m)^n = a^{m \cdot n}$$

examples: $(3^2)^4 = 3^{2 \cdot 4} = 3^8$
$(x^3)^6 = x^{3 \cdot 6} = x^{18}$

5. Third Law of Exponents:

$$(a \cdot b)^n = a^n \cdot b^n$$

examples: $(2 \cdot 5)^3 = 2^3 \cdot 5^3$

$(3xy^2)^4 = 3^4 \cdot x^4 \cdot (y^2)^4 = 3^4 x^4 y^8$

---

6. Fourth Law of Exponents:

$$\frac{a^m}{a^n} = a^{m-n} \qquad \text{if } m \geq n$$

$$\frac{a^m}{a^n} = \frac{1}{a^{n-m}} \qquad \text{if } n > m$$

---

examples: $\dfrac{2^9}{2^3} = 2^{9-3} = 2^6$

$$\frac{x^7}{x} = x^{7-1} = x^6$$

$$\frac{y^3}{y^5} = \frac{1}{y^{5-3}} = \frac{1}{y^2}$$

---

7. Fifth Law of Exponents:

$$\left(\frac{a}{b}\right)^n = \frac{a^n}{b^n}$$

---

examples: $\left(\dfrac{2}{3}\right)^4 = \dfrac{2^4}{3^4} = \dfrac{16}{81}$

$$\left(\frac{x}{y}\right)^3 = \frac{x^3}{y^3}$$

$$\left(\frac{3x^2}{z}\right)^2 = \frac{(3x^3)^2}{z^2} = \frac{3^2(x^3)^2}{z^2} = \frac{9x^6}{z^2}$$

12.1 Working with Negative Exponents

So far, we have used only whole numbers $(0, 1, 2, 3, 4, \ldots)$ as exponents, and you may have wondered if negative integers $(\ldots, -4, -3, -2, -1)$ could ever appear as exponents. The answer is yes, indeed, but let's see how to give meaning to expressions such as

$$2^{-1} \qquad \text{or} \qquad x^{-2} \qquad \text{or} \qquad y^{-3}$$

However we decide to define negative exponents, we must agree that we do not wish to lose all the laws of exponents that we have already learned. We would like negative exponents to continue to obey *all* those laws. For instance, we would like

$$2^1 \cdot 2^{-1} = 2^{1+(-1)} = 2^0 = 1$$
$$x^2 \cdot x^{-2} = x^{2+(-2)} = x^0 = 1$$

because of the first law of exponents. Let's examine the first example more closely.

$$2^1 \cdot 2^{-1} = 1$$

Here we are saying that 2 multiplied times 2^{-1} is equal to 1. We would like to give some sensible meaning to the expression 2^{-1}. Can you think of any number that would make

$$2 \cdot ? = 1$$

a true statement? Of course, the only possible replacement for ? is $\frac{1}{2}$, because

$$2 \cdot \frac{1}{2} = 1$$

It seems, then, that if

$$2 \cdot 2^{-1} = 1$$

and

$$2 \cdot \frac{1}{2} = 1$$

it is reasonable to conclude that

$$2^{-1} = \frac{1}{2}$$

Let's look at the second example in which

$$x^2 \cdot x^{-2} = 1$$

Again, we must think of a replacement for ? in the statement

$$x^2 \cdot ? = 1$$

The only possible replacement is $\frac{1}{x^2}$, since

$$\frac{x^2}{1} \cdot \frac{1}{x^2} = 1$$

Again, since we know that

$$x^2 \cdot x^{-2} = 1$$

and

$$x^2 \cdot \frac{1}{x^2} = 1$$

we must agree that

$$x^{-2} = \frac{1}{x^2}$$

Indeed, these two examples give us a good clue for the definition of a negative exponent. We have found that

$$2^{-1} = \frac{1}{2} = \frac{1}{2^1}$$

$$x^{-2} = \frac{1}{x^2}$$

How do you suppose that we evaluate the following?

$$y^{-3} = ?$$
$$x^{-5} = ?$$
$$b^{-8} = ?$$

If you have followed the pattern, you should agree that the following interpretations seem reasonable.

$$y^{-3} = \frac{1}{y^3}$$

$$x^{-5} = \frac{1}{x^5}$$

$$z^{-8} = \frac{1}{z^8}$$

In fact, we are ready now to state the general definition of a base raised to a negative power.

> **Definition**
>
> $a^{-n} = \dfrac{1}{a^n}$ (provided that $a \neq 0$)

Let's practice simplifying some expressions by rewriting them using positive exponents only.

Example 1. Simplify x^{-1}.

   Solution. $x^{-1} = \dfrac{1}{x^1} = \dfrac{1}{x}$.

Example 2. Simplify y^{-4}.

   Solution. $y^{-4} = \dfrac{1}{y^4}$.

Example 3. Simplify 2^{-5}.

   Solution

$$2^{-5} = \frac{1}{2^5}$$
$$= \frac{1}{32}$$

Example 4. Simplify $(-3)^{-2}$.

   Solution

$$(-3)^{-2} = \frac{1}{(-3)^2}$$
$$= \frac{1}{(-3)(-3)}$$
$$= \frac{1}{9}$$

Example 5. Simplify $(-2)^{-3}$.

   Solution

$$(-2)^{-3} = \frac{1}{(-2)^3}$$
$$= \frac{1}{(-2)(-2)(-2)}$$
$$= \frac{1}{-8}$$
$$= \frac{-1}{8}$$

Example 6. Simplify $\left(\dfrac{2}{3}\right)^{-1}$.

$$\left(\frac{2}{3}\right)^{-1} = \frac{1}{\left(\frac{2}{3}\right)^1}$$

$$= \frac{1}{\frac{2}{3}}$$

$$= 1 \div \frac{2}{3}$$

$$= \frac{1}{1} \cdot \frac{3}{2}$$

$$= \frac{3}{2}$$

Example 7. Simplify $(2x)^{-1}$.

Solution

$$(2x)^{-1} = \frac{1}{(2x)^1}$$

$$= \frac{1}{2x}$$

Example 8. Simplify $2x^{-1}$.

Solution

$$2x^{-1} = 2 \cdot \frac{1}{x}$$

$$= \frac{2}{x}$$

We should notice two important facts from these examples.

1. A negative exponent has *no* effect on the sign of the base. The negative exponent merely says "rewrite the expression as 1 divided by the base raised to the opposite (positive) power." The sign of the answer is determined *after* using the negative exponent to rewrite the expression with a positive exponent. So

$$(-1)^{-4} = \frac{1}{(-1)^4} = \frac{1}{1} = 1 \qquad \text{because } (-1)^4 = 1$$

$$(-1)^{-5} = \frac{1}{(-1)^5} = \frac{1}{-1} = -1 \qquad \text{because } (-1)^5 = -1$$

2. A negative exponent applies only to the base to which it is attached, unless parentheses indicate otherwise. You should recall that this was true for positive exponents also. So

$$3 \cdot 5^2 \quad = 3 \cdot 25 = 75$$

$$(3 \cdot 5)^2 = 15^2 \quad = 225$$

$$2x^{-3} \quad = 2 \cdot \frac{1}{x^3} = \frac{2}{x^3}$$

$$(2x)^{-3} \quad = \frac{1}{(2x)^3} \quad = \frac{1}{8x^3}$$

Let's show that negative exponents do indeed obey all our laws of exponents.

Consider the following product:

$$x^{-2} \cdot x^5$$

Let's first approach this problem using the definition of a negative exponent.
   Approach 1:

$$x^{-2} \cdot x^5 = \frac{1}{x^2} \cdot \frac{x^5}{1}$$

$$= \frac{x^5}{x^2} = x^3$$

Now let's use the First Law of Exponents.
   Approach 2:

$$x^{-2} \cdot x^5 = x^{-2+5} = x^3$$

We see that our answers are identical, so we conclude that

Negative exponents obey the First Law of Exponents.

And we can use that law to simplify products. Let's agree to use positive exponents in our answers.

Example 9. Simplify $x^2 \cdot x^{-7}$.

   Solution. $x^2 \cdot x^{-7} = x^{2+(-7)} = x^{-5} = \dfrac{1}{x^5}$.

Example 10. Simplify $a^{-3} \cdot a^{-4}$.

   Solution. $a^{-3} \cdot a^{-4} = a^{-3+(-4)} = a^{-7} = \dfrac{1}{a^7}$.

Example 11. Simplify $(2x^{-3})(5x^{-1})$.

   Solution

$$(2x^{-3})(5x^{-1}) = 2 \cdot 5 \cdot x^{-3} \cdot x^{-1}$$
$$= 10x^{-3+(-1)}$$
$$= 10x^{-4}$$
$$= 10 \cdot \frac{1}{x^4}$$
$$= \frac{10}{x^4}$$

Example 12. Simplify $(3a^{-2})(-4a^2)$.

   Solution

$$(3a^{-2})(-4a^2) = 3(-4)a^{-2} \cdot a^2$$
$$= -12a^{-2+2}$$
$$= -12a^0$$
$$= -12 \cdot 1$$
$$= -12$$

Example 13. Simplify $x^{-3} \cdot x^5 \cdot x^{-1}$.

Solution

$$x^{-3} \cdot x^5 \cdot x^{-1} = x^{-3+5+(-1)}$$
$$= x^1$$
$$= x$$

Trial Run   *Simplify. Write answers using positive exponents only.*

_____ **1.** $(5)^{-2}$

_____ **2.** $(-2)^{-3}$

_____ **3.** $\left(\dfrac{3}{4}\right)^{-1}$

_____ **4.** $(3a)^{-1}$

_____ **5.** $5x^{-2}$

_____ **6.** $a^4 \cdot a^{-1}$

_____ **7.** $x^3 \cdot x^{-5}$

_____ **8.** $(3x^{-6})(4x^2)$

_____ **9.** $(-5x^2)(5x^{-2})$

ANSWERS

1. $\dfrac{1}{25}$.   **2.** $\dfrac{-1}{8}$.   **3.** $\dfrac{4}{3}$.   **4.** $\dfrac{1}{3a}$.   **5.** $\dfrac{5}{x^2}$.   **6.** a^3.   **7.** $\dfrac{1}{x^2}$.   **8.** $\dfrac{12}{x^4}$.   **9.** -25.

What about the Second Law of Exponents? Will negative exponents obey this law also?

> **Second Law of Exponents**
>
> $$(a^m)^n = a^{m \cdot n}$$

Let's try a problem using two approaches again.

$$(x^4)^{-2}$$

Approach 1:

$$(x^4)^{-2} = \frac{1}{(x^4)^2} = \frac{1}{x^{4 \cdot 2}} = \frac{1}{x^8}$$

Approach 2:

$$(x^4)^{-2} = x^{4(-2)} = x^{-8} = \frac{1}{x^8}$$

Perhaps another problem will convince us completely.

$$(x^{-2})^{-1}$$

Approach 1:

$$(x^{-2})^{-1} = \left(\frac{1}{x^2}\right)^{-1} = \frac{1}{\frac{1}{x^2}}$$

$$= 1 \div \frac{1}{x^2}$$

$$= \frac{1}{1} \cdot \frac{x^2}{1}$$

$$= x^2$$

Approach 2:

$$(x^{-2})^{-1} = x^{-2(-1)} = x^2$$

In each example, both approaches yielded the same answer. Especially in the second example, the approach using the second law of exponents was quicker than the other approach. It is very handy to use the fact that

> Negative exponents obey the Second Law of Exponents.

Example 14. Simplify $(x^{-3})^2$.

Solution. $(x^{-3})^2 = x^{-3 \cdot 2} = x^{-6} = \frac{1}{x^6}$.

Example 15. Simplify $(x^{-5})^{-2}$.

Solution. $(x^{-5})^{-2} = x^{-5(-2)} = x^{10}$.

Example 16. Simplify $(x^{10})^{-1} \cdot (y^{-4})^{-2}$.

Solution

$$(x^{10})^{-1} \cdot (y^{-4})^{-2} = x^{10(-1)}y^{-4(-2)}$$

$$= x^{-10}y^8$$

$$= \frac{1}{x^{10}} \cdot \frac{y^8}{1}$$

$$= \frac{y^8}{x^{10}}$$

Example 17. Simplify $(x^{-2})^0$.

Solution. $(x^{-2})^0 = x^{-2(0)} = x^0 = 1$.

Example 18. Simplify $x^5(x^{-3})^2$.

Solution

$$x^5(x^{-3})^2 = x^5 \cdot x^{-3(2)}$$

$$= x^5 \cdot x^{-6}$$

$$= x^{5+(-6)}$$

$$= x^{-1}$$

$$= \frac{1}{x}$$

Example 19. Simplify $(x^{-1})^9 \cdot (x^2)^{-3}$.

Solution

$$(x^{-1})^9 \cdot (x^2)^{-3} = x^{-1(9)} \cdot x^{2(-3)}$$
$$= x^{-9} \cdot x^{-6}$$
$$= x^{-9+(-6)}$$
$$= x^{-15}$$
$$= \frac{1}{x^{15}}$$

We must now investigate the Third Law of Exponents to see whether negative exponents obey it also.

> **Third Law of Exponents**
>
> $$(a \cdot b)^n = a^n \cdot b^n$$

Let us try our two approaches to the problem

$$(2x)^{-4}$$

Approach 1:

$$(2x)^{-4} = \frac{1}{(2x)^4} = \frac{1}{2^4 x^4} = \frac{1}{16x^4}$$

Approach 2:

$$(2x)^{-4} = 2^{-4}x^{-4} = \frac{1}{2^4} \cdot \frac{1}{x^4} = \frac{1}{16x^4}$$

And another problem to simplify using both approaches,

$$(x^{-2}y^{-3})^{-1}$$

Approach 1:

$$(x^{-2}y^{-3})^{-1} = \left(\frac{1}{x^2} \cdot \frac{1}{y^3}\right)^{-1}$$
$$= \left(\frac{1}{x^2 y^3}\right)^{-1}$$
$$= \frac{1}{\frac{1}{x^2 y^3}}$$
$$= 1 \div \frac{1}{x^2 y^3}$$
$$= \frac{1}{1} \cdot \frac{x^2 y^3}{1}$$
$$= x^2 y^3$$

Approach 2:

$$(x^{-2}y^{-3})^{-1} = x^{-2(-1)}y^{-3(-1)}$$
$$= x^2 y^3$$

Again, both approaches to each example yielded the same answer, but Approach 2 using the Third Law of Exponents certainly seemed simpler. We shall find it very useful to use the fact that

> Negative exponents obey the Third Law of Exponents.

Example 20. Simplify $(5x)^{-2}$.

Solution

$$(5x)^{-2} = 5^{-2}x^{-2} = \frac{1}{5^2} \cdot \frac{1}{x^2}$$

$$= \frac{1}{25} \cdot \frac{1}{x^2}$$

$$= \frac{1}{25x^2}$$

Example 21. Simplify $(3x^{-1})^4$.

Solution

$$(3x^{-1})^4 = 3^4(x^{-1})^4 = 3^4 x^{-1 \cdot 4}$$

$$= 3^4 x^{-4}$$

$$= 81x^{-4}$$

$$= 81 \cdot \frac{1}{x^4}$$

$$= \frac{81}{x^4}$$

Example 22. Simplify $(2^{-1}x^{-3}y^{-5})^{-1}$.

Solution

$$(2^{-1}x^{-3}y^{-5})^{-1} = (2^{-1})^{-1}(x^{-3})^{-1}(y^{-5})^{-1}$$
$$= 2^{-1(-1)}x^{-3(-1)}y^{-5(-1)}$$
$$= 2^1 x^3 y^5$$
$$= 2x^3 y^5$$

Note in this example that two steps could have been skipped if you realize that you must multiply the exponent on each factor within the parentheses times the exponent on the entire quantity. You may be able to do this computation in your head if you are careful.

Example 23. Simplify $(3^{-2}x^4 z^{-1})^{-2}$.

Solution

$$(3^{-2}x^4 z^{-1})^{-2} = 3^4 x^{-8} z^2$$

$$= 81 \cdot \frac{1}{x^8} \cdot z^2$$

$$= \frac{81z^2}{x^8}$$

Example 24. $(2^{-1}x^2)^{-4}(2x^{-2})^{-3}$.

Solution

$$(2^{-1}x^2)^{-4}(2x^{-2})^{-3} = 2^4 x^{-8} \cdot 2^{-3} x^6$$

$$= 2^4 \cdot 2^{-3} \cdot x^{-8} \cdot x^6$$

$$= 2^{4+(-3)} \cdot x^{-8+6}$$

$$= 2^1 x^{-2}$$

$$= 2 \cdot \frac{1}{x^2}$$

$$= \frac{2}{x^2}$$

Trial Run *Simplify, writing answers with positive exponents.*

_____ **1.** $(x^{-4})^2$

_____ **2.** $(a^{-3})^{-4}$

_____ **3.** $(y^5)^{-1}(y^{-3})^{-2}$

_____ **4.** $3x^{-2}(x^{-1})^3$

_____ **5.** $(3x)^{-2}$

_____ **6.** $(2a^{-2})^3$

_____ **7.** $(3^{-1}y^4)^{-3}$

_____ **8.** $(2a^2b^{-3})^2$

ANSWERS

1. $\dfrac{1}{x^8}$. **2.** a^{12}. **3.** y. **4.** $\dfrac{3}{x^5}$. **5.** $\dfrac{1}{9x^2}$. **6.** $\dfrac{8}{a^6}$. **7.** $\dfrac{27}{y^{12}}$. **8.** $\dfrac{4a^4}{b^6}$.

The Fourth Law of Exponents involves division. Now that we are allowed to use negative exponents, we need only one version of that law.

> **Fourth Law of Exponents**
>
> $$\frac{a^m}{a^n} = a^{m-n}$$

Earlier, we said that

$$\frac{x^5}{x^2} = x^{5-2} = x^3$$

$$\frac{x^6}{x^9} = \frac{1}{x^{9-6}} = \frac{1}{x^3}$$

Let's see why our new version for this law works whether the exponent in the denominator is less than or greater than the exponent in the numerator. Consider the problem

$$\frac{x^6}{x^9}$$

According to our new version of the Fourth Law, we have

$$\frac{x^6}{x^9} = x^{6-9} = x^{-3} = \frac{1}{x^3}$$

which is the same answer obtained earlier.

But will this law work when the exponents in the original problem are negative? Let's tackle a problem using two approaches. Simplify

$$\frac{x^{-3}}{x^{-5}}$$

Approach 1:

$$\frac{x^{-3}}{x^{-5}} = \frac{\dfrac{1}{x^3}}{\dfrac{1}{x^5}} = \frac{1}{x^3} \div \frac{1}{x^5} = \frac{1}{x^3} \cdot \frac{x^5}{1}$$

$$= \frac{x^5}{x^3} = x^{5-3} = x^2$$

Approach 2:

$$\frac{x^{-3}}{x^{-5}} = x^{-3-(-5)} = x^{-3+5} = x^2$$

Our conclusion?

> Negative exponents obey the Fourth Law of Exponents.

Example 25. Simplify $\dfrac{x^9}{x^{-6}}$.

Solution

$$\frac{x^9}{x^{-6}} = x^{9-(-6)}$$
$$= x^{9+6}$$
$$= x^{15}$$

Example 26. Simplify $\dfrac{10x^{-8}}{2x^{-2}}$.

Solution

$$\frac{10x^{-8}}{2x^{-2}} = \frac{10}{2} \cdot x^{-8-(-2)}$$
$$= 5x^{-8+2}$$
$$= 5x^{-6}$$
$$= 5 \cdot \frac{1}{x^6}$$
$$= \frac{5}{x^6}$$

Example 27. Simplify $\dfrac{2x^{-3}y^{-4}}{6x^8y^{-5}}$.

Solution

$$\frac{2x^{-3}y^{-4}}{6x^8y^{-5}} = \frac{2}{6} \cdot x^{-3-8} \cdot y^{-4-(-5)}$$
$$= \frac{1}{3} \cdot x^{-11}y^{-4+5}$$
$$= \frac{1}{3} \cdot \frac{1}{x^{11}} \cdot y$$
$$= \frac{y}{3x^{11}}$$

Example 28. $\dfrac{(-2x^{-3})(5x^{-5})}{30x^{-10}}$.

Solution

$$\dfrac{(-2x^{-3})(5x^{-5})}{30x^{-10}} = \dfrac{-2 \cdot 5 \cdot x^{-3} \cdot x^{-5}}{30x^{-10}}$$

$$= \dfrac{-10x^{-3-5}}{30x^{-10}}$$

$$= \dfrac{-1}{3} \cdot \dfrac{x^{-8}}{x^{-10}}$$

$$= \dfrac{-1}{3} x^{-8-(-10)}$$

$$= \dfrac{-1}{3} x^{-8+10}$$

$$= \dfrac{-1}{3} x^{2}$$

$$= \dfrac{-x^{2}}{3}$$

Trial Run   *Simplify, writing answers with positive exponents.*

_____ **1.** $\dfrac{x^{10}}{x^{-6}}$

_____ **2.** $\dfrac{a^{-4}}{a^{4}}$

_____ **3.** $\dfrac{12y^{-7}}{6y^{-3}}$

_____ **4.** $\dfrac{3a^{-2}b^{-3}}{9a^{3}b^{-6}}$

_____ **5.** $\dfrac{(-3x^{-2})(2x^{-3})}{9x^{-11}}$

_____ **6.** $\dfrac{(2x^{3})^{-2}(x^{2})^{-1}}{(4x^{10})^{-1}}$

ANSWERS

1. x^{16}.   **2.** $\dfrac{1}{a^{8}}$.   **3.** $\dfrac{2}{y^{4}}$.   **4.** $\dfrac{b^{3}}{3a^{5}}$.   **5.** $\dfrac{-2x^{6}}{3}$.   **6.** x^{2}.

The Fifth Law of Exponents provides a rule for raising a quotient (or fraction) to a power.

┌─────────────────────────────────┐

Fifth Law of Exponents

$$\left(\dfrac{a}{b}\right)^{n} = \dfrac{a^{n}}{b^{n}}$$

└─────────────────────────────────┘

To see whether negative exponents obey this law, we shall use two approaches to simplify

$$\left(\frac{x}{3}\right)^{-2}$$

Approach 1:

$$\left(\frac{x}{3}\right)^{-2} = \frac{1}{\left(\frac{x}{3}\right)^2} = \frac{1}{\frac{x^2}{3^2}} = \frac{1}{\frac{x^2}{9}}$$

$$= 1 \div \frac{x^2}{9} = \frac{1}{1} \cdot \frac{9}{x^2} = \frac{9}{x^2}$$

Approach 2:

$$\left(\frac{x}{3}\right)^{-2} = \frac{x^{-2}}{3^{-2}} = \frac{\frac{1}{x^2}}{\frac{1}{3^2}}$$

$$= \frac{1}{x^2} \div \frac{1}{9} = \frac{1}{x^2} \cdot \frac{9}{1} = \frac{9}{x^2}$$

Indeed our answers agree, so we conclude that

> Negative exponents obey the Fifth Law of Exponents.

The examples that follow will emphasize how this rule makes simplification much quicker.

Example 29. Simplify $\left(\frac{x^{-1}}{y^{-3}}\right)^{-2}$.

Solution. $\left(\frac{x^{-1}}{y^{-3}}\right)^{-2} = \frac{x^{-1(-2)}}{y^{-3(-2)}} = \frac{x^2}{y^6}$.

Example 30. Simplify $\left(\frac{2a^2}{b^{-5}}\right)^{-4}$.

Solution

$$\left(\frac{2a^2}{b^{-5}}\right)^{-4} = \frac{2^{-4}a^{2(-4)}}{b^{-5(-4)}}$$

$$= \frac{2^{-4}a^{-8}}{b^{20}}$$

$$= \frac{1}{2^4} \cdot \frac{1}{a^8} \cdot \frac{1}{b^{20}}$$

$$= \frac{1}{16a^8b^{20}}$$

Example 31. Simplify $\left(\frac{6x^2y^{-3}}{-2x^5y^{-9}}\right)^{-4}$.

Solution

$$\left(\frac{6x^2y^{-3}}{-2x^5y^{-9}}\right)^{-4} = \left[\frac{6}{-2}x^{2-5}y^{-3-(-9)}\right]^{-4}$$

$$= (-3x^{-3}y^6)^{-4}$$

$$= (-3)^{-4}x^{-3(-4)}y^{6(-4)}$$

$$= \frac{1}{(-3)^4} \cdot x^{12} \cdot y^{-24}$$

$$= \frac{1}{81} \cdot x^{12} \cdot \frac{1}{y^{24}}$$

$$= \frac{x^{12}}{81y^{24}}$$

Trial Run   *Simplify, writing answers with positive exponents.*

_____ 1. $\left(\dfrac{x}{y}\right)^{-1}$

_____ 2. $\left(\dfrac{a^{-2}}{b^3}\right)^{-3}$

_____ 3. $\left(\dfrac{x^{-4}}{y^{-2}}\right)^{-1}$

_____ 4. $\left(\dfrac{3x^2}{y^{-3}}\right)^{-2}$

_____ 5. $\left(\dfrac{8a^2b^{-3}}{-4a^3b^{-7}}\right)^{-2}$

ANSWERS

1. $\dfrac{y}{x}$.   2. a^6b^9.   3. $\dfrac{x^4}{y^2}$.   4. $\dfrac{1}{9x^4y^6}$.   5. $\dfrac{a^2}{4b^8}$.

We have discovered that negative exponents obey all the laws of exponents used earlier. The important fact to remember in simplifying expressions involving any kind of exponents is to *use the laws of exponents to simplify the expression first*. When you have reached the end of your simplification, *then* you may make use of the fact that $a^{-n} = \dfrac{1}{a^n}$ if you wish to write your answer with positive exponents only.

EXERCISE SET 12.1

Simplify using the laws of exponents. Give answers with positive exponents.

_____ **1.** $(2x^2)^3$ _____ **2.** $(-3x^4)^2$

_____ **3.** $\dfrac{a^5}{a^2}$ _____ **4.** $\dfrac{a^9}{a^3}$

_____ **5.** $\dfrac{5y^4}{y^7}$ _____ **6.** $\dfrac{4y^5}{y^9}$

_____ **7.** $\left(\dfrac{-3a^2}{b}\right)^3$ _____ **8.** $\left(\dfrac{-2a^3}{b}\right)^5$

_____ **9.** $(3x)^{-2}$ _____ **10.** $(2x)^{-4}$

_____ **11.** $(-5)^{-3}$ _____ **12.** $(-4)^{-2}$

_____ **13.** $\left(\dfrac{4}{5}\right)^{-2}$ _____ **14.** $\left(\dfrac{2}{3}\right)^{-3}$

_____ **15.** $(6x)^{-2}$ _____ **16.** $(2x)^{-3}$

_____ **17.** $6x^{-2}$ _____ **18.** $2x^{-3}$

_____ **19.** $4y^7 \cdot y^{-7}$ _____ **20.** $5y^9y^{-9}$

_____ **21.** $a^8 \cdot a^{-11}$ _____ **22.** $a^{-9} \cdot a^6$

_____ **23.** $(2x^{-3})(-5x^7)$ _____ **24.** $(4x^{-2})(-7x^5)$

_____ **25.** $(-5x^{-1})(3x^{-2})$ _____ **26.** $(-8x^{-2})(5x^{-3})$

_____ **27.** $(a^{-6})^2$ _____ **28.** $(a^{-7})^3$

_____ **29.** $(y^9)^{-1} \cdot (y^{-5})^{-2}$ _____ **30.** $(y^{-3})^{-2} \cdot (y^5)^{-1}$

_____ **31.** $9x^{-3}(x^{-7})^0$ _____ **32.** $8x^{-7}(x^{-11})^0$

_____ **33.** $(4x)^{-3}$ _____ **34.** $(9x)^{-2}$

_____ **35.** $(3a^{-4})^3$ _____ **36.** $(6a^{-5})^2$

_____ 37. $(2^3y^{-5})^{-2}$ _____ 38. $(3y^{-7})^{-3}$

_____ 39. $(5a^3b^{-5})^2$ _____ 40. $(6a^{-3}b^7)^2$

_____ 41. $(9x^{-5}y^{-3})^{-2}$ _____ 42. $(11x^{-4}y^{-5})^{-2}$

_____ 43. $(7^{-1}x^3)^{-2} \cdot (7^{-1}x^2)^3$ _____ 44. $(5^{-3}x^4)^{-1} \cdot (5^{-1}x^2)^2$

_____ 45. $\dfrac{x^{13}}{x^{-2}}$ _____ 46. $\dfrac{x^9}{x^{-3}}$

_____ 47. $\dfrac{9a^{-5}}{15a^5}$ _____ 48. $\dfrac{8a^{-6}}{12a^6}$

_____ 49. $\dfrac{7a^{-4}b^{-5}}{21a^3b^7}$ _____ 50. $\dfrac{8a^{-9}b^{-2}}{32a^6b^9}$

_____ 51. $\dfrac{(-4x^{-3})(3x^{-7})}{6x^{-12}}$ _____ 52. $\dfrac{(-9x^{-7})(-2x^{-4})}{-6x^{-8}}$

_____ 53. $\dfrac{(3x^2)^{-2} \cdot (x^5)^{-1}}{(9x^8)^{-1}}$ _____ 54. $\dfrac{(2x^3)^{-3} \cdot (x^4)^{-1}}{(8x^{12})^{-1}}$

_____ 55. $\left(\dfrac{2x}{3}\right)^{-3}$ _____ 56. $\left(\dfrac{3x}{4}\right)^{-2}$

_____ 57. $\left(\dfrac{a^{-1}}{b^2}\right)^{-2}$ _____ 58. $\left(\dfrac{a^{-2}}{b^3}\right)^{-1}$

_____ 59. $\left(\dfrac{x^{-4}}{y^{-2}}\right)^{-3}$ _____ 60. $\left(\dfrac{x^{-5}}{y^{-3}}\right)^{-2}$

_____ 61. $\left(\dfrac{2a^5b^{-2}}{-10a^4b^2}\right)^{-2}$ _____ 62. $\left(\dfrac{3a^{10}b^{-1}}{-9a^8b}\right)^{-2}$

_____ 63. $\left(\dfrac{16x^{-5}y^0}{4x^{-5}y^{-3}}\right)^{-2}$ _____ 64. $\left(\dfrac{15x^0y^{-7}}{5x^{-2}y^{-7}}\right)^{-3}$

CHECKUP 12.1

Simplify and give answers with positive exponents.

_____ **1.** $(2x)^{-5}$

_____ **2.** $a^7 \cdot a^{-3}$

_____ **3.** $3y^{-2}$

_____ **4.** $2x^{-8}(x^{-4})^{-2}$

_____ **5.** $\dfrac{-12x^5}{4x^{-2}}$

_____ **6.** $\dfrac{9a^{-3}}{18a^{-5}}$

_____ **7.** $\left(\dfrac{7x}{5}\right)^{-2}$

_____ **8.** $\left(\dfrac{3x^{-2}}{4y^{-4}}\right)^{-1}$

_____ **9.** $\dfrac{(-2x^{-4})(15x^6)}{6x^{-2}}$

_____ **10.** $\dfrac{9x^{-4}y^{-5}}{3x^4y^{-5}}$

12.2   Working with Square Roots

We have talked often about squares of numbers and variables in our work with polynomials and factoring. Recall that we learned to recognize the "difference of two squares" in Chapter 5. In order to factor expressions such as

$$x^2 - 9$$
$$y^2 - 16$$
$$a^2 - 1$$

we asked ourselves whether each of these terms was the square of some variable or some number. We agreed that

x^2 was a square	because	$x^2 = x \cdot x$
9 was a square	because	$9 = 3 \cdot 3$
y^2 was a square	because	$y^2 = y \cdot y$
16 was a square	because	$16 = 4 \cdot 4$
a^2 was a square	because	$a^2 = a \cdot a$
1 was a square	because	$1 = 1 \cdot 1$

12.2-1   Finding Square Roots

In determining whether a certain number was a square, we searched for a number that could be multiplied times itself to give the original number. Such a number is called the **square root** of the original number; so we may say that

the square root of 9 is 3 because $3^2 = 9$

the square root of 16 is 4 because $4^2 = 16$

the square root of 1 is 1 because $1^2 = 1$

You may be thinking that 3 is not the only square root of 9 because you know that

$$(-3)(-3) = 9$$
$$(-3)^2 = 9$$

and indeed you are correct. To avoid confusion, mathematicians have agreed that

> The **principal square root** of a positive number is always the *positive* square root.

Let's practice finding some principal square roots. From now on, we shall use the phrase square root to mean the principal (positive) square root.

What is the square root of 4? The square root of 4 is 2 because $2^2 = 2 \cdot 2 = 4$.

What is the square root of 0? The square root of 0 is 0 because $0^2 = 0 \cdot 0 = 0$. We note here that

> The square root of zero is zero.

What is the square root of -9? We must find a number that will multiply times itself to give an answer of -9. Will 3 work? No, because $3 \cdot 3 = 9$. Will -3 work? No, because $(-3)(-3) = +9$. Indeed, there is no possible square root for -9.

As a matter of fact, *using the numbers with which we are familiar so far,*

> It is impossible to find the square root of a negative number.

Mathematicians have invented a symbol to indicate that we are to find the principal square root of a number. That symbol is called a **radical** and it looks like this: $\sqrt{}$. The expression $\sqrt{9}$ means "the principal square root of 9." We may write

$$\sqrt{9} = 3 \qquad \text{because} \qquad 3^2 = 9$$
$$\sqrt{25} = 5 \qquad \text{because} \qquad 5^2 = 25$$
$$\sqrt{81} = 9 \qquad \text{because} \qquad 9^2 = 81$$
$$\sqrt{1} = 1 \qquad \text{because} \qquad 1^2 = 1$$
$$\sqrt{0} = 0 \qquad \text{because} \qquad 0^2 = 0$$

In general, we agree that if a is nonnegative and b is nonnegative, then to say that

> $$\sqrt{a} = b \qquad \text{means} \qquad b^2 = a$$

In this radical statement, a is called the **radicand** and b is called the **root**.

Example 1. Find $\sqrt{36}$.

Solution. $\sqrt{36} = 6$ because $6^2 = 36$.

Example 2. Find $\sqrt{\dfrac{4}{9}}$.

Solution. $\sqrt{\dfrac{4}{9}} = \dfrac{2}{3}$ because $\left(\dfrac{2}{3}\right)^2 = \dfrac{2}{3} \cdot \dfrac{2}{3} = \dfrac{4}{9}$.

Example 3. Find $\sqrt{-16}$.

Solution. $\sqrt{-16}$. We know of no number whose square is -16.

It is also worthwhile to note what happens when we square a square root. Can you compute $(\sqrt{9})^2$?

$$(\sqrt{9})^2 = \sqrt{9} \cdot \sqrt{9} = 3 \cdot 3 = 9$$

Similarly,

$$(\sqrt{36})^2 = \sqrt{36} \cdot \sqrt{36} = 6 \cdot 6 = 36$$

In general, then, we should agree that

> $$(\sqrt{a})^2 = a \qquad \text{(provided that } a \text{ is nonnegative)}$$

and this rule allows us to compute squares of square roots without computing the square root first.

$$(\sqrt{81})^2 = 81$$

$$(\sqrt{1})^2 = 1$$
$$(\sqrt{5})^2 = 5$$
$$(\sqrt{71})^2 = 71$$

To express the square root of a variable expression, it is absolutely necessary that the variable expression represents a positive quantity or zero. Why? Because we have no way to find the square root of a negative quantity.

> Throughout this chapter, we assume that variable expressions under square root radicals do *not* represent negative quantities.

Once we have agreed on this very important condition, we may simplify some radicals containing variables. Recall our two rules for working with square roots:

> **1.** $\sqrt{a} = b$ means $b^2 = a$
>
> **2.** $(\sqrt{a})^2 = a$

Let us try simplifying some radical expressions.

$$\sqrt{x^2} = x \qquad \text{because} \qquad (x)^2 = x^2$$

$$\sqrt{y^4} = y^2 \qquad \text{because} \qquad (y^2)^2 = y^4$$

$$\sqrt{\frac{x^2}{4}} = \frac{x}{2} \qquad \text{because} \qquad \left(\frac{x}{2}\right)^2 = \frac{x}{2} \cdot \frac{x}{2} = \frac{x^2}{4}$$

$$(\sqrt{x^2})^2 = x^2$$
$$(\sqrt{y})^2 = y$$
$$(\sqrt{5x})^2 = 5x$$

So far, we have practiced finding square roots that were rational numbers. Recall that rational numbers are numbers that can be expressed as fractions of integers.

$$\sqrt{9} = 3 \qquad \text{is rational because} \qquad 3 = \frac{3}{1}$$

$$\sqrt{\frac{4}{9}} = \frac{2}{3} \qquad \text{is rational because} \qquad \frac{2}{3} = \frac{2}{3}$$

We do not have to look very far in our list of natural numbers (1, 2, 3, 4, . . .) before we find a number whose square root is *not* rational. For instance

$$\sqrt{2} \text{ is not rational}$$

You could hunt forever and never find a rational number that can be multiplied times itself to give an answer of 2. This does *not* mean that $\sqrt{2}$ does not exist. It merely means that $\sqrt{2}$ is not rational; $\sqrt{2}$ cannot be expressed as a fraction of integers.

A number that is *not* rational is called an **irrational** number. Its value can be approximated, but it cannot be expressed exactly as a fraction of integers.

$$\sqrt{2} \text{ is irrational}$$
$$\sqrt{3} \text{ is irrational}$$
$$\sqrt{4} \text{ is rational} \qquad \text{because} \qquad \sqrt{4} = 2$$
$$\sqrt{5} \text{ is irrational}$$

We may approximate irrational square roots using the symbol $\doteq$ to mean "is approximately equal to."

$$\sqrt{2} \doteq 1.414$$

$$\sqrt{3} \doteq 1.732$$

$$\sqrt{5} \doteq 2.236$$

However, you should realize that these values are not exact, since

$$(1.414)^2 = 1.999396 \neq 2$$

$$(1.732)^2 = 2.999824 \neq 3$$

$$(2.236)^2 = 4.999696 \neq 5$$

In order to represent *exact* values of square roots that are irrational numbers, we must leave them in radical form.

Let's look at some square roots and decide whether they represent rational or irrational numbers. We shall simplify the roots that are rational.

$$\sqrt{7} \text{ is irrational}$$

$$\sqrt{9} \text{ is rational;} \quad \sqrt{9} = 3$$

$$\sqrt{10} \text{ is irrational}$$

$$\sqrt{17} \text{ is irrational}$$

$$\sqrt{25} \text{ is rational;} \quad \sqrt{25} = 5$$

Now that we have discussed rational numbers *and* irrational numbers, we can define a new set of numbers formed by putting these two sets together. This set is called the set of **real numbers** and we shall be using real numbers throughout the remainder of this chapter. Any number that is rational or that is irrational belongs in the set of real numbers.

Notice that this set contains *all* the numbers that we have discussed in this book. All the number properties that we have used will continue to work for the set of real numbers.

Earlier, we mentioned that it was "impossible" to find the square root of a negative number. Another way to express this fact is to say that

---

The square root of a negative number is *not* a real number.

For example, to say that it is impossible to find the square root of -9 actually means that

$$\sqrt{-9} \text{ is not a real number}$$

In a future algebra course, you will learn that there is another set of numbers to which $\sqrt{-9}$ belongs. This set is called the set of **complex numbers**, but we will not be using those numbers in this course.

See if you agree which of the following numbers are real.

$$\sqrt{3} \text{ is real}$$

$$\sqrt{\frac{7}{4}} \text{ is real}$$

$$\sqrt{0} \text{ is real}$$

$$\sqrt{53} \text{ is real}$$

$$\sqrt{-53} \text{ is } not \text{ real}$$

Trial Run *Simplify the radical expressions.*

_____ **1.** $\sqrt{64}$

_____ **2.** $\sqrt{\dfrac{a^2}{9}}$

_____ **3.** $(\sqrt{25})^2$

Decide if the expressions represent rational or irrational numbers.

_____ **4.** $\sqrt{49}$

_____ **5.** $\sqrt{13}$

_____ **6.** $\sqrt{0}$

ANSWERS

1. 8. **2.** $\dfrac{a}{3}$. **3.** 25. **4.** Rational. **5.** Irrational. **6.** Rational.

12.2-2 Simplifying Radical Expressions

We know that we may simplify

$$\sqrt{36} = 6$$

but suppose that we approach this radical in another way. Notice that

$$\begin{aligned}
\sqrt{36} &= \sqrt{4 \cdot 9} \\
&= \sqrt{4} \cdot \sqrt{9} \\
&= 2 \cdot 3 \\
&= 6
\end{aligned}$$

and we have arrived at the same answer. In this second approach, we made use of the fact that the square root of a product is equal to the product of the square roots. Is this always true? Let's try another example.
We know

$$\sqrt{144} = 12 \qquad \text{because} \qquad 12^2 = 144$$

But

$$\begin{aligned}
\sqrt{144} &= \sqrt{9 \cdot 16} \\
&= \sqrt{9} \cdot \sqrt{16} \\
&= 3 \cdot 4 \\
\sqrt{144} &= 12
\end{aligned}$$

Although we have verified this rule for only two examples, it is indeed true that

---

The square root of a product is equal to the product of the square roots.

In other words, we may say that

$$\boxed{\sqrt{a \cdot b} = \sqrt{a} \cdot \sqrt{b} \qquad \text{(provided that } a \text{ and } b \text{ are not negative)}}$$

We shall make use of this rule in simplifying radical expressions. For instance, to simplify

$$\sqrt{4x^2}$$

we may rewrite our radical as

$$\sqrt{4x^2} = \sqrt{4 \cdot x^2} = \sqrt{4} \cdot \sqrt{x^2} = 2x$$

Similarly,

$$\begin{aligned}
\sqrt{9a^2b^4} &= \sqrt{9 \cdot a^2 \cdot b^2 \cdot b^2} \\
&= \sqrt{9} \cdot \sqrt{a^2} \cdot \sqrt{b^2} \cdot \sqrt{b^2} \\
&= 3 \cdot a \cdot b \cdot b \\
&= 3ab^2
\end{aligned}$$

Sometimes a radical will not disappear completely in the simplification process. The best that we can do is to try rewriting the radicand as a product of factors, some of which are perfect squares. The factors whose roots are irrational numbers must then be left under the radical, as the next example illustrates.

Simplify $\sqrt{12}$. Our aim is to rewrite 12 as a product of factors. However, we should be careful about the factors we choose, since we would like at least one of those factors to be a perfect square. Since $12 = 4 \cdot 3$ and 4 is a perfect square, we may write

$$\begin{aligned}
\sqrt{12} &= \sqrt{4 \cdot 3} \\
&= \sqrt{4} \cdot \sqrt{3} \\
&= 2 \cdot \sqrt{3} \\
&= 2\sqrt{3}
\end{aligned}$$

Since $\sqrt{3}$ is irrational, it cannot be simplified. We have now written $\sqrt{12}$ in simplest radical form as $2\sqrt{3}$.

Let's try writing $\sqrt{50x^3}$ in simplest radical form. Remember, we would like to rewrite the radicand $50x^3$ as a product of factors, some of which are perfect squares.

$$\begin{aligned}
\sqrt{50x^3} &= \sqrt{25 \cdot 2 \cdot x^2 \cdot x} \\
&= \sqrt{25} \cdot \sqrt{2} \cdot \sqrt{x^2} \cdot \sqrt{x} \\
&= 5 \cdot \sqrt{2} \cdot x \cdot \sqrt{x} \\
&= 5 \cdot x \cdot \sqrt{2} \cdot \sqrt{x} \\
&= 5x\sqrt{2x}
\end{aligned}$$

To simplify the radical $\sqrt{700x^5y^7}$ we must again rewrite the radicand as a product of factors, some of which are perfect squares.

$$\begin{aligned}
\sqrt{700x^5y^7} &= \sqrt{100 \cdot 7 \cdot x^2 \cdot x^2 \cdot x \cdot y^2 \cdot y^2 \cdot y^2 \cdot y} \\
&= \sqrt{100} \cdot \sqrt{7} \cdot \sqrt{x^2} \cdot \sqrt{x^2} \cdot \sqrt{x} \cdot \sqrt{y^2} \cdot \sqrt{y^2} \cdot \sqrt{y^2} \cdot \sqrt{y} \\
&= 10 \cdot \sqrt{7} \cdot x \cdot x \cdot \sqrt{x} \cdot y \cdot y \cdot y \cdot \sqrt{y} \\
&= 10 \cdot x \cdot x \cdot y \cdot y \cdot y \cdot \sqrt{7} \cdot \sqrt{x} \cdot \sqrt{y} \\
&= 10x^2y^3\sqrt{7xy}
\end{aligned}$$

A square root radical is said to be in **simplest form** when the radicand contains no perfect square factors. From the examples we have worked, we see that if the exponent on a variable factor is greater than 1, some simplification is necessary.

Example 4. Simplify $\sqrt{162x^3y^2}$.

Solution

$$\sqrt{162x^3y^2} = \sqrt{81 \cdot 2 \cdot x^2 \cdot x \cdot y^2}$$
$$= \sqrt{81} \cdot \sqrt{2} \cdot \sqrt{x^2} \cdot \sqrt{x} \cdot \sqrt{y^2}$$
$$= 9xy\sqrt{2x}$$

Example 5. Simplify $\sqrt{108ab^4}$.

Solution

$$\sqrt{108ab^4} = \sqrt{36 \cdot 3 \cdot a \cdot b^2 \cdot b^2}$$
$$= \sqrt{36} \cdot \sqrt{3} \cdot \sqrt{a} \cdot \sqrt{b^2} \cdot \sqrt{b^2}$$
$$= 6 \cdot b \cdot b \sqrt{3a}$$
$$= 6b^2\sqrt{3a}$$

We decided earlier that

$$\sqrt{\frac{4}{9}} = \frac{2}{3} \qquad \text{because} \qquad \left(\frac{2}{3}\right)^2 = \frac{4}{9}$$

Another approach to this problem might have been to consider the numerator and denominator separately. Let's see if we obtain the same square root when we consider

$$\sqrt{\frac{4}{9}} \quad \text{as} \quad \frac{\sqrt{4}}{\sqrt{9}}$$

We know $\sqrt{4} = 2$ and $\sqrt{9} = 3$, so

$$\frac{\sqrt{4}}{\sqrt{9}} = \frac{2}{3}$$

Therefore, this approach to the problem seems to give the same result and we can write

$$\sqrt{\frac{4}{9}} = \frac{\sqrt{4}}{\sqrt{9}} = \frac{2}{3}$$

It seems, then, that we may conclude that

> The square root of a quotient is the quotient of the square roots.

In other words, we may note in general that

> $$\sqrt{\frac{a}{b}} = \frac{\sqrt{a}}{\sqrt{b}} \qquad \text{(provided that } a \geq 0 \text{ and } b > 0\text{)}$$

This rule gives us another tool for simplifying radicals, as the following problems will illustrate.

To simplify $\sqrt{\dfrac{25x^2}{49y^2}}$ we may write

$$\sqrt{\frac{25x^2}{49y^2}} = \frac{\sqrt{25x^2}}{\sqrt{49y^2}}$$
$$= \frac{5x}{7y}$$

Similarly, we may simplify

$$\sqrt{\frac{32x^3}{y^4}} = \frac{\sqrt{32x^3}}{\sqrt{y^4}} = \frac{\sqrt{16 \cdot 2 \cdot x^2 \cdot x}}{\sqrt{y^2 \cdot y^2}}$$

$$= \frac{4x\sqrt{2x}}{y \cdot y}$$

$$= \frac{4x\sqrt{2x}}{y^2}$$

This rule for quotients can also be used in reverse. That is, we may sometimes wish to make use of the fact that

$$\frac{\sqrt{a}}{\sqrt{b}} = \sqrt{\frac{a}{b}}$$

Suppose, for instance, that we wish to simplify

$$\frac{\sqrt{50x^3}}{\sqrt{2x}}$$

Writing this quotient of radicals as a single radical, we have

$$\frac{\sqrt{50x^3}}{\sqrt{2x}} = \sqrt{\frac{50x^3}{2x}}$$

Now we may use our division rules to simplify the radicand before we take the square root.

$$\sqrt{\frac{50x^3}{2x}} = \sqrt{25x^2}$$

$$= 5x$$

Similar steps allow us to simplify the radical expression.

$$\frac{\sqrt{363x^5y}}{\sqrt{3xy^3}} = \sqrt{\frac{363x^5y}{3xy^3}}$$

$$= \sqrt{\frac{121x^4}{y^2}}$$

$$= \frac{\sqrt{121x^4}}{\sqrt{y^2}}$$

$$= \frac{11x^2}{y}$$

In general, we note that a quotient of two radicals should be written as a single radical whenever the numerator and denominator contain common factors. Once the numerator and denominator have been rid of common factors, the fraction should be considered as a quotient of two separate radicals.

Example 6. Simplify $\sqrt{\dfrac{9x^2}{16}}$.

Solution

$$\sqrt{\frac{9x^2}{16}} = \frac{\sqrt{9x^2}}{\sqrt{16}}$$

$$= \frac{3x}{4}$$

Example 7. Simplify $\sqrt{\dfrac{45a^3}{4b^2}}$.

Solution

$$\sqrt{\frac{45a^3}{4b^2}} = \frac{\sqrt{45a^3}}{\sqrt{4b^2}}$$

$$= \frac{\sqrt{9 \cdot 5 \cdot a^2 \cdot a}}{\sqrt{4 \cdot b^2}}$$

$$= \frac{3a\sqrt{5a}}{2b}$$

Example 8. Simplify $\sqrt{\dfrac{12a^3b^3}{600ab}}$.

Solution

$$\sqrt{\frac{12a^3b^3}{600ab}} = \sqrt{\frac{2a^2b^2}{100}}$$

$$= \frac{\sqrt{2a^2b^2}}{\sqrt{100}}$$

$$= \frac{ab\sqrt{2}}{10}$$

Example 9. Simplify $\dfrac{\sqrt{20x^3y^4}}{\sqrt{125xy}}$.

Solution

$$\frac{\sqrt{20x^3y^4}}{\sqrt{125xy}} = \sqrt{\frac{20x^3y^4}{125xy}}$$

$$= \sqrt{\frac{4x^2y^3}{25}}$$

$$= \frac{\sqrt{4x^2y^3}}{\sqrt{25}}$$

$$= \frac{\sqrt{4 \cdot x^2 \cdot y^2 \cdot y}}{\sqrt{25}}$$

$$= \frac{2xy\sqrt{y}}{5}$$

Example 10. Simplify $\dfrac{\sqrt{18a^7}}{\sqrt{6a^6}}$.

Solution

$$\frac{\sqrt{18a^7}}{\sqrt{6a^6}} = \sqrt{\frac{18a^7}{6a^6}}$$

$$= \sqrt{3a}$$

Trial Run   *Write the following radicals in simplest form.*

_____ 1. $\sqrt{16x^2}$

_____ 2. $\sqrt{18}$

_____ 3. $\sqrt{27a^3}$

_____ 4. $\sqrt{20x^3y^4}$

_____ 5. $\sqrt{\dfrac{9x^2}{49y^2}}$

_____ 6. $\sqrt{\dfrac{50x^3}{y^4}}$

_____ 7. $\dfrac{\sqrt{2x^5y}}{\sqrt{98xy}}$

_____ 8. $\dfrac{\sqrt{169a^4b^2}}{\sqrt{36c^2}}$

ANSWERS

1. $4x$.   **2.** $3\sqrt{2}$.   **3.** $3a\sqrt{3a}$.   **4.** $2xy^2\sqrt{5x}$.   **5.** $\dfrac{3x}{7y}$.   **6.** $\dfrac{5x\sqrt{2x}}{y^2}$.   **7.** $\dfrac{x^2}{7}$.

8. $\dfrac{13a^2b}{6c}$.

EXERCISE SET 12.2

Decide if the following expressions represent rational or irrational numbers.

_____ 1. $\sqrt{25}$

_____ 2. $\sqrt{49}$

_____ 3. $\sqrt{5}$

_____ 4. $\sqrt{3}$

_____ 5. $\sqrt{1}$

_____ 6. $\sqrt{0}$

_____ 7. $\sqrt{8}$

_____ 8. $\sqrt{12}$

_____ 9. $\sqrt{\dfrac{25}{36}}$

_____ 10. $\sqrt{\dfrac{1}{4}}$

_____ 11. $\sqrt{65}$

_____ 12. $\sqrt{35}$

_____ 13. $\sqrt{\dfrac{2}{3}}$

_____ 14. $\sqrt{\dfrac{5}{7}}$

_____ 15. $\sqrt{225}$

_____ 16. $\sqrt{196}$

Simplify the following radical expressions.

_____ 17. $\sqrt{144}$

_____ 18. $\sqrt{169}$

_____ 19. $\sqrt{a^4}$

_____ 20. $\sqrt{a^6}$

_____ 21. $(\sqrt{8})^2$

_____ 22. $(\sqrt{10})^2$

_____ 23. $(\sqrt{x^5})^2$

_____ 24. $(\sqrt{a^3})^2$

_____ 25. $\sqrt{25b^2}$

_____ 26. $\sqrt{64c^2}$

_____ 27. $\sqrt{\dfrac{x^2}{36}}$

_____ 28. $\sqrt{\dfrac{y^2}{49}}$

_____ 29. $\sqrt{45}$

_____ 30. $\sqrt{40}$

_____ 31. $\sqrt{600}$

_____ 32. $\sqrt{128}$

_____ 33. $\sqrt{75}$

_____ 34. $\sqrt{108}$

_____ 35. $\sqrt{44}$ _____ 36. $\sqrt{80}$

_____ 37. $\sqrt{x^2}$ _____ 38. $\sqrt{x^4}$

_____ 39. $\sqrt{y^5}$ _____ 40. $\sqrt{y^3}$

_____ 41. $\sqrt{x^2y^3}$ _____ 42. $\sqrt{x^3y^2}$

_____ 43. $\sqrt{32a^3}$ _____ 44. $\sqrt{18b^3}$

_____ 45. $\sqrt{4x^2y}$ _____ 46. $\sqrt{9xy^2}$

_____ 47. $\sqrt{81a^5b^4}$ _____ 48. $\sqrt{100a^3b^6}$

_____ 49. $\sqrt{192x^2y^4z^3}$ _____ 50. $\sqrt{243x^4y^5z^2}$

_____ 51. $\sqrt{180a^7b^3c}$ _____ 52. $\sqrt{98a^5b^4c}$

_____ 53. $\sqrt{\dfrac{4x^2}{25}}$ _____ 54. $\sqrt{\dfrac{9x^2}{121}}$

_____ 55. $\sqrt{\dfrac{32a^3}{b^4}}$ _____ 56. $\sqrt{\dfrac{75a}{b^2}}$

_____ 57. $\sqrt{\dfrac{40x^3}{9y^2}}$ _____ 58. $\sqrt{\dfrac{27x^5}{4y^2}}$

_____ 59. $\sqrt{\dfrac{30a^3b^3}{500ab}}$ _____ 60. $\sqrt{\dfrac{35a^5b^5}{28a^3b^3}}$

_____ 61. $\dfrac{\sqrt{75x^3}}{\sqrt{3x}}$ _____ 62. $\dfrac{\sqrt{80x^5}}{\sqrt{5x}}$

_____ 63. $\dfrac{\sqrt{15x^7}}{\sqrt{5x^3}}$ _____ 64. $\dfrac{\sqrt{56x^5}}{\sqrt{7x^4}}$

_____ 65. $\dfrac{\sqrt{3xy^3}}{\sqrt{363xy^2}}$ _____ 66. $\dfrac{\sqrt{5xy^2}}{\sqrt{720y}}$

CHECKUP 12.2

Simplify the following radical expressions.

_____ 1. $\sqrt{49}$

_____ 2. $\sqrt{9a^2}$

_____ 3. $(\sqrt{12})^2$

_____ 4. $\sqrt{80}$

_____ 5. $\sqrt{75y^4}$

_____ 6. $\sqrt{45a^3b^2}$

_____ 7. $\sqrt{\dfrac{25x^2}{36y^2}}$

_____ 8. $\sqrt{\dfrac{32x^5}{y^6}}$

_____ 9. $\dfrac{\sqrt{50a^5b^3}}{\sqrt{2ab}}$

_____ 10. $\dfrac{\sqrt{15a^5}}{\sqrt{5a}}$

12.3 Operating with Radical Expressions

So far, we have spent time recognizing radical expressions and learning how to simplify them. As with any kind of algebraic expressions, we must now discuss the basic operations of addition, subtraction, and multiplication.

12.3-1 Adding and Subtracting Radical Expressions

First we turn our attention to learning how to compute sums and differences of radical expressions such as

$$2\sqrt{3} + 5\sqrt{3}$$
$$17\sqrt{x} - 12\sqrt{x} + 4\sqrt{x}$$
$$\sqrt{18a} - \sqrt{8a} + \sqrt{2a}$$
$$\sqrt{\frac{75x}{9}} + \sqrt{300x} - 2\sqrt{48x}$$

Our first problem seems simple enough.

$$2\sqrt{3} + 5\sqrt{3}$$

Here we are being asked to add 2 square roots of 3 and 5 square roots of 3. From our knowledge of combining like terms, it seems logical to conclude that the sum will be 7 square roots of 3.

$$2\sqrt{3} + 5\sqrt{3} = 7\sqrt{3}$$

The distributive property can be used to verify that this result is reasonable.

$$2\sqrt{3} + 5\sqrt{3} = (2 + 5)\sqrt{3}$$
$$= 7\sqrt{3}$$

Let's try the same approach to our second problem.

$$17\sqrt{x} - 12\sqrt{x} + 4\sqrt{x} = (17 - 12 + 4)\sqrt{x}$$
$$= 9\sqrt{x}$$

The key fact to realize here is that in both our problems, the *radicals matched exactly*. As in our work with combining like terms, the rule for combining radical expressions demands that the radicals be *exactly alike*. If they are exactly alike, we combine **like radicals** by combining their numerical coefficients and keeping the radical part.

> To combine like radicals, we combine the numerical coefficients and keep the radical part.

Let's take a look at the third problem.

$$\sqrt{18a} - \sqrt{8a} + \sqrt{2a}$$

Notice that the radicals are all square roots, but the radicands are all *different*. Before we decide that these radicals cannot be combined, we must try to simplify each radical completely.

$$\sqrt{18a} = \sqrt{9 \cdot 2 \cdot a} = 3\sqrt{2a}$$
$$\sqrt{8a} = \sqrt{4 \cdot 2 \cdot a} = 2\sqrt{2a}$$
$$\sqrt{2a} = 1\sqrt{2a}$$

Now we may rewrite our problem as follows:

$$\sqrt{18a} - \sqrt{8a} + \sqrt{2a} = 3\sqrt{2a} - 2\sqrt{2a} + 1\sqrt{2a}$$

and our radicals do indeed match; so

$$3\sqrt{2a} - 2\sqrt{2a} + 1\sqrt{2a} = (3 - 2 + 1)\sqrt{2a}$$
$$= 2\sqrt{2a}$$

Remembering to simplify all radicals before trying to combine, let's tackle the last problem.

$$\sqrt{\frac{75x}{9}} + \sqrt{300x} - 2\sqrt{48x}$$

First, we must try to simplify each radical expression.

$$\sqrt{\frac{75x}{9}} = \frac{\sqrt{75x}}{\sqrt{9}} = \frac{\sqrt{25 \cdot 3 \cdot x}}{\sqrt{9}} = \frac{5\sqrt{3x}}{3}$$
$$\sqrt{300x} = \sqrt{100 \cdot 3 \cdot x} = 10\sqrt{3x}$$
$$2\sqrt{48x} = 2\sqrt{16 \cdot 3 \cdot x} = 2 \cdot 4\sqrt{3x} = 8\sqrt{3x}$$

So

$$\sqrt{\frac{75x}{9}} + \sqrt{300x} - 2\sqrt{48x} = \frac{5\sqrt{3x}}{3} + 10\sqrt{3x} - 8\sqrt{3x}$$

and we must use a common denominator of 3 to combine these terms.

$$\frac{5\sqrt{3x}}{3} + 10\sqrt{3x} - 8\sqrt{3x} = \frac{5\sqrt{3x}}{3} + \frac{3(10\sqrt{3x})}{3} - \frac{3(8\sqrt{3x})}{3}$$
$$= \frac{5\sqrt{3x}}{3} + \frac{30\sqrt{3x}}{3} - \frac{24\sqrt{3x}}{3}$$
$$= \frac{5\sqrt{3x} + 30\sqrt{3x} - 24\sqrt{3x}}{3}$$
$$= \frac{(5 + 30 - 24)\sqrt{3x}}{3}$$
$$= \frac{11\sqrt{3x}}{3}$$

Let's emphasize again the similarity between combining like radicals and combining like terms. In order to combine radicals, their radicands must match exactly. We shall work one more problem in detail.

$$5x\sqrt{7x} + 3\sqrt{28x^3} - 4\sqrt{7y}$$
$$= 5x\sqrt{7x} + 3\sqrt{4 \cdot 7 \cdot x^2 \cdot x} - 4\sqrt{7y}$$
$$= 5x\sqrt{7x} + 3 \cdot 2 \cdot x\sqrt{7x} - 4\sqrt{7y}$$
$$= 5x\sqrt{7x} + 6x\sqrt{7x} - 4\sqrt{7y}$$
$$= 11x\sqrt{7x} - 4\sqrt{7y}$$

There is no way to combine these last two terms since their radical parts are different. The final answer must be written

$$11x\sqrt{7x} - 4\sqrt{7y}$$

Example 1. Simplify $3\sqrt{7} + 6\sqrt{7} - 10\sqrt{7}$.

Solution

$$3\sqrt{7} + 6\sqrt{7} - 10\sqrt{7} = (3 + 6 - 10)\sqrt{7}$$
$$= -1\sqrt{7}$$
$$= -\sqrt{7}$$

Example 2. Simplify $\sqrt{50a} - 6\sqrt{32a}$.

Solution

$$
\begin{aligned}
\sqrt{50a} - 6\sqrt{32a} &= \sqrt{25 \cdot 2 \cdot a} - 6\sqrt{16 \cdot 2 \cdot a} \\
&= 5\sqrt{2a} - 6 \cdot 4\sqrt{2a} \\
&= 5\sqrt{2a} - 24\sqrt{2a} \\
&= (5 - 24)\sqrt{2a} \\
&= -19\sqrt{2a}
\end{aligned}
$$

Example 3. Simplify $6 + \sqrt{27} + \sqrt{48}$.

Solution

$$
\begin{aligned}
6 + \sqrt{27} + \sqrt{48} &= 6 + \sqrt{9 \cdot 3} + \sqrt{16 \cdot 3} \\
&= 6 + 3\sqrt{3} + 4\sqrt{3} \\
&= 6 + (3 + 4)\sqrt{3} \\
&= 6 + 7\sqrt{3}
\end{aligned}
$$

Example 4. Simplify $\sqrt{\dfrac{75}{3}} + \sqrt{500} - \sqrt{125}$.

Solution

$$
\begin{aligned}
\sqrt{\frac{75}{3}} + \sqrt{500} - \sqrt{125} &= \sqrt{\frac{75}{3}} + \sqrt{100 \cdot 5} - \sqrt{25 \cdot 5} \\
&= \sqrt{25} + 10\sqrt{5} - 5\sqrt{5} \\
&= 5 + (10 - 5)\sqrt{5} \\
&= 5 + 5\sqrt{5}
\end{aligned}
$$

Example 5. Simplify $\dfrac{\sqrt{3x}}{\sqrt{75}} - 4\sqrt{x} + 5\sqrt{9x}$.

Solution

$$
\begin{aligned}
\frac{\sqrt{3x}}{\sqrt{75}} - 4\sqrt{x} + 5\sqrt{9x} &= \sqrt{\frac{3x}{75}} - 4\sqrt{x} + 5\sqrt{9x} \\
&= \sqrt{\frac{x}{25}} - 4\sqrt{x} + 5 \cdot 3\sqrt{x} \\
&= \frac{\sqrt{x}}{5} - 4\sqrt{x} + 15\sqrt{x} \\
&= \frac{\sqrt{x}}{5} - \frac{5(4\sqrt{x})}{5} + \frac{5(15\sqrt{x})}{5} \\
&= \frac{\sqrt{x}}{5} - \frac{20\sqrt{x}}{5} + \frac{75\sqrt{x}}{5} \\
&= \frac{\sqrt{x} - 20\sqrt{x} + 75\sqrt{x}}{5} \\
&= \frac{(1 - 20 + 75)\sqrt{x}}{5} \\
&= \frac{56\sqrt{x}}{5}
\end{aligned}
$$

Trial Run *Simplify.*

_____ 1. $2\sqrt{3} + 5\sqrt{3} - 9\sqrt{3}$

_____ 2. $8\sqrt{x} - 7\sqrt{x} + \sqrt{x}$

_____ 3. $\sqrt{32a} - 5\sqrt{18a}$

_____ 4. $11 + \sqrt{75} - \sqrt{12}$

_____ 5. $\dfrac{\sqrt{8}}{\sqrt{2}} + \sqrt{50} - \sqrt{72}$

_____ 6. $\dfrac{\sqrt{4x}}{\sqrt{36}} - 5\sqrt{x} + \sqrt{16x}$

ANSWERS

1. $-2\sqrt{3}$.   2. $2\sqrt{x}$.   3. $-11\sqrt{2a}$.   4. $11 + 3\sqrt{3}$.   5. $2 - \sqrt{2}$.   6. $\dfrac{-2\sqrt{x}}{3}$.

12.3-2 Multiplying Radical Expressions

The methods used for multiplying radical expressions are similar to those used for multiplying polynomials. As you remember, multiplying polynomials required use of the associative and distributive properties and the rules for combining like terms.

We discussed multiplying radicals in our work with simplifying radicals, but a few examples are included here for review.

Example 6. Multiply $\sqrt{2} \cdot \sqrt{5} \cdot \sqrt{10}$.

Solution. $\sqrt{2} \cdot \sqrt{5} \cdot \sqrt{10} = \sqrt{2 \cdot 5 \cdot 10} = \sqrt{100} = 10$.

Example 7. Multiply $\sqrt{3a} \cdot \sqrt{3a}$.

Solution. $\sqrt{3a} \cdot \sqrt{3a} = \sqrt{3a \cdot 3a} = \sqrt{(3a)^2} = 3a$.

Example 8. Multiply $\sqrt{x + 1} \cdot \sqrt{x + 1}$.

Solution. $\sqrt{x + 1} \cdot \sqrt{x + 1} = \sqrt{(x + 1)^2} = x + 1$.

Examples 7 and 8 illustrate a fact that shall be very useful to us shortly.

$$\sqrt{a} \cdot \sqrt{a} = \sqrt{a^2} = a \qquad \text{(provided that } a \text{ is not negative)}$$

Example 9. Multiply $\sqrt{2x - 1} \cdot \sqrt{2x - 1}$.

Solution. $\sqrt{2x - 1} \cdot \sqrt{2x - 1} = \sqrt{(2x - 1)^2} = 2x - 1$.

Example 10. Multiply $5\sqrt{2a} \cdot \sqrt{8a}$.

Solution

$$5\sqrt{2a}\sqrt{8a} = 5\sqrt{2a \cdot 8a}$$
$$= 5\sqrt{16a^2}$$
$$= 5 \cdot 4 \cdot a = 20a$$

Example 11. Multiply $(-3\sqrt{x})(-6\sqrt{x})$.

Solution

$$(-3\sqrt{x})(-6\sqrt{x}) = (-3)(-6)(\sqrt{x})(\sqrt{x})$$
$$= 18\sqrt{x^2}$$
$$= 18x$$

To multiply a single radical times a sum, we must use the distributive property and then simplify. For instance,

Example 12. Multiply $\sqrt{3}(5 + \sqrt{3})$.

Solution

$$\sqrt{3}(5 + \sqrt{3}) = 5 \cdot \sqrt{3} + \sqrt{3} \cdot \sqrt{3}$$
$$= 5\sqrt{3} + \sqrt{3^2}$$
$$= 5\sqrt{3} + 3$$

Example 13. Multiply $2\sqrt{2}(\sqrt{3} + \sqrt{8})$.

Solution

$$2\sqrt{2}(\sqrt{3} + \sqrt{8}) = 2\sqrt{2} \cdot \sqrt{3} + 2\sqrt{2} \cdot \sqrt{8}$$
$$= 2\sqrt{2 \cdot 3} + 2\sqrt{2 \cdot 8}$$
$$= 2\sqrt{6} + 2\sqrt{16}$$
$$= 2\sqrt{6} + 2 \cdot 4$$
$$= 2\sqrt{6} + 8$$

Example 14. Multiply $5(\sqrt{x} - 2\sqrt{3} + 4)$.

Solution

$$5(\sqrt{x} - 2\sqrt{3} + 4) = 5\sqrt{x} - 5 \cdot 2\sqrt{3} + 20$$
$$= 5\sqrt{x} - 10\sqrt{3} + 20$$

Example 15. Simplify $\sqrt{x}(\sqrt{x} + 3) - 6(\sqrt{x} + 3)$.

Solution

$$\sqrt{x}(\sqrt{x} + 3) - 6(\sqrt{x} + 3) = \sqrt{x} \cdot \sqrt{x} + 3 \cdot \sqrt{x} - 6 \cdot \sqrt{x} - 6 \cdot 3$$
$$= x + 3\sqrt{x} - 6\sqrt{x} - 18$$
$$= x + (3 - 6)\sqrt{x} - 18$$
$$= x - 3\sqrt{x} - 18$$

Example 16. Simplify $\sqrt{3}(\sqrt{3} - 5) + 5(\sqrt{3} - 5)$.

Solution

$$\sqrt{3}(\sqrt{3} - 5) + 5(\sqrt{3} - 5) = \sqrt{3} \cdot \sqrt{3} - 5 \cdot \sqrt{3} + 5 \cdot \sqrt{3} - 5 \cdot 5$$
$$= 3 - 5\sqrt{3} + 5\sqrt{3} - 25$$
$$= 3 - 25 = -22$$

Trial Run *Multiply and give answers in simplest form.*

_____ 1. $\sqrt{2} \cdot \sqrt{5}$

_____ 2. $\sqrt{5a} \cdot \sqrt{5a}$

_____ **3.** $4\sqrt{3a} \cdot \sqrt{12a}$

_____ **4.** $(-2\sqrt{x})(5\sqrt{3x})$

_____ **5.** $\sqrt{2}(3 - \sqrt{2})$

_____ **6.** $3\sqrt{2}(\sqrt{5} + \sqrt{18})$

_____ **7.** $\sqrt{x}(\sqrt{x} - 3)$

_____ **8.** $\sqrt{x}(\sqrt{x} - 4) + 4(\sqrt{x} - 4)$

ANSWERS
1. $\sqrt{10}$.　**2.** $5a$.　**3.** $24a$.　**4.** $-10x\sqrt{3}$.　**5.** $3\sqrt{2} - 2$.　**6.** $3\sqrt{10} + 18$.
7. $x - 3\sqrt{x}$.　**8.** $x - 16$.

Let's spend a moment reviewing the multiplication of binomials to prepare ourselves for multiplying sums and differences of radicals. Recall how we used the FOIL method to find products of binomials.

$$(x + 3)(x + 7) = x^2 + 7x + 3x + 21$$

$$= x^2 + 10x + 21$$

$$(x - 6)(x + 6) = x^2 + 6x - 6x - 36$$

$$= x^2 - 36$$

We can use the same method to multiply sums and/or differences of radicals. Let's try $(2 + \sqrt{3})(4 + \sqrt{2})$.

$$(2 + \sqrt{3})(4 + \sqrt{2}) = 2 \cdot 4 + 2 \cdot \sqrt{2} + 4 \cdot \sqrt{3} + \sqrt{3} \cdot \sqrt{2}$$

$$= 8 + 2\sqrt{2} + 4\sqrt{3} + \sqrt{6}$$

Since there are no like radicals in our answer, we cannot combine any terms. Let's look at another product. Multiply $(5 + \sqrt{3})(7 - 2\sqrt{3})$.

$$(5 + \sqrt{3})(7 - 2\sqrt{3}) = 5 \cdot 7 - 10\sqrt{3} + 7\sqrt{3} - 2\sqrt{3} \cdot \sqrt{3}$$

$$= 35 - 10\sqrt{3} + 7\sqrt{3} - 2 \cdot 3$$
$$= 35 - 10\sqrt{3} + 7\sqrt{3} - 6$$
$$= 35 - 6 - 10\sqrt{3} + 7\sqrt{3}$$
$$= 29 - 3\sqrt{3}$$

Notice that in this example we were able to combine like terms and simplify our answer. Let's try another and multiply $(\sqrt{2} - \sqrt{3})(5\sqrt{2} - \sqrt{3})$.

$$(\sqrt{2} - \sqrt{3})(5\sqrt{2} - \sqrt{3}) = 5\sqrt{2} \cdot \sqrt{2} - \sqrt{2} \cdot \sqrt{3} - 5\sqrt{3} \cdot \sqrt{2} + \sqrt{3} \cdot \sqrt{3}$$

$$= 5 \cdot 2 - \sqrt{6} - 5\sqrt{6} + 3$$
$$= 10 - \sqrt{6} - 5\sqrt{6} + 3$$
$$= 10 + 3 - \sqrt{6} - 5\sqrt{6}$$
$$= 13 - 6\sqrt{6}$$

And one more product, $(\sqrt{5} + 2\sqrt{7})(\sqrt{5} - 2\sqrt{7})$.

$$(\sqrt{5} + 2\sqrt{7})(\sqrt{5} - 2\sqrt{7}) = \sqrt{5} \cdot \sqrt{5} - 2\sqrt{5} \cdot \sqrt{7} + 2\sqrt{5} \cdot \sqrt{7} - 4\sqrt{7} \cdot \sqrt{7}$$

$$= 5 - 2\sqrt{35} + 2\sqrt{35} - 4 \cdot 7$$
$$= 5 - 2\sqrt{35} + 2\sqrt{35} - 28$$
$$= 5 - 28 - 2\sqrt{35} + 2\sqrt{35}$$
$$= -23 + 0$$
$$= -23$$

This example is especially interesting because we multiplied two irrational quantities together and our product was a *rational* number. Notice that our original quantities were the *sum and difference of the same two terms.* Let's look at another such product.

Multiply $(\sqrt{3} + \sqrt{2})(\sqrt{3} - \sqrt{2})$.

$$(\sqrt{3} + \sqrt{2})(\sqrt{3} - \sqrt{2}) = \sqrt{3} \cdot \sqrt{3} - \sqrt{3} \cdot \sqrt{2} + \sqrt{2} \cdot \sqrt{3} - \sqrt{2} \cdot \sqrt{2}$$

$$= 3 - \sqrt{6} + \sqrt{6} - 2$$
$$= 3 - 2 - \sqrt{6} + \sqrt{6}$$
$$= 1$$

Notice again that our answer was rational; there were no radicals left after we combined like terms. This special product will be important to us later in the chapter. For now, just be sure that you can multiply sums and/or differences of radicals.

Example 17. Multiply $(2 - \sqrt{3})(\sqrt{5} - \sqrt{3})$.

Solution

$$(2 - \sqrt{3})(\sqrt{5} - \sqrt{3}) = 2 \cdot \sqrt{5} - 2 \cdot \sqrt{3} - \sqrt{3} \cdot \sqrt{5} + \sqrt{3} \cdot \sqrt{3}$$
$$= 2\sqrt{5} - 2\sqrt{3} - \sqrt{15} + 3$$

Example 18. Multiply $(7\sqrt{2} - \sqrt{11})(3\sqrt{2} + 2\sqrt{11})$.

Solution

$$(7\sqrt{2} - \sqrt{11})(3\sqrt{2} + 2\sqrt{11}) = 21 \cdot \sqrt{2}\sqrt{2} + 14 \cdot \sqrt{2} \cdot \sqrt{11} - 3 \cdot \sqrt{11} \cdot \sqrt{2}$$
$$-2 \cdot \sqrt{11} \cdot \sqrt{11}$$

$$= 21 \cdot 2 + 14\sqrt{22} - 3\sqrt{22} - 2 \cdot 11$$
$$= 42 + 14\sqrt{22} - 3\sqrt{22} - 22$$
$$= 20 + 11\sqrt{22}$$

Example 19. Multiply $(6 - \sqrt{17})(6 + \sqrt{17})$.

Solution

$$\begin{aligned}(6 - \sqrt{17})(6 + \sqrt{17}) &= 6 \cdot 6 + 6 \cdot \sqrt{17} - 6 \cdot \sqrt{17} - \sqrt{17} \cdot \sqrt{17}\\ &= 36 + 6\sqrt{17} - 6\sqrt{17} - 17\\ &= 36 - 17\\ &= 19\end{aligned}$$

Example 20. Multiply $(\sqrt{x} + 2)(\sqrt{x} - 4)$.

Solution

$$\begin{aligned}(\sqrt{x} + 2)(\sqrt{x} - 4) &= \sqrt{x} \cdot \sqrt{x} - 4 \cdot \sqrt{x} + 2 \cdot \sqrt{x} - 2 \cdot 4\\ &= \sqrt{x^2} - 4\sqrt{x} + 2\sqrt{x} - 8\\ &= x - 2\sqrt{x} - 8\end{aligned}$$

Example 21. Multiply $(5 + \sqrt{x})(3 + 4\sqrt{x})$.

Solution

$$\begin{aligned}(5 + \sqrt{x})(3 + 4\sqrt{x}) &= 5 \cdot 3 + 5 \cdot 4\sqrt{x} + 3\sqrt{x} + 4\sqrt{x} \cdot \sqrt{x}\\ &= 15 + 20\sqrt{x} + 3\sqrt{x} + 4\sqrt{x^2}\\ &= 15 + 23\sqrt{x} + 4x\end{aligned}$$

Example 22. Multiply $(\sqrt{y} + 7)(\sqrt{y} - 7)$.

Solution

$$\begin{aligned}(\sqrt{y} + 7)(\sqrt{y} - 7) &= \sqrt{y} \cdot \sqrt{y} - 7 \cdot \sqrt{y} + 7 \cdot \sqrt{y} - 7 \cdot 7\\ &= \sqrt{y^2} - 7\sqrt{y} + 7\sqrt{y} - 49\\ &= y - 49\end{aligned}$$

Trial Run *Multiply.*

_____ **1.** $(1 + \sqrt{2})(2 - \sqrt{3})$

_____ **2.** $(5 - 2\sqrt{5})(2 + \sqrt{5})$

_____ **3.** $(2 - 3\sqrt{7})(2 + 3\sqrt{7})$

_____ **4.** $(\sqrt{5} - 3)(\sqrt{5} + 3)$

_____ **5.** $(4\sqrt{2} - \sqrt{3})(5\sqrt{2} - 2\sqrt{3})$

_____ **6.** $(\sqrt{x} - 5)(\sqrt{x} - 5)$

ANSWERS
1. $2 - \sqrt{3} + 2\sqrt{2} - \sqrt{6}$.   **2.** $\sqrt{5}$.   **3.** -59.   **4.** -4.   **5.** $46 - 13\sqrt{6}$.
6. $x - 10\sqrt{x} + 25$.

EXERCISE SET 12.3

Simplify each expression.

_____ 1. $2\sqrt{3} + \sqrt{3}$

_____ 2. $5\sqrt{6} - \sqrt{6}$

_____ 3. $5\sqrt{x} - 2\sqrt{x} + 8\sqrt{x}$

_____ 4. $9\sqrt{x} - 5\sqrt{x} - 6\sqrt{x}$

_____ 5. $\sqrt{27} - \sqrt{300}$

_____ 6. $\sqrt{32} - \sqrt{98}$

_____ 7. $\sqrt{54a} + \sqrt{24a}$

_____ 8. $\sqrt{125a} + \sqrt{80a}$

_____ 9. $\sqrt{81} - \sqrt{12} - \sqrt{48}$

_____ 10. $\sqrt{49} - \sqrt{45} - \sqrt{180}$

_____ 11. $4\sqrt{12x} + 3\sqrt{75x}$

_____ 12. $5\sqrt{72x} + 3\sqrt{18x}$

_____ 13. $5\sqrt{28} - 4\sqrt{50}$

_____ 14. $4\sqrt{27} - 3\sqrt{20}$

_____ 15. $\dfrac{\sqrt{288}}{\sqrt{4}} + 3\sqrt{50} - \sqrt{200}$

_____ 16. $\dfrac{\sqrt{360}}{\sqrt{8}} - 2\sqrt{405} + \sqrt{80}$

_____ 17. $\sqrt{75x^2} - \sqrt{300x^2}$

_____ 18. $\sqrt{72x^2} - \sqrt{32x^2}$

_____ 19. $\dfrac{\sqrt{9a}}{\sqrt{25}} - 5\sqrt{a} + \sqrt{4a}$

_____ 20. $\dfrac{\sqrt{16a}}{\sqrt{9}} - 2\sqrt{a} + \sqrt{49a}$

_____ 21. $\sqrt{3} \cdot \sqrt{7}$

_____ 22. $\sqrt{5} \cdot \sqrt{11}$

_____ 23. $\sqrt{2} \cdot \sqrt{12} \cdot \sqrt{6}$

_____ 24. $\sqrt{4} \cdot \sqrt{8} \cdot \sqrt{2}$

_____ 25. $\sqrt{11a} \cdot \sqrt{11a}$

_____ 26. $\sqrt{17a} \cdot \sqrt{17a}$

_____ 27. $\sqrt{x-1} \cdot \sqrt{x-1}$

_____ 28. $\sqrt{x+5} \cdot \sqrt{x+5}$

_____ 29. $\sqrt{15a} \cdot \sqrt{3a}$

_____ 30. $\sqrt{14a} \cdot \sqrt{2a}$

_____ 31. $3\sqrt{50} \cdot \sqrt{2}$

_____ 32. $5\sqrt{27} \cdot \sqrt{3}$

_____ 33. $(5\sqrt{18x})(-2\sqrt{8x})$

_____ 34. $(-2\sqrt{27x})(4\sqrt{3x})$

_____ 35. $(\sqrt{5x})(3\sqrt{10x})$

_____ 36. $(5\sqrt{3x})(\sqrt{21x})$

_____ 37. $\sqrt{5}(2 - \sqrt{5})$

_____ 38. $\sqrt{11}(2 - \sqrt{11})$

_____ 39. $2\sqrt{3}(\sqrt{2} + \sqrt{12})$

_____ 40. $3\sqrt{5}(\sqrt{2} + \sqrt{20})$

_____ 41. $-3(\sqrt{x} - 2\sqrt{5} + \sqrt{6})$

_____ 42. $-5(\sqrt{x} - 4\sqrt{2} + \sqrt{7})$

_____ 43. $\sqrt{x}(\sqrt{x} - 16)$

_____ 44. $\sqrt{y}(\sqrt{y} - 9)$

_____ 45. $\sqrt{6}(\sqrt{6} - 5) - 5(\sqrt{6} - 5)$

_____ 46. $\sqrt{2}(\sqrt{2} - 11) - 11(\sqrt{2} - 11)$

_____ 47. $\sqrt{x}(\sqrt{x} - 5) + 5 + 5(\sqrt{x} - 5)$

_____ 48. $\sqrt{a}(\sqrt{a} - 7) + 7(\sqrt{a} - 7)$

_____ 49. $4(3 - \sqrt{11}) - \sqrt{11}(3 - \sqrt{11})$

_____ 50. $2(7 - \sqrt{5}) + \sqrt{5}(7 - \sqrt{5})$

_____ 51. $2\sqrt{x}(\sqrt{x} - 1) - 3(\sqrt{x} + 1)$

_____ 52. $3\sqrt{x}(\sqrt{x} - 2) - 5(\sqrt{x} + 2)$

_____ 53. $(5 - \sqrt{2})(1 - \sqrt{3})$

_____ 54. $(4 - \sqrt{3})(1 + \sqrt{5})$

_____ 55. $(7 - 2\sqrt{5})(3 + \sqrt{5})$

_____ 56. $(4 - \sqrt{3})(9 + 2\sqrt{3})$

_____ 57. $(2 - \sqrt{3x})(4 + \sqrt{3x})$

_____ 58. $(5 - \sqrt{2x})(3 + \sqrt{2x})$

_____ 59. $(\sqrt{3} - \sqrt{7})(\sqrt{3} + \sqrt{7})$

_____ 60. $(\sqrt{10} - \sqrt{5})(\sqrt{10} + \sqrt{5})$

_____ 61. $(\sqrt{x} - 10)(\sqrt{x} + 10)$

_____ 62. $(\sqrt{x} - 9)(\sqrt{x} + 9)$

_____ 63. $(5\sqrt{2} - \sqrt{7})(3\sqrt{2} - 4\sqrt{7})$

_____ 64. $(2\sqrt{3} - \sqrt{5})(\sqrt{3} - 7\sqrt{5})$

_____ 65. $(\sqrt{y} - \sqrt{7})^2$

_____ 66. $(\sqrt{y} - \sqrt{11})^2$

_____ 67. $(a\sqrt{2} - 5)^2$

_____ 68. $(a\sqrt{3} - 4)^2$

CHECKUP 12.3

Simplify each expression.

_____ 1. $\sqrt{24x} + \sqrt{54x}$

_____ 2. $\sqrt{12} + \sqrt{75} - \sqrt{48}$

_____ 3. $\dfrac{\sqrt{108}}{\sqrt{4}} - 5\sqrt{12} + \sqrt{300}$

_____ 4. $\sqrt{15a} \cdot \sqrt{3a}$

_____ 5. $(5\sqrt{6})(-2\sqrt{21})$

_____ 6. $\sqrt{5}(7 - \sqrt{10})$

_____ 7. $2\sqrt{2}(\sqrt{3} + \sqrt{6})$

_____ 8. $\sqrt{2}(\sqrt{5} - 1) - 7(\sqrt{2} + 2)$

_____ 9. $(3 + 2\sqrt{5})(4 - 3\sqrt{5})$

_____ 10. $(\sqrt{x} - \sqrt{3})(\sqrt{x} + \sqrt{3})$

12.4 Rationalizing Denominators

Mathematicians have agreed that a fraction containing any radicals in the denominator is not in simplest form. In other words, it is not acceptable to leave irrational numbers in the denominator of a fraction. Fractions such as

$$\frac{1}{\sqrt{3}} \quad \text{or} \quad \frac{2\sqrt{2}}{\sqrt{5}} \quad \text{or} \quad \frac{3}{1 + \sqrt{11}} \quad \text{or} \quad \frac{3 + \sqrt{7}}{5 - \sqrt{10}}$$

are not in simplest form because each one contains a radical in the denominator.

We would like to change each irrational denominator to a *rational* number. Recall that we are allowed to multiply the numerator and denominator of a fraction by the same number. Let's see if we can choose such a number for the fraction

$$\frac{1}{\sqrt{3}}$$

We would like to multiply the denominator ($\sqrt{3}$) by some number that will leave us with no radical in the denominator. Remembering that $\sqrt{3} \cdot \sqrt{3} = 3$, let's multiply numerator and denominator by $\sqrt{3}$ and simplify.

$$\frac{1}{\sqrt{3}} \cdot \frac{\sqrt{3}}{\sqrt{3}} = \frac{1 \cdot \sqrt{3}}{\sqrt{3} \cdot \sqrt{3}} = \frac{\sqrt{3}}{3}$$

We notice that our denominator is now a rational number. Indeed, the numerator now contains a radical, but that is acceptable.

This process is called **rationalizing the denominator** because it involves changing the denominator into a rational number. Let's try to rationalize the denominator in our next fraction.

$$\frac{2\sqrt{2}}{\sqrt{5}}$$

We concentrate on the denominator and use the fact that $\sqrt{5} \cdot \sqrt{5} = 5$ to choose $\sqrt{5}$ as our rationalizing factor. We multiply numerator and denominator by $\sqrt{5}$.

$$\frac{2\sqrt{2}}{\sqrt{5}} \cdot \frac{\sqrt{5}}{\sqrt{5}} = \frac{2\sqrt{2} \cdot \sqrt{5}}{\sqrt{5} \cdot \sqrt{5}}$$
$$= \frac{2\sqrt{10}}{5}$$

> To rationalize a denominator that is a single term containing a radical ($\sqrt{a}$), we multiply numerator and denominator by that radical ($\sqrt{a}$).

Example 1. Rationalize the denominator for $\dfrac{4}{\sqrt{2}}$.

Solution

$$\frac{4}{\sqrt{2}} = \frac{4}{\sqrt{2}} \cdot \frac{\sqrt{2}}{\sqrt{2}} = \frac{4 \cdot \sqrt{2}}{\sqrt{2} \cdot \sqrt{2}}$$
$$= \frac{4\sqrt{2}}{2}$$
$$= 2\sqrt{2}$$

Notice that we reduced our final fraction in the usual way, dividing numerator and denominator by the common factor 2.

Example 2. Rationalize the denominator for $\sqrt{\dfrac{5}{3}}$.

Solution

$$\sqrt{\frac{5}{3}} = \frac{\sqrt{5}}{\sqrt{3}} = \frac{\sqrt{5}}{\sqrt{3}} \cdot \frac{\sqrt{3}}{\sqrt{3}} = \frac{\sqrt{5} \cdot \sqrt{3}}{\sqrt{3} \cdot \sqrt{3}}$$

$$= \frac{\sqrt{15}}{3}$$

Example 3. Rationalize the denominator for $\dfrac{1 + \sqrt{3}}{\sqrt{7}}$.

Solution

$$\frac{1 + \sqrt{3}}{\sqrt{7}} = \frac{1 + \sqrt{3}}{\sqrt{7}} \cdot \frac{\sqrt{7}}{\sqrt{7}} = \frac{(1 + \sqrt{3})\sqrt{7}}{\sqrt{7} \cdot \sqrt{7}}$$

$$= \frac{\sqrt{7} + \sqrt{3} \cdot \sqrt{7}}{7}$$

$$= \frac{\sqrt{7} + \sqrt{21}}{7}$$

Example 4. Rationalize the denominator for $\dfrac{\sqrt{6} - 2}{\sqrt{6}}$.

Solution

$$\frac{\sqrt{6} - 2}{\sqrt{6}} = \frac{\sqrt{6} - 2}{\sqrt{6}} \cdot \frac{\sqrt{6}}{\sqrt{6}} = \frac{(\sqrt{6} - 2)\sqrt{6}}{\sqrt{6} \cdot \sqrt{6}}$$

$$= \frac{\sqrt{6} \cdot \sqrt{6} - 2 \cdot \sqrt{6}}{6}$$

$$= \frac{6 - 2\sqrt{6}}{6}$$

$$= \frac{2(3 - \sqrt{6})}{6}$$

$$= \frac{3 - \sqrt{6}}{3}$$

Trial Run   *Rationalize the denominator and simplify.*

_____ 1. $\dfrac{1}{\sqrt{5}}$

_____ 2. $\dfrac{5}{\sqrt{3}}$

_____ 3. $\sqrt{\dfrac{7}{2}}$

_____ 4. $\dfrac{1 + \sqrt{2}}{\sqrt{3}}$

_____ 5. $\dfrac{\sqrt{10} - 4}{\sqrt{2}}$

To rationalize denominators containing *two* terms requires a different kind of rationalizing factor. Can you see why? Suppose that we were to try to rationalize the denominator of the fraction $\dfrac{3}{1+\sqrt{11}}$ by multiplying numerator and denominator by $\sqrt{11}$.

$$\frac{3}{1+\sqrt{11}} \cdot \frac{\sqrt{11}}{\sqrt{11}} = \frac{3\cdot\sqrt{11}}{(1+\sqrt{11})\sqrt{11}} = \frac{3\sqrt{11}}{1\cdot\sqrt{11}+\sqrt{11}\cdot\sqrt{11}}$$

$$= \frac{3\sqrt{11}}{1\cdot\sqrt{11}+\sqrt{11}\cdot\sqrt{11}}$$

$$= \frac{3\sqrt{11}}{\sqrt{11}+11}$$

Is this new fraction in better form than the original fraction? Not at all, since the denominator still contains a radical.

Recall from our multiplication examples what happens when we multiply the sum and difference of the same two terms.

$$(1+\sqrt{11})(1-\sqrt{11}) = 1 - 1\cdot\sqrt{11}+1\cdot\sqrt{11}-\sqrt{11}\cdot\sqrt{11}$$
$$= 1 - \sqrt{11}+\sqrt{11}-11$$
$$= 1 - 11 = -10$$

Multiplying $1+\sqrt{11}$ by $1-\sqrt{11}$ yielded a product (-10) that was a *rational* number! Since this is the result we are seeking, it seems sensible to rationalize the denominator of the fraction

$$\frac{3}{1+\sqrt{11}}$$

by multiplying numerator and denominator by $1-\sqrt{11}$.

$$\frac{3}{1+\sqrt{11}} \cdot \frac{1-\sqrt{11}}{1-\sqrt{11}} = \frac{3(1-\sqrt{11})}{(1+\sqrt{11})(1-\sqrt{11})}$$
$$= \frac{3-3\sqrt{11}}{-10}$$

Let's rationalize the denominator of the fraction

$$\frac{\sqrt{3}}{5-\sqrt{2}}$$

Since our denominator is a *difference* of two terms $(5-\sqrt{2})$, we should multiply numerator and denominator by the *sum* of the same two terms $(5+\sqrt{2})$.

$$\frac{\sqrt{3}}{5-\sqrt{2}} \cdot \frac{5+\sqrt{2}}{5+\sqrt{2}} = \frac{\sqrt{3}(5+\sqrt{2})}{(5-\sqrt{2})(5+\sqrt{2})}$$
$$= \frac{5\sqrt{3}+\sqrt{3}\cdot\sqrt{2}}{25+5\sqrt{2}-5\sqrt{2}-\sqrt{2}\cdot\sqrt{2}}$$
$$= \frac{5\sqrt{3}+\sqrt{6}}{25-2}$$
$$= \frac{5\sqrt{3}+\sqrt{6}}{23}$$

To rationalize a denominator containing a sum (or difference) of two terms, multiply the numerator and denominator by the difference (or sum) of the same two terms.

Example 5. Rationalize the denominator for $\dfrac{3 + \sqrt{7}}{5 - \sqrt{10}}$.

Solution

$$\frac{3 + \sqrt{7}}{5 - \sqrt{10}} = \frac{3 + \sqrt{7}}{5 - \sqrt{10}} \cdot \frac{5 + \sqrt{10}}{5 + \sqrt{10}}$$

$$= \frac{(3 + \sqrt{7})(5 + \sqrt{10})}{(5 - \sqrt{10})(5 + \sqrt{10})}$$

$$= \frac{15 + 3\sqrt{10} + 5\sqrt{7} + \sqrt{7} \cdot \sqrt{10}}{25 + 5\sqrt{10} - 5\sqrt{10} - 10}$$

$$= \frac{15 + 3\sqrt{10} + 5\sqrt{7} + \sqrt{70}}{25 - 10}$$

$$= \frac{15 + 3\sqrt{10} + 5\sqrt{7} + \sqrt{70}}{15}$$

Example 6. Rationalize the denominator for $\dfrac{\sqrt{3}}{9 + \sqrt{3}}$.

Solution

$$\frac{\sqrt{3}}{9 + \sqrt{3}} = \frac{\sqrt{3}}{9 + \sqrt{3}} \cdot \frac{9 - \sqrt{3}}{9 - \sqrt{3}} = \frac{\sqrt{3}(9 - \sqrt{3})}{(9 + \sqrt{3})(9 - \sqrt{3})}$$

$$= \frac{9\sqrt{3} - \sqrt{3} \cdot \sqrt{3}}{81 - 9\sqrt{3} + 9\sqrt{3} - 3}$$

$$= \frac{9\sqrt{3} - 3}{81 - 3}$$

$$= \frac{9\sqrt{3} - 3}{78}$$

$$= \frac{3(3\sqrt{3} - 1)}{3 \cdot 26}$$

$$= \frac{3\sqrt{3} - 1}{26}$$

Example 7. Rationalize the denominator for $\dfrac{8}{\sqrt{2} - \sqrt{3}}$.

Solution

$$\frac{8}{\sqrt{2} - \sqrt{3}} = \frac{8}{\sqrt{2} - \sqrt{3}} \cdot \frac{\sqrt{2} + \sqrt{3}}{\sqrt{2} + \sqrt{3}}$$

$$= \frac{8(\sqrt{2} + \sqrt{3})}{(\sqrt{2} - \sqrt{3})(\sqrt{2} + \sqrt{3})}$$

$$= \frac{8\sqrt{2} + 8\sqrt{3}}{2 + \sqrt{2} \cdot \sqrt{3} - \sqrt{3} \cdot \sqrt{2} - 3}$$

$$= \frac{8\sqrt{2} + 8\sqrt{3}}{2 - 3}$$

$$= \frac{8\sqrt{2} + 8\sqrt{3}}{-1}$$

Letting the numerator "carry the sign," we have

$$= -1(8\sqrt{2} + 8\sqrt{3})$$

$$= -8\sqrt{2} - 8\sqrt{3}$$

Trial Run *Rationalize the denominators and give answers in simplest form.*

———— **1.** $\dfrac{2}{7 + \sqrt{5}}$

———— **2.** $\dfrac{-3}{\sqrt{2} + 1}$

———— **3.** $\dfrac{4 - \sqrt{6}}{3 - \sqrt{2}}$

———— **4.** $\dfrac{\sqrt{5}}{4 - \sqrt{5}}$

———— **5.** $\dfrac{3 - \sqrt{5}}{3 + \sqrt{5}}$

ANSWERS

1. $\dfrac{7 - \sqrt{5}}{22}$.   **2.** $-3\sqrt{2} + 3$.   **3.** $\dfrac{12 + 4\sqrt{2} - 3\sqrt{6} - 2\sqrt{3}}{7}$.   **4.** $\dfrac{4\sqrt{5} + 5}{11}$.

5. $\dfrac{7 - 3\sqrt{5}}{2}$.

EXERCISE SET 12.4

Rationalize the denominators. Write your answers in simplest form.

_____ 1. $\dfrac{3}{\sqrt{3}}$

_____ 2. $\dfrac{5}{\sqrt{5}}$

_____ 3. $\dfrac{2}{\sqrt{7}}$

_____ 4. $\dfrac{5}{\sqrt{6}}$

_____ 5. $\dfrac{4}{\sqrt{2}}$

_____ 6. $\dfrac{9}{\sqrt{3}}$

_____ 7. $\dfrac{12}{\sqrt{10}}$

_____ 8. $\dfrac{21}{\sqrt{6}}$

_____ 9. $\dfrac{9}{\sqrt{32}}$

_____ 10. $\dfrac{8}{\sqrt{27}}$

_____ 11. $\dfrac{\sqrt{5}}{\sqrt{8}}$

_____ 12. $\dfrac{\sqrt{7}}{\sqrt{12}}$

_____ 13. $\dfrac{\sqrt{10}}{\sqrt{5}}$

_____ 14. $\dfrac{\sqrt{6}}{\sqrt{2}}$

_____ 15. $\sqrt{\dfrac{1}{2}}$

_____ 16. $\sqrt{\dfrac{1}{3}}$

_____ 17. $\sqrt{\dfrac{4}{5}}$

_____ 18. $\sqrt{\dfrac{9}{7}}$

_____ 19. $\sqrt{\dfrac{3}{x}}$

_____ 20. $\sqrt{\dfrac{2}{x}}$

_____ 21. $\dfrac{1 + \sqrt{2}}{\sqrt{5}}$

_____ 22. $\dfrac{2 + \sqrt{3}}{\sqrt{2}}$

_____ 23. $\dfrac{\sqrt{8} - 6}{\sqrt{2}}$

_____ 24. $\dfrac{\sqrt{10} - 5}{\sqrt{5}}$

_____ 25. $\dfrac{\sqrt{2} + \sqrt{5}}{\sqrt{10}}$

_____ 26. $\dfrac{\sqrt{3} + \sqrt{2}}{\sqrt{6}}$

_____ 27. $\dfrac{1}{2 + \sqrt{3}}$

_____ 28. $\dfrac{1}{5 - \sqrt{2}}$

_____ 29. $\dfrac{-12}{3 - \sqrt{6}}$

_____ 30. $\dfrac{-6}{2 + \sqrt{7}}$

_____ 31. $\dfrac{\sqrt{2}}{1 - \sqrt{2}}$

_____ 32. $\dfrac{\sqrt{3}}{2 - \sqrt{3}}$

_____ 33. $\dfrac{\sqrt{8}}{\sqrt{3} + 1}$

_____ 34. $\dfrac{\sqrt{12}}{\sqrt{5} - 1}$

_____ 35. $\dfrac{2\sqrt{3}}{\sqrt{5} + \sqrt{2}}$

_____ 36. $\dfrac{2\sqrt{5}}{\sqrt{11} - \sqrt{3}}$

_____ 37. $\dfrac{2 + \sqrt{2}}{\sqrt{2} + 1}$

_____ 38. $\dfrac{3 + \sqrt{5}}{\sqrt{5} - 2}$

_____ 39. $\dfrac{\sqrt{6} + 2}{\sqrt{6} - 2}$

_____ 40. $\dfrac{\sqrt{7} + 1}{\sqrt{7} - 1}$

Name _____ Date _____

CHECKUP 12.4

Rationalize the denominators. Write your answers in simplest form.

_____ 1. $\dfrac{6}{\sqrt{6}}$

_____ 2. $\dfrac{20}{\sqrt{10}}$

_____ 3. $\dfrac{7}{\sqrt{12}}$

_____ 4. $\dfrac{\sqrt{6}}{\sqrt{15}}$

_____ 5. $\sqrt{\dfrac{8}{3}}$

_____ 6. $\dfrac{\sqrt{12}-10}{\sqrt{2}}$

_____ 7. $\dfrac{\sqrt{5}+\sqrt{15}}{\sqrt{20}}$

_____ 8. $\dfrac{\sqrt{3}}{6-\sqrt{3}}$

_____ 9. $\dfrac{5+\sqrt{11}}{\sqrt{11}-3}$

_____ 10. $\dfrac{\sqrt{7}-2}{\sqrt{7}+2}$

12.5 Solving More Quadratic Equations

Recall from Chapter 6 that a **quadratic equation** is an equation of the form

$$ax^2 + bx + c = 0 \qquad (a \neq 0)$$

In that chapter we learned to solve such equations by factoring and setting each factor equal to zero. For instance, we solved

$$x^2 + 3x - 4 = 0$$
$$(x + 4)(x - 1) = 0$$

$$x + 4 = 0 \qquad \text{or} \qquad x - 1 = 0$$
$$x = -4 \qquad \text{or} \qquad x = 1$$

and

$$x^2 - 9 = 0$$
$$(x + 3)(x - 3) = 0$$

$$x + 3 = 0 \qquad \text{or} \qquad x - 3 = 0$$
$$x = -3 \qquad \text{or} \qquad x = 3$$

Every quadratic equation in Chapter 6 could be solved by means of factoring to yield rational solutions. Unfortunately, it is *not* true that *all* quadratic equations have solutions that are rational numbers. We now turn our attention to methods that can be used for solving *any* quadratic equations.

12.5-1 Solving Quadratic Equations of the Form $ax^2 = c$

Many quadratic equations do not contain a first-degree term. For instance

$$x^2 - 2 = 0$$
$$x^2 = 7$$

$$3x^2 - 5 = 0$$
$$x^2 + 1 = 0$$

are all examples of quadratic (second-degree) equations containing no first-degree term. Let's consider each one in turn.

To solve $x^2 - 2 = 0$ we would like to factor the left-hand side, but no rational factors will "work." Suppose that we rewrite our equation by adding 2 to both sides.

$$x^2 = 2$$

Now we are looking for a number x, whose square is the number 2. Certainly, $x = \sqrt{2}$ is a possible solution, since

$$x^2 = (\sqrt{2})^2 = \sqrt{2} \cdot \sqrt{2} = 2$$

but another solution is $x = -\sqrt{2}$ because

$$x^2 = (-\sqrt{2})^2 = (-\sqrt{2})(-\sqrt{2}) = 2$$

Our solutions, then, must be

$$x = \sqrt{2} \qquad \text{or} \qquad x = -\sqrt{2}$$

As before, we discover that a second-degree equation has two solutions.

We can use similar reasoning to solve the next problem,

$$x^2 = 7$$

We are looking for a number x whose square is 7. You should agree that the solutions are

$$x = \sqrt{7} \qquad \text{because} \qquad (\sqrt{7})^2 = \sqrt{7} \cdot \sqrt{7} = 7$$

or $\qquad x = -\sqrt{7} \qquad$ because $\qquad (-\sqrt{7})^2 = (-\sqrt{7})(-\sqrt{7}) = 7$

So our solutions for $x^2 = 7$ are $x = \sqrt{7}$ or $x = -\sqrt{7}$.

Let's solve

$$3x^2 - 5 = 0$$

We should be sure that the left-hand side does not factor; it does not. Then we proceed as before, isolating the x^2 term.

$$3x^2 - 5 = 0$$
$$3x^2 = 5$$
$$\frac{3x^2}{3} = \frac{5}{3}$$
$$x^2 = \frac{5}{3}$$

Our solutions must be
$$x = \sqrt{\frac{5}{3}} \qquad \text{or} \qquad x = -\sqrt{\frac{5}{3}}$$

But wait a minute! These answers contain radicals in the denominators that must be rationalized.

$$x = \sqrt{\frac{5}{3}} = \frac{\sqrt{5}}{\sqrt{3}} = \frac{\sqrt{5}}{\sqrt{3}} \cdot \frac{\sqrt{3}}{\sqrt{3}} = \frac{\sqrt{15}}{3}$$

$$x = -\sqrt{\frac{5}{3}} = -\frac{\sqrt{5}}{\sqrt{3}} \cdot \frac{\sqrt{3}}{\sqrt{3}} = \frac{-\sqrt{15}}{3}$$

So our solutions for $3x^2 - 5 = 0$ are

$$x = \frac{\sqrt{15}}{3} \qquad \text{or} \qquad x = \frac{-\sqrt{15}}{3}$$

Now let's tackle the last problem,

$$x^2 + 1 = 0$$

Since the left-hand side cannot be factored, we isolate the x^2 term by subtracting 1 from both sides.

$$x^2 = -1$$

We are seeking a number x whose square is -1. Is this possible? Is $x = -1$ a solution?

$$(-1)^2 = (-1)(-1) = 1 \neq -1$$

In fact, we know that the *square of any real number is always positive*. There is no real solution for the problem

$$x^2 = -1$$

Indeed, there will never be a real solution for a problem in which x^2 equals a *negative* number. Keep that in mind.

Let's summarize what we have discovered about solving quadratic equations of the form $x^2 = c$.

If $x^2 = c$ and c is a nonnegative constant, then the solutions are $x = \sqrt{c}$ or $x = -\sqrt{c}$. If $x^2 = c$ and c is a negative constant, then there is no real solution.

Example 1. Solve $x^2 - 23 = 0$.

Solution

$$x^2 - 23 = 0$$
$$x^2 = 23$$

so

$$x = \sqrt{23} \quad \text{or} \quad x = -\sqrt{23}$$

Example 2. Solve $2x^2 = 22$.

Solution

$$2x^2 = 22$$
$$\frac{2x^2}{2} = \frac{22}{2}$$
$$x^2 = 11$$

so

$$x = \sqrt{11} \quad \text{or} \quad x = -\sqrt{11}$$

Example 3. Solve $2x^2 - 11 = 2$.

Solution

$$2x^2 - 11 = 2$$
$$2x^2 - 11 + 11 = 2 + 11$$
$$2x^2 = 13$$
$$\frac{2x^2}{2} = \frac{13}{2}$$
$$x = \sqrt{\frac{13}{2}} = \frac{\sqrt{13}}{\sqrt{2}} \cdot \frac{\sqrt{2}}{\sqrt{2}} = \frac{\sqrt{26}}{2}$$

or

$$x = -\sqrt{\frac{13}{2}} = \frac{-\sqrt{13}}{\sqrt{2}} \cdot \frac{\sqrt{2}}{\sqrt{2}} = \frac{-\sqrt{26}}{2}$$

so

$$x = \frac{\sqrt{26}}{2} \quad \text{or} \quad x = \frac{-\sqrt{26}}{2}$$

Example 4. Solve $x^2 + 7 = 5$.

Solution

$$x^2 + 7 = 5$$
$$x^2 + 7 - 7 = 5 - 7$$
$$x^2 = -2$$

so there is no real solution.

Trial Run   *Solve.*

———— **1.** $x^2 - 21 = 0$

———— **2.** $3x^2 = 15$

_____ **3.** $2x^2 - 16 = 0$

_____ **4.** $2x^2 - 9 = 5$

_____ **5.** $x^2 + 7 = 3$

_____ **6.** $15 = 5x^2 - 20$

ANSWERS
1. $x = \sqrt{21}$ or $x = -\sqrt{21}$.　**2.** $x = \sqrt{5}$ or $x = -\sqrt{5}$.　**3.** $x = 2\sqrt{2}$ or $x = -2\sqrt{2}$.
4. $x = \sqrt{7}$ or $x = -\sqrt{7}$.　**5.** No real solution.　**6.** $x = \sqrt{7}$ or $x = -\sqrt{7}$.

12.5-2 Using the Quadratic Formula

To solve a quadratic equation of the form

$$ax^2 + bx + c = 0$$

we have learned to factor the left-hand side of the equation and set each factor equal to zero. This technique works as long as the left-hand side can be factored, but suppose that it cannot be factored by our usual methods. For instance, suppose that we wish to solve the equations

$$x^2 + 5x + 1 = 0$$
$$2x^2 - 7x + 2 = 0$$
$$x^2 - x - 1 = 0$$

A quick check reveals that none of these second-degree polynomials can be factored. Fortunately, mathematicians have found a formula that allows us to solve such equations. The formula for solving quadratic equations is called the **quadratic formula**, and it works in the following way.

Quadratic Formula.　If $ax^2 + bx + c = 0$, then the solutions can be found by

$$x = \frac{-b + \sqrt{b^2 - 4ac}}{2a} \qquad \text{or} \qquad x = \frac{-b - \sqrt{b^2 - 4ac}}{2a}$$

As a matter of fact, this quadratic formula can be used to solve *any* quadratic equation—even those that can be solved by factoring. Let's practice with an equation whose solutions can also be found by factoring; then we can be truly convinced that the formula "works."

To solve the equation $x^2 + 2x - 3 = 0$, we can use either of our approaches.

1. Using factoring,

$$x^2 + 2x - 3 = 0$$
$$(x + 3)(x - 1) = 0$$

$$x + 3 = 0 \qquad \text{or} \qquad x - 1 = 0$$
$$x = -3 \qquad \text{or} \qquad x = 1$$

2. Using the quadratic formula, our equation

$$x^2 + 2x - 3 = 0$$

fits into the pattern of a quadratic equation

$$ax^2 + bx + c = 0$$

and we carefully note that, here,

$$a = 1$$
$$b = 2$$
$$c = -3$$

According to the quadratic formula, the solutions for our equation must be

$$x = \frac{-b + \sqrt{b^2 - 4ac}}{2a} \quad \text{or} \quad x = \frac{-b - \sqrt{b^2 - 4ac}}{2a}$$

To find our solution, we must substitute our values for a, b, and c into the quadratic formula.

$$x = \frac{-2 + \sqrt{2^2 - 4(1)(-3)}}{2(1)} = \frac{-2 + \sqrt{4 + 12}}{2}$$

$$= \frac{-2 + \sqrt{16}}{2} = \frac{-2 + 4}{2} = \frac{2}{2} = 1$$

$$x = 1$$

or

$$x = \frac{-2 - \sqrt{2^2 - 4(1)(-3)}}{2(1)} = \frac{-2 - \sqrt{4 + 12}}{2}$$

$$= \frac{-2 - \sqrt{16}}{2} = \frac{-2 - 4}{2} = \frac{-6}{2} = -3$$

$$x = -3$$

Our solutions, therefore, are

$$x = 1 \quad \text{or} \quad x = -3$$

Notice that both approaches yielded the same pair of solutions. This should convince you that the quadratic formula will "work" for *any* quadratic equation. Surely you agree, however, that factoring was a quicker method to use in this example. We shall always solve by factoring when possible. Now we turn our attention to quadratic equations that cannot be solved by factoring and see how the quadratic formula allows us to find solutions.

Example 5. Solve $x^2 + 5x + 1 = 0$.

Solution. $x^2 + 5x + 1 = 0$. Here we note that

$$a = 1$$
$$b = 5$$
$$c = 1$$

Then we substitute into the quadratic formula.

$$x = \frac{-b + \sqrt{b^2 - 4ac}}{2a} = \frac{-5 + \sqrt{5^2 - 4(1)(1)}}{2(1)}$$

$$= \frac{-5 + \sqrt{25 - 4}}{2}$$

$$= \frac{-5 + \sqrt{21}}{2}$$

or

$$x = \frac{-b - \sqrt{b^2 - 4ac}}{2a} = \frac{-5 - \sqrt{5^2 - 4(1)(1)}}{2(1)}$$

$$= \frac{-5 - \sqrt{25 - 4}}{2}$$

$$= \frac{-5 - \sqrt{21}}{2}$$

So our two solutions are

$$x = \frac{-5 + \sqrt{21}}{2} \qquad \text{or} \qquad x = \frac{-5 - \sqrt{21}}{2}$$

Before continuing let's make a couple of observations. Notice that these solutions are *irrational* numbers. Notice also that they are very similar looking; they differ only because of the sign in front of the radical. We could save ourselves some writing by stating both solutions in the single formula

$$x = \frac{-b \pm \sqrt{b^2 - 4ac}}{2a}$$

where the symbol "$\pm$" means use the $+$ sign to find one solution and the $-$ sign to find the other solution. Let's try this single formula in the next example.

Example 6. Solve $2x^2 - 7x + 2 = 0$.

Solution. $2x^2 - 7x + 2 = 0$. Here we note that

$$a = 2$$
$$b = -7$$
$$c = 2$$

Then we substitute into the formula

$$x = \frac{-b \pm \sqrt{b^2 - 4ac}}{2a}$$

$$x = \frac{-(-7) \pm \sqrt{(-7)^2 - 4(2)(2)}}{2(2)}$$

$$x = \frac{7 \pm \sqrt{49 - 16}}{4}$$

$$x = \frac{7 \pm \sqrt{33}}{4}$$

So our two solutions are

$$x = \frac{7 + \sqrt{33}}{4} \qquad \text{or} \qquad x = \frac{7 - \sqrt{33}}{4}$$

Example 7. Solve $x^2 + 2x - 1 = 0$.

Solution. $x^2 + 2x - 1 = 0$. We note that

$$a = 1$$
$$b = 2$$
$$c = -1$$

Then we substitute into the quadratic formula.

$$x = \frac{-b \pm \sqrt{b^2 - 4ac}}{2a}$$

$$x = \frac{-2 \pm \sqrt{2^2 - 4(1)(-1)}}{2(1)}$$

$$x = \frac{-2 \pm \sqrt{4 + 4}}{2}$$

$$x = \frac{-2 \pm \sqrt{8}}{2}$$

$$x = \frac{-2 \pm \sqrt{4 \cdot 2}}{2}$$

$$x = \frac{-2 \pm 2\sqrt{2}}{2}$$

$$x = \frac{2(-1 \pm \sqrt{2})}{2}$$

$$x = -1 \pm \sqrt{2}$$

So our solutions are

$$x = -1 + \sqrt{2} \quad \text{or} \quad x = -1 - \sqrt{2}$$

Notice that we simplified our radical and reduced our fraction by the usual methods in this example.

Example 8. Solve $x^2 + x + 3 = 0$.

Solution. $x^2 + x + 3 = 0$. We note here that

$$a = 1$$
$$b = 1$$
$$c = 3$$

Then we substitute into the quadratic formula.

$$x = \frac{-b \pm \sqrt{b^2 - 4ac}}{2a}$$

$$x = \frac{-1 \pm \sqrt{1^2 - 4(1)(3)}}{2(1)}$$

$$x = \frac{-1 \pm \sqrt{1 - 12}}{2}$$

$$x = \frac{-1 \pm \sqrt{-11}}{2}$$

We are in trouble here because we know that it is impossible to find the square root of a negative number. We conclude that the equation

$$x^2 + x + 3 = 0$$

has *no real solutions.*

Trial Run   *Write the following equations in the form $ax^2 + bx + c = 0$ and give the values of a, b, and c for each equation.*

_____ **1.** $2x^2 = 3x - 2$

_____ **2.** $2x^2 - 5x + 2 = 7x - 5$

Use the quadratic formula to solve each equation.

_____ **3.** $x^2 + x - 3 = 0$

_____ **4.** $2x^2 - 3x - 1 = 0$

_____ **5.** $2x^2 + 4x = -1$

_____ **6.** $x^2 = x - 1$

ANSWERS

1. $2x^2 - 3x + 2 = 0, a = 2, b = -3, c = 2.$

2. $2x^2 - 12x + 7 = 0, a = 2, b = -12, c = 7.$

3. $\dfrac{-1 \pm \sqrt{13}}{2}.$ **4.** $\dfrac{3 \pm \sqrt{17}}{4}.$ **5.** $\dfrac{-2 \pm \sqrt{2}}{2}.$ **6.** No real solution.

EXERCISE SET 12.5

Solve each of the following quadratic equations. Write all answers in simplest radical form and reduce all answers to lowest terms.

_____ 1. $x^2 = 12$ _____ 2. $x^2 = 32$

_____ 3. $9a^2 = 5$ _____ 4. $4a^2 = 7$

_____ 5. $5x^2 - 1 = 0$ _____ 6. $6x^2 - 5 = 0$

_____ 7. $6y^2 - 15 = 0$ _____ 8. $8y^2 - 28 = 0$

_____ 9. $x^2 + 7 = 0$ _____ 10. $x^2 + 13 = 0$

_____ 11. $3y^2 - 8 = 7$ _____ 12. $5y^2 - 17 = 8$

_____ 13. $9 = 11 - a^2$ _____ 14. $8 = 13 - a^2$

_____ 15. $4x^2 + 3 = 3$ _____ 16. $6x^2 + 8 = 8$

_____ 17. $a^2 - 80 = 0$ _____ 18. $a^2 - 75 = 0$

_____ 19. $x^2 + x - 3 = 0$ _____ 20. $x^2 + 6x + 1 = 0$

_____ 21. $a^2 = 5a + 5$ _____ 22. $a^2 = 2a + 2$

_____ 23. $x^2 - 3 = 4x$ _____ 24. $x^2 - 2 = 3x$

_____ 25. $y^2 = 2y + 1$ _____ 26. $y^2 = 4y + 2$

_____ 27. $a^2 + 7a = 1$ _____ 28. $a^2 + 5a = 4$

_____ 29. $3x^2 - 2x - 2 = 0$ _____ 30. $5x^2 - 3x - 1 = 0$

_____ 31. $2x^2 + 9x = -11$ _____ 32. $3x^2 + 5x = -5$

_____ 33. $4 - 6y = -y^2$ _____ 34. $4z - 1 = -z^2$

_____ 35. $2z = 15 - 2z^2$ _____ 36. $7z = 6z^2 - 1$

CHECKUP 12.5

Solve.

_____ 1. $x^2 - 2 = 0$

_____ 2. $4a^2 = 20$

_____ 3. $y^2 - 32 = 0$

_____ 4. $9x^2 - 80 = 100$

_____ 5. $9 - 3x^2 = 9$

_____ 6. $x^2 + 4x - 4 = 0$

_____ 7. $x^2 = 3x - 1$

_____ 8. $3x^2 - 6x + 2 = 0$

_____ 9. $2x^2 - 5x = 1$

_____ 10. $2x^2 = 3x - 6$

SUMMARY

Our work in this chapter dealt with three major topics: negative exponents, square roots, and solving quadratic equations.

After making the following definition for a negative exponent

$$a^{-n} = \frac{1}{a^n} \qquad \text{(where } a \neq 0\text{)}$$

we discovered that all our laws of exponents continued to "work" for negative exponents and we practiced using those laws to simplify expressions containing negative exponents.

We agreed to use the radical symbol $\sqrt{a}$ to represent the principal (nonnegative) square root of a. If a is nonnegative and b is nonnegative, we decided that

$$\sqrt{a} = b \qquad \text{means that} \qquad b^2 = a$$

We noted that if a is a perfect square number, $\sqrt{a}$ will be a rational number; if not, $\sqrt{a}$ will be an irrational number. If a number is either rational or irrational, it is called a real number.

After developing the basic properties of radicals, we practiced adding and multiplying radicals and rationalizing denominators. Our knowledge of irrational numbers allowed us to solve quadratic equations without factoring. We agreed that if

$$x^2 = c$$

and c is a nonnegative constant, then the solutions are

$$x = \sqrt{c} \qquad \text{or} \qquad x = -\sqrt{c}$$

and in general, if

$$ax^2 + bx + c = 0$$

then the solutions are given by the quadratic formula

$$x = \frac{-b \pm \sqrt{b^2 - 4ac}}{2a}$$

REVIEW EXERCISES

Simplify and write the answer with positive exponents.

SECTION 12.1

_____ **1.** $(5x)^{-2}$

_____ **2.** $4y^{-3}$

_____ **3.** $\dfrac{-35x^6}{5x^{-3}}$

_____ **4.** $\dfrac{13a^{-2}}{52a^{-5}}$

_____ **5.** $\left(\dfrac{3x}{2}\right)^{-4}$

_____ **6.** $\left[\dfrac{-2y^{-3}}{5y^{-4}}\right]^{-1}$

_____ **7.** $\dfrac{(-2x^{-3})^2(12x^7)}{16x^{-3}}$

_____ **8.** $\dfrac{25x^{-3}y^{-4}}{5x^4y^{-4}}$

Simplify the following radical expressions.

SECTION 12.2

_____ **9.** $\sqrt{144}$

_____ **10.** $\sqrt{64a^2}$

_____ **11.** $(\sqrt{13})^2$

_____ **12.** $\sqrt{45y^4}$

_____ **13.** $\sqrt{48a^3b^2}$

_____ **14.** $\sqrt{\dfrac{27x^5}{y^4}}$

_____ **15.** $\dfrac{\sqrt{98a^7b^3}}{\sqrt{2ab}}$

_____ **16.** $\dfrac{\sqrt{63a^7}}{\sqrt{7a}}$

SECTION 12.3

_____ **17.** $\sqrt{20x} + \sqrt{80x} - \sqrt{45x}$

_____ **18.** $\sqrt{21a} \cdot \sqrt{7a}$

_____ **19.** $(3\sqrt{10})(-4\sqrt{15})$

_____ **20.** $\sqrt{2}\,(3 - \sqrt{14})$

_____ **21.** $5\sqrt{3}\,(\sqrt{2} + \sqrt{8})$

_____ **22.** $\sqrt{5}\,(\sqrt{3} - 2) - 4(\sqrt{5} + 3)$

_____ **23.** $(4 - 3\sqrt{6})(1 + 2\sqrt{6})$

_____ **24.** $(\sqrt{a} - \sqrt{5})(\sqrt{a} + \sqrt{5})$

Rationalize the denominators. Write the answer in simplest form.

SECTION 12.4

_____ **25.** $\dfrac{15}{\sqrt{3}}$

_____ **26.** $\dfrac{\sqrt{15}}{\sqrt{12}}$

_____ **27.** $\sqrt{\dfrac{7}{5}}$

_____ **28.** $\dfrac{\sqrt{12} - 21}{\sqrt{3}}$

_____ **29.** $\dfrac{\sqrt{7} + \sqrt{21}}{\sqrt{14}}$

_____ **30.** $\dfrac{\sqrt{11}}{5 - \sqrt{11}}$

_____ **31.** $\dfrac{4 + \sqrt{3}}{\sqrt{3} - 2}$

_____ **32.** $\dfrac{\sqrt{13} - 2}{\sqrt{13} + 2}$

Solve the equations.

_____ **33.** $y^2 - 48 = 0$ _____ **34.** $9a^2 = 18$

_____ **35.** $7 - 4x^2 = 7$ _____ **36.** $x^2 + 6x - 3 = 0$

_____ **37.** $x^2 = 4x - 1$ _____ **38.** $5x^2 - 10x + 2 = 0$

_____ **39.** $2x^2 - 3x = 1$ _____ **40.** $3x^2 = 4x + 2$

Answers to Odd-Numbered Exercises, Checkups, and Review Exercises

CHAPTER 1

Exercise 1.1 (page 17)

1.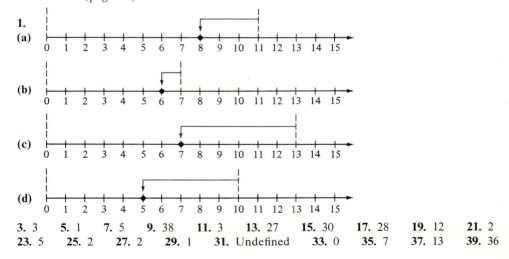

3. $A(2)$; $B(6)$; $C(9)$; $D(12)$ **5.** 6   **7.** 48   **9.** 24   **11.** 160   **13.** 14   **15.** 65
17. 47   **19.** 64   **21.** 28   **23.** 60   **25.** 51   **27.** 15   **29.** 10   **31.** 7   **33.** 0
35. $20 + 6$ **37.** $(3 + 1)$ **39.** $(7 + 2)$ **41.** $1 + (3 + 2)$ **43.** 9   **45.** 0   **47.** $13 \cdot 8$
49. $(4 \cdot 2)$ **51.** $(9 \cdot 8)$ **53.** $(7 \cdot 6)5$ **55.** $3 \cdot 2$ **57.** $9 \cdot 6$ **59.** $7 \cdot 3 + 7 \cdot 5$ **61.** $8(\$18)$
63. $5 \cdot 4$ **65.** $2(\$600) + \$100$

Checkup 1.1 (page 19)

1. 12   **2.** 96   **3.** 42   **4.** 27   **5.** 32   **6.** 104   **7.** $13 + 5$ **8.** $3 \cdot (5 \cdot 2)$
9. $3 \cdot 7 + 3 \cdot 5$ **10.** $110 + 8$

Exercise 1.2 (page 35)

1.
(a)

(b)

(c)

(d)

3. 3   **5.** 1   **7.** 5   **9.** 38   **11.** 3   **13.** 27   **15.** 30   **17.** 28   **19.** 12   **21.** 2
23. 5   **25.** 2   **27.** 2   **29.** 1   **31.** Undefined   **33.** 0   **35.** 7   **37.** 13   **39.** 36

41. 24   **43.** 7   **45.** 24   **47.** 6   **49.** 14   **51.** Undefined   **53.** 1   **55.** 40   **57.** 28
59. 0   **61.** $45 \cdot 100$ **63.** $235 - 190$ **65.** $2 + 8(\$1)$ — incorrect
$2 + 1(8-1)$

Checkup 1.2 (*page 37*)
1. 16   **2.** 15   **3.** 40   **4.** 5   **5.** 4   **6.** 13   **7.** Undefined   **8.** 90   **9.** 0
10. $\dfrac{\$625{,}000 - \$125{,}000}{3}$

Exercise 1.3 (*page 47*)
1.

3. $A(-5); \; B(-2); \; C(0); \; D(3)$

5.
(a)

(b)

(c)

(d)

7. (a) $5 < 12$ **(b)** $^-4 > {}^-10$ **(c)** $8 > {}^-11$ **(d)** $^-3 < 7$
9. (a) $15 > 9$ **(b)** $^-3 > {}^-8$ **(c)** $5 > {}^-16$ **(d)** $12 > {}^-8$ **11.** 22   **13.** $^-12$ **15.** 12
17. $^-6$ **19.** $^-7$ **21.** 13   **23.** 20   **25.** 8   **27.** $^-17$ **29.** $^-14$ **31.** 10   **33.** $^-5$
35. $^-12$ **37.** 24   **39.** 18   **41.** 3   **43.** 65   **45.** 3   **47.** 3   **49.** 3   **51.** Undefined
53. $^-7$ **55.** 0   **57.** $^-8$ **59.** 18   **61.** 32   **63.** 100   **65.** 6

Checkup 1.3 (*page 49*)
1. $^-10$ **2.** $^-5$ **3.** 8   **4.** $^-1$ **5.** $^-12$ **6.** 15   **7.** 24   **8.** $^-9$ **9.** 1   **10.** 70

Exercise 1.4 (*page 55*)
1. $12 + {}^-9 = 3$ **3.** $11 + {}^-27 = {}^-16$ **5.** $^-15 + {}^-7 = {}^-22$ **7.** $12 + 3 = 15$
9. $^-9 + 7 = {}^-2$ **11.** 10   **13.** 25   **15.** $^-11$ **17.** $^-9$ **19.** 3   **21.** $^-12$ **23.** $^-10$
25. $^-6$ **27.** 0   **29.** 1   **31.** 15   **33.** $^-2$ **35.** $^-6$ **37.** 40   **39.** 63   **41.** 15
43. 90   **45.** 10   **47.** 2   **49.** 2   **51.** Undefined   **53.** $^-21$ **55.** 0   **57.** $^-16$ **59.** 8
61. 70   **63.** 30   **65.** 5

Checkup 1.4 (*page 57*)
1. $^-18$ **2.** 21   **3.** $^-22$ **4.** $^-10$ **5.** $^-23$ **6.** 32   **7.** 66   **8.** 100   **9.** 5   **10.** 60

Exercise 1.5 (*page 69*)
1. $^-30$ **3.** 42   **5.** $^-24$ **7.** 150   **9.** 0   **11.** 45   **13.** $^-80$ **15.** 37   **17.** $^-4$
19. $^-2$ **21.** 5   **23.** $^-2$ **25.** $^-7$ **27.** $^-9$ **29.** $^-12$ **31.** Undefined   **33.** $^-1$
35. 1   **37.** $^-4$ **39.** $^-7$ **41.** 2   **43.** $^-13$ **45.** $^-3$ **47.** 51   **49.** $^-200$ **51.** 1
53. $^-47$ **55.** $2(^-5) + 7 + {}^-3 = {}^-6$. She owes him \$6.   **57.** $\dfrac{2(^-3) + 2 \cdot 4 + 2 \cdot 0 + 5}{7} = 1°\,\text{F}$
59. $16 \cdot 10 + 32 \cdot 11 = 512$

Checkup 1.5 (*page 71*)
1. $^-30$ **2.** 48   **3.** $^-5$ **4.** 10   **5.** 30   **6.** $^-10$ **7.** $^-2$ **8.** $^-2$ **9.** 50
10. $16 + 7(^-3) = {}^-5°$

Review Exercises (*page 75*)
1.

2. $(^-6)$; $B(^-1)$; $C(4)$; $D(9)$ **3.** 8   **4.** $(2 \cdot 4)$ **5.** $^-2$ **6.** $3 + (0 + ^-1)$ **7.** $2 \cdot 5$
8. $2 \cdot 8 + 2(^-3)$ **9.** 57   **10.** 72   **11.** 65   **12.** 24   **13.** 29   **14.** 6   **15.** 8   **16.** 35
17. 21   **18.** $^-18$ **19.** 3   **20.** $^-5$ **21.** $^-20$ **22.** 0   **23.** 38   **24.** $^-16$ **25.** $^-23$
26. 105   **27.** 3   **28.** 16   **29.** $^-4$ **30.** 0   **31.** 1   **32.** $^-2$ **33.** $5 \cdot 7 + 3 \cdot 5 = \$50$
34. $8 \cdot 4 - 6 = \$26$ **35.** $2 \cdot 4 + 15 + 10(^-2) = \$3$

CHAPTER 2

Exercise 2.1 (page 85)

1. $10 + y$ **3.** ^-3m **5.** $a - 6$ **7.** $\dfrac{t}{^-15}$ **9.** $x + ^-9$ **11.** $32q$ **13.** $n - 6$ **15.** $\dfrac{2x}{7}$

17. $6k + 9$ **19.** $\dfrac{p}{7} - 5$ **21.** $9 - 4x$ **23.** $^-4(x + 12)$ **25.** 15; 5; 9   **27.** 7; $^-6$; 0

29. 6; $^-3$; 0   **31.** 5; undefined; $^-9$ **33.** 42; $^-49$; 0   **35.** $^-32$; 20; 0   **37.** 10; 6; 5
39. $^-4$; 0; 14   **41.** 1; $^-1$; 0   **43.** 9; $^-39$; 0   **45.** $2y$; \$20; \$50   **47.** $20 - x$; 12; 7
49. $200 - 3n$; 182; 110

Checkup 2.1 (page 87)

1. $^-3 + x$ **2.** $y - 11$ **3.** ^-6m **4.** $2b + 5$ **5.** $\dfrac{y}{^-2} - 6$ **6.** $^-3$; 24; 0

7. $^-4$; undefined; 3   **8.** $^-21$; 6; 0   **9.** 1; $^-4$; 0   **10.** $14x$; 70; 168

Exercise 2.2 (page 95)
1. $3x$, $^-9$ **3.** $2a$, ^-7b, 9   **5.** 1, ^-5m, $6m$ **7.** $13x$, ^-5y, $2z$, $^-7$
9. $^-19$ (constant); $4x$ (variable)   **11.** $2x$, $3y$ (variable)   **13.** 5 (constant); ^-4a, ^-7b (variable)
15. 1, $^-4$ (constant); $7m$, ^-5n, $14p$ (variable)   **17.** 3 is coefficient of x.
19. 4 is coefficient of x; $^-12$ is coefficient of y.   **21.** 3 is coefficient of a; 4 is coefficient of b.
23. 3 is coefficient of m; $^-1$ is coefficient of n; $^-2$ is coefficient of p.   **25.** $8a$, ^-3a
27. $3h$ and $5h$; $6k$ and ^-2k **29.** $5a$ and $3a$; $^-6$ and $^-9$
31. ^-2x and $3x$; ^-2y and $4y$; 1 and $^-7$ **33.** ^-2a **35.** $3x$ **37.** $9 - 5y$ **39.** $15m + 3$
41. $7h + k - 4$ **43.** $11x + 5y + 7z$ **45.** $^-8a + 10b + 7$ **47.** $5m - 11n + 4p$
49. $^-v - 6w$

Checkup 2.2 (page 96)
1. 3, ^-7x, $2y$ **2.** 11, $^-13$ (constant); ^-4a, $7a$ (variable)
3. 3 is coefficient of x; $^-1$ is coefficient of y; $^-4$ is coefficient of z.
4. $5x$ and ^-7x; ^-2y and $2y$; 4 and $^-1$ **5.** ^-4a **6.** $4x$ **7.** $8y + 5$ **8.** $m + 2$
9. $^-x + 7y - 6$ **10.** $^-5b + 2c$

Exercise 2.3 (page 107)
1. $8x$ **3.** ^-15x **5.** ^-24m **7.** $6x$ **9.** ^-24y **11.** $30k$ **13.** ^-5x **15.** $5x$ **17.** a
19. m **21.** ^-5y **23.** ^-n **25.** $11x$ **27.** $7x$ **29.** $4a - 4b$ **31.** $-3x - 6y$
33. $21x - 14y$ **35.** $^-20a + 28b$ **37.** $^-18m - 27n$ **39.** $3x - 12y$ **41.** $-2x + 5y$
43. $3a + 7b - 5$ **45.** $x + 1$ **47.** $7a + 5b$ **49.** ^-6m **51.** $9x - 4y$ **53.** $11m - 20$
55. $^-7x + 10y$ **57.** $4a + 12$ **59.** $10x - 15y$ **61.** $^-38a + 6b$ **63.** ^-20x **65.** $x + 4$

Checkup 2.3 (page 109)
1. ^-32x **2.** $40y$ **3.** ^-4x **4.** ^-a **5.** ^-5x **6.** $^-10x + 15y$ **7.** $2x - 8$
8. $12a - 15b$ **9.** $^-6a + 6$ **10.** $49n$

Review Exercises (page 113)
1. $6x$ **2.** $y + 9$ **3.** $3m + 10$ **4.** $23 - 4k$ **5.** $9(x + 7)$ **6.** 4; $^-7$; $^-21$
7. 4; undefined; $^-8$ **8.** $^-6$; 0; 33   **9.** 3; $^-9$; 11   **10.** 0; $^-12$; $^-4$ **11.** $4h$; 12

12. $\dfrac{x}{3}$; \$70,000   **13.** $200 + 20x$; \$260   **14.** $^-12$ (constant); $3x$ (variable)

15. ^-2x, $7y$ (variable)   **16.** 2 (constant); $4a$, ^-3b (variable)
17. $^-2$, $^-9$ (constant); $3m$, $4n$, ^-p (variable)   **18.** 3 is coefficient of x.
19. 7 is coefficient of x; $^-2$ is coefficient of y.   **20.** $^-3$ is coefficient of a; 1 is coefficient of b.
21. 1 is coefficient of m; $^-1$ is coefficient of n; 2 is coefficient of p.   **22.** ^-2x **23.** $5m + 2$
24. $^-7a - 13b + 3$ **25.** $4x + 5z$ **26.** ^-20x **27.** $24y$ **28.** ^-6x **29.** a **30.** ^-3x
31. $28a - 35b$ **32.** $^-7x + 2y$ **33.** $^-4x + 7$ **34.** ^-16m **35.** $^-9a - 6$ **36.** $20x - 50y$
37. $15x + 3$

CHAPTER 3

Exercise 3.1 (page 127)
1. Yes   **3.** No   **5.** Yes   **7.** No   **9.** Yes   **11.** No   **13.** $x = 7$ **15.** $x = 12$
17. $y = -2$ **19.** $y = -8$ **21.** $n = 5$ **23.** $a = 10$ **25.** $k = 6$ **27.** $x = 0$
29. $m = 5$ **31.** $y = -8$ **33.** $a = -7$ **35.** $x = 0$ **37.** $x = 7$ **39.** $x = -9$
41. $x = -17$ **43.** $x = 6$ **45.** $x = 0$ **47.** $x = -6$ **49.** $x = 12$ **51.** $x = -21$
53. $x = -56$ **55.** $x = 0$ **57.** $x = -40$ **59.** $x = 63$

Checkup 3.1 (page 129)
1. Yes   **2.** $x = -1$ **3.** $x = 7$ **4.** $x = 4$ **5.** $x = 0$ **6.** $x = 9$ **7.** $x = -6$
8. $x = 0$ **9.** $x = 24$ **10.** $x = -60$

Exercise 3.2 (page 137)
1. $x = 7$ **3.** $x = 10$ **5.** $y = -5$ **7.** $x = -1$ **9.** $a = -9$ **11.** $m = -6$
13. $x = -13$ **15.** $x = -3$ **17.** $x = 13$ **19.** $y = 0$ **21.** $a = 8$ **23.** $x = 12$
25. $a = -40$ **27.** $y = 16$ **29.** $x = -4$ **31.** $x = 10$ **33.** $x = 6$ **35.** $x = 0$
37. $x = -10$ **39.** $x = 5$ **41.** $x = 8$ **43.** $x = -3$ **45.** $x = 17$ **47.** $x = 1$
49. $x = 5$

Checkup 3.2 (page 138)
1. $x = 5$ **2.** $x = -1$ **3.** $x = -4$ **4.** $y = -6$ **5.** $x = 25$ **6.** $x = -3$ **7.** $x = 4$
8. $a = 0$ **9.** $x = -8$ **10.** $x = 6$

Exercise 3.3 (page 147)
1. $x = 4$ **3.** $x = 7$ **5.** $x = -2$ **7.** $x = 2$ **9.** $x = -4$ **11.** $x = 3$ **13.** $x = 3$
15. $x = -4$ **17.** $x = 0$ **19.** $x = -4$ **21.** $x = -11$ **23.** $x = 3$ **25.** $x = 9$
27. $x = -3$ **29.** $x = 3$ **31.** Lyle 7 pounds, Lydia 5 pounds   **33.** 37   **35.** 9
37. 115   **39.** 15 by 20 feet

Checkup 3.3 (page 149)
1. $x = 3$ **2.** $x = -5$ **3.** $y = 2$ **4.** $x = 4$ **5.** $a = -3$ **6.** $y = 8$ **7.** 9   **8.** \$20

Review Exercises (page 153)
1. Yes   **2.** No   **3.** No   **4.** Yes   **5.** $x = 10$ **6.** $y = -14$ **7.** $m = 4$
8. $a = -12$ **9.** $x = 8$ **10.** $x = -18$ **11.** $x = -7$ **12.** $x = 0$ **13.** $x = -3$
14. $y = -14$ **15.** $x = 9$ **16.** $x = 1$ **17.** $x = -3$ **18.** $x = 4$ **19.** $x = -6$
20. $x = 4$ **21.** $x = 4$ **22.** $x = -10$ **23.** $y = 5$ **24.** $x = 4$ **25.** $a = 2$
26. $x = -19$ **27.** $x = -3$ **28.** $x = 0$ **29.** $x = 11$ **30.** $x = -3$ **31.** 6 and 7
32. 4 by 12 feet   **33.** 4

CHAPTER 4

Exercise 4.1 (page 163)
1. Monomial   **3.** Binomial   **5.** Trinomial   **7.** Binomial   **9.** Binomial   **11.** Trinomial
13. Polynomial of 4 terms   **15.** $3x^2y^3$ **17.** $-a^4$ **19.** $(-2x)^3$ **21.** $3x^3 - 4y^4$
23. $8 \cdot a \cdot a \cdot a$ **25.** $27 \cdot a \cdot a \cdot a$ **27.** $-1 \cdot x \cdot x \cdot x \cdot x$ **29.** $3 \cdot x \cdot x \cdot y \cdot y \cdot y$ **31.** -27
33. -16 **35.** 4   **37.** 64   **39.** 1   **41.** $4x - 17y$ **43.** $4x^2 - 15x - 2$ **45.** $-y^2 + 1$
47. 1   **49.** $-y^2 - 3$

Checkup 4.1 (page 164)
1. Binomial   **2.** Trinomial   **3.** $-3x^4$ **4.** $2a^2 - b^4$ **5.** $-8 \cdot x \cdot x \cdot x$
6. $5 \cdot x \cdot x \cdot x \cdot y \cdot y$ **7.** 16   **8.** -27 **9.** $-2x^2 + 13x - 6$ **10.** $y^2 - y - 3$

Exercise 4.2 (page 171)
1. x^5 **3.** y^6 **5.** x^3y^2 **7.** a^7 **9.** y^9 **11.** $10x^5$ **13.** $-20x^8$ **15.** $10a^{12}$
17. $-9y^4$ **19.** $-6m^8$ **21.** $-12x^7y^5$ **23.** $6x^7y^4$ **25.** $42x^4y^4$ **27.** $35x^3y^4z^5$
29. $-24x^5yz^2$ **31.** a^6 **33.** x^5y^5 **35.** $49x^2$ **37.** $9y^4$ **39.** $-64x^3$ **41.** $25a^5$
43. x^6y^3 **45.** $81x^4y^6$ **47.** $3a^{16}b^{10}$ **49.** x^4y^{12} **51.** $20x^7$ **53.** $-200y^{23}$
55. $-24a^4b^7$ **57.** $11x^2y^6$ **59.** $-9a^6b^{12}$

Checkup 4.2 (*page 172*)
1. x^8 **2.** y^6 **3.** $21a^7$ **4.** $-36x^{10}$ **5.** $10x^3y^7$ **6.** $14x^5y^4$ **7.** $-12x^3y^7z^7$ **8.** $9x^2$
9. $-8y^3$ **10.** $54a^8b^3$

Exercise 4.3 (*page 175*)
1. $10x + 15$ **3.** $-14x^2 + 28$ **5.** $6x^2 - 12x$ **7.** $2y^3 - 5y^2$ **9.** $-a^4 + 2a^2$
11. $36x^4 - 16x^2$ **13.** $-21a^5 + 9a^4$ **15.** $10x^2y + 4xy^2$ **17.** $-11x^3y^2 + 4x^2y^3$
19. $14a^4b^2 - 2a^2b^4$ **21.** $2m^3 - 10m^2 + 12m$ **23.** $15x^4y^4 - 5x^3y^3 - 20x^2y^2$
25. $-3a^2b + 2a^3b^2 - 4a^4b^3$ **27.** $32m^4n - 8m^3n^2 + 4m^2n^3 - 12mn^4$ **29.** $4x^3y$
31. $x^4 + 2x^2 + 24$ **33.** $9m^2 - n^2$ **35.** $4ax - 5axy$ **37.** $7x^2y - 10xy^2$
39. $-13x^3y^7 - 5x^3y^3$

Checkup 4.3 (*page 176*)
1. $10x^2 - 20x$ **2.** $3y^3 - 2y^2$ **3.** $-2a^4 + 14a^3$ **4.** $15x^2y - 6xy^2$
5. $16m^3 - 28m^2 + 8m$ **6.** $12x^4y^4 - 18x^3y^3 + 24x^2y^2$
7. $-8m^4n + 2m^3n^2 - 4m^2n^3 + 2mn^4$ **8.** $-x^3y$ **9.** $15x^2 + xy + y^2$ **10.** $7ax$

Exercise 4.4 (*page 183*)
1. $x^2 + x - 6$ **3.** $12y^2 - 23y + 5$ **5.** $-6a^2 - 11a + 35$ **7.** $12x^2 + 8x - 15$
9. $18 - 13y + 2y^2$ **11.** $2a^2 - 11ab + 5b^2$ **13.** $2x^2 - 5xy - 3y^2$
15. $-2x^2 + 5xy - 3y^2$ **17.** $a^2 + 6ab + 9b^2$ **19.** $4 - 25x^2$ **21.** $16x^2 - 8xy + y^2$
23. $ac + ad - 2bc - 2bd$ **25.** $a^3 - 3a^2 - 4a + 12$ **27.** $6a^4 - 31a^2 + 5$
29. $25a^4 - 20a^2 + 4$ **31.** $a^6 - 2a^3 - 3$ **33.** $x^2y^2 - 5xy - 50$ **35.** $a^2b^2 - 4c^2$
37. $9y^2z^2 - 12yz + 4$ **39.** $36a^4 - b^2$ **41.** $x^4 + 4x^2y^2 + 4y^4$ **43.** $x^3 - 5x^2 + 7x - 3$
45. $b^3 - 125$ **47.** $x^4 + 3x^3 - 2x^2 - 12x - 8$ **49.** $6x^3 + x^2 + x + 2$
51. $3x^3 - 21x^2 + 30x$ **53.** $y^4 - y^3 - 56y^2$ **55.** $a^3 - 4a^2b - 7ab^2 + 10b^3$
57. $x^3 - 4x^2 - 3x + 18$ **59.** $y^3 - 3y^2 - 24y + 80$ **61.** $a^3 + 6a^2 + 12a + 8$
63. $27x^3 - 27x^2y + 9xy^2 - y^3$

Checkup 4.4 (*page 184*)
1. $x^2 + 3x - 10$ **2.** $2y^2 - 15y + 18$ **3.** $25 - 20y + 4y^2$ **4.** $6a^2 - 13ab - 5b^2$
5. $x^4 - 25$ **6.** $-2x^2 - xy + 3y^2$ **7.** $a^3 - 6a^2 + 7a - 42$ **8.** $x^2y^2 - xy - 6$
9. $x^3 + 8$ **10.** $2x^3 + 4x^2 - 30x$

Exercise 4.5 (*page 189*)
1. $x^2 + 12x$ **3.** $x^2 - 7x$ **5.** $5x^2 + 20x + 25$ **7.** $x^2 + x$ **9.** $x^2 + 5x - 45$
11. $2x^2 + 4x$ **13.** $x^2 + 4x$ **15.** $\frac{1}{2}h^2 + \frac{5}{2}h$ **17.** $x^2 + 15x$ **19.** $5x^2 + 12x$

Checkup 4.5 (*page 190*)
1. $x^2 - 3x$ **2.** $7x^2 - 3x$ **3.** $2x^2 + 1$ **4.** $2x^2 + 14x + 49$ **5.** $2h^2$

Review Exercises (*page 193*)
1. **(a)** Binomial   **(b)** Binomial   **(c)** Monomial   **2.** **(a)** $-2x^2y^3$ **(b)** $(3a)^3$ **(c)** $5m^2 - 3n^3$
3. **(a)** $2 \cdot x \cdot x \cdot x \cdot x$ **(b)** $-125 \cdot a \cdot a \cdot a$ **(c)** $-3 \cdot m \cdot m \cdot m \cdot n \cdot n$ **4.** **(a)** -27 **(b)** -2 **(c)** 1
5. $-7a - 2b$ **6.** $-3x^2 - 4x - 14$ **7.** x^{10} **8.** $-6a^8$ **9.** $-18x^5y^8$ **10.** m^6
11. $81y^4$ **12.** $x^6y^4z^2$ **13.** $3a^{13}b^6$ **14.** $-15x^4y^6$ **15.** $40x - 24$ **16.** $21y^2 + 14y$
17. $3m^4 - 12m^3n - 15m^2n^2$ **18.** $35x^2y - 55xy^2$ **19.** $-6a^2b + 3a^3b^2 - 4a^4b^3$
20. $2x^2y + xy^2$ **21.** $x^2 + x - 72$ **22.** $12y^2 - 5y - 2$ **23.** $6a^2 - 11ab + 3b^2$
24. $x^2 - 10xy + 25y^2$ **25.** $m^4 + 2m^2 - 15$ **26.** $4x^2y^2 - 4xy + 1$ **27.** $y^3 - 4y^2 - y + 4$
28. $2x^3 - 18x$ **29.** $a^3 - 4a^2 + a + 6$ **30.** $y^3 - y^2 - 8y + 12$
31. $4x^2 + 27x + 64$ **32.** $x^2 - 10x$

CHAPTER 5

Exercise 5.1 (*page 201*)
1. $7(x + 2)$ **3.** $7(3x - 4)$ **5.** $6(7 - y)$ **7.** $6(a^2 - 6)$ **9.** $-6(2x - 3y)$
11. $15(2x^2 + y^2)$ **13.** $11(a^3 + 4b^3)$ **15.** $7(x^2 - 3x - 1)$ **17.** $5(2y^2 + 3y - 7)$
19. $7(4 - a + 5a^2)$ **21.** $10(5x^2 - xy + 2y^2)$ **23.** $2(x^3 - 3x^2 + 2x + 4)$ **25.** $x(x + 1)$
27. $-y(y^2 - y + 2)$ **29.** $2b(b^2 - 3b + 2)$ **31.** $b^3(7 + 2b^2 - 3b^4)$
33. $3x(x^2 - 2xy + 4y^2)$ **35.** $-a^2b(a^2 - 6ab + 5b^2)$ **37.** $2xy(2x^2y^2 - 10xy + 1)$

39. $x^3y^2z(1 - 2xy + x^2y^2)$ **41.** $9x(7x + 8)$ **43.** $20(5a^3 - 3b^3)$ **45.** $16b^3(1 + 4b)$
47. $6a(a^2 - 7a - 13)$ **49.** $5xy(x^2 - 16y^2)$ **51.** $-7b^2(1 + 6b - 5b^2)$
53. $xy(9x^2 - 12xy + 20y^2)$ **55.** $8a^2(9a^2 - 6a + 1)$ **57.** $4xy(3x^2 - 2xy - y^2)$
59. $9xyz(x^3 + 2x^2y - 3xy^2 - 5y^3)$

Checkup 5.1 (page 202)
1. $3(x + 5)$ **2.** $7(5 - 2y)$ **3.** $-x(x - 1)$ **4.** $9(x^2 - 2x + 3)$ **5.** $5(3x^2 - 2xy + 5y^2)$
6. $8a(a^2 - 2a - 5)$ **7.** $2xy(5x^2 - 13xy + 7y^2)$ **8.** $3ab(3a^2b^2 - 9ab + 1)$
9. $-8m(m^2 + 3)$ **10.** $x^2yz(x^2 - xy + y^2)$

Exercise 5.2 (page 207)
1. $(x + 5)(x - 5)$ **3.** $(3y + 1)(3y - 1)$ **5.** $(7 + a)(7 - a)$ **7.** $(2a + 11)(2a - 11)$
9. $(5b + 13)(5b - 13)$ **11.** $(x + y)(x - y)$ **13.** $(a + 6b)(a - 6b)$
15. $(2x + 15y)(2x - 15y)$ **17.** $(mn + 12)(mn - 12)$ **19.** $(3ab + c)(3ab - c)$
21. $(5xy + 4z)(5xy - 4z)$ **23.** $(x^2 + 9)(x + 3)(x - 3)$ **25.** $(5x^2 + 11)(5x^2 - 11)$
27. $(15 + a^2b^2)(15 - a^2b^2)$ **29.** $3(x + 5)(x - 5)$ **31.** $11(x + 2y)(x - 2y)$
33. $y(4 + 3x)(4 - 3x)$ **35.** $6m(m + 5)(m - 5)$ **37.** $b(7a + 2b)(7a - 2b)$
39. $2y^2(x + 6y)(x - 6y)$ **41.** $3(m^2 + 9)(m + 3)(m - 3)$ **43.** $2xy(x + 2y)(x - 2y)$
45. $3xy(3xy + 2)(3xy - 2)$ **47.** $a^2(a^2 + 16)(a + 4)(a - 4)$ **49.** $z^2(17xy + 1)(17xy - 1)$

Checkup 5.2 (page 208)
1. $(x + 8)(x - 8)$ **2.** $(5 + y)(5 - y)$ **3.** $(3a + 10)(3a - 10)$ **4.** $(x + 7y)(x - 7y)$
5. $(4x^2 + 9)(2x + 3)(2x - 3)$ **6.** $3(x + 5y)(x - 5y)$ **7.** $7y^2(x + 2y)(x - 2y)$
8. $5(m^2 + 9)(m + 3)(m - 3)$ **9.** $4xy(x + 10y)(x - 10y)$ **10.** $3a(1 + 9a)(1 - 9a)$

Exercise 5.3 (page 217)
1. $(x - 3)(x - 4)$ **3.** $(x + 5)(x + 4)$ **5.** $(x - 12)(x - 12)$ **7.** $(a + 7)(a + 2)$ *error should be* –
9. $(4 + x)(4 + x)$ **11.** $(y - 6)(y + 5)$ **13.** $(x - 7)(x + 5)$ **15.** $(m + 8)(m + 4)$
17. $(x + 9)(x - 8)$ **19.** $(b - 11)(b + 3)$ **21.** $(x - 2y)(x - 2y)$ **23.** $(a + 6b)(a + 7b)$
25. $(m + 3n)(m - 2n)$ **27.** $(x - 5y)(x - 5y)$ **29.** $(x + 11y)(x + 11y)$
31. $(13 - m)(3 - m)$ **33.** $(xy + 5)(xy - 3)$ **35.** $(ab + 6c)(ab - c)$
37. $(x^2 - 3)(x^2 - 2)$ **39.** $3(x - 3)(x + 2)$ **41.** $5a(a - 4)(a + 3)$
43. $-2y(y - 5)(y + 2)$ **45.** $7(x - 2y)(x - y)$ **47.** $-10a(b^2 - 3)(b^2 - 3)$
49. $4(m^2 + 3n^2)(m^2 - 2n^2)$ **51.** $xy(x - 10y)(x - 8y)$ **53.** $2a^2b(a - 16b)(a + 2b)$
55. $-y^2(y + 19)(y - 2)$ **57.** $abc(a - 9b)(a + 7b)$

Checkup 5.3 (page 218)
1. $(x - 8)(x - 7)$ **2.** $(x + 3)(x + 2)$ **3.** $(a + 8)(a - 3)$ **4.** $(x - 2y)(x - 2y)$
5. $(6 + x)(2 + x)$ **6.** $(xy + 7z)(xy - z)$ **7.** $-3(x - 5)(x - 4)$ **8.** $x^2(a - 4b)(a + 2b)$
9. $-xy(x + 7y)(x + 4y)$ **10.** $3y(x - 3y)(x + y)$

Exercise 5.4 (page 227)
1. $(2x + 1)(x + 3)$ **3.** $(5x + 2)(x + 1)$ **5.** $(3a - 2)(a - 5)$ **7.** $(5x - 2)(2x - 3)$
9. $(5y + 4)(2y - 3)$ **11.** $(4m - 3)(2m + 5)$ **13.** $(3x + 8y)(2x - 7y)$
15. $(9x - 2y)(9x - 2y)$ **17.** $(13a - 4b)(3a - 2b)$ **19.** $(11 + 6a)(11 + 6a)$
21. $5(3c - d)(2c - 9d)$ **23.** $-2(4y - 3)(2y + 5)$ **25.** $x^2(3x - 7)(3x - 4)$ **27.** Prime
29. $a^2b^2(4b - 1)(b - 6)$ **31.** $-9(2x^2 - x + 3)$ **33.** $2xy(3x + 7)(4x - 5)$
35. $(3m - 5n)(4m + 11n)$ **37.** $x^2(2x^2 + 3y^2)(2x^2 + 3y^2)$ **39.** Prime   **41.** $(5 + c)(a + b)$
43. $(x - 5)(x^2 + 6)$ **45.** $(2x + 3y)(7a - 5)$ **47.** $(x - 2)(x + 5)(x - 5)$
49. $(7a - 4b)(10a + 3)$ **51.** $3x^2(x^2 - 5y^2)(x^2 - 5y^2)$ **53.** $(x^2 - 5)(x + 2)(x - 2)$

Checkup 5.4 (page 228)
1. $(3x + 1)(x + 5)$ **2.** $(2a - 7)(a - 3)$ **3.** $(2a - 5b)(2a + 3b)$ **4.** $3(8a - 7)(a + 1)$
5. Prime   **6.** $-4(2x - 1)(x + 6)$ **7.** $2x^2(3x - 5y)(3x - 5y)$ **8.** $(3x^2 - 4y^2)(x^2 + 2y^2)$
9. $(m + n)(5 + p)$ **10.** $(x + y)(3a - 2b)$

Review Exercises (page 231)
1. $5(2x - 3)$ **2.** $3x(5x - 4)$ **3.** $y^2(2y - 5)$ **4.** $4(3x^2 + 2y)(3x^2 - 2y)$
5. $2xy(5x + 2y)$ **6.** $2m(m - 3)(m - 2)$ **7.** $-a^2b(3 - 2ab + 4a^2b^2)$ **8.** $(x + 7)(x - 7)$
9. $(5 + y)(5 - y)$ **10.** $(a + 8b)(a - 8b)$ **11.** $(xy + 11)(xy - 11)$ **12.** $3(m + 4)(m - 4)$
13. $b(8a + 5b)(8a - 5b)$ **14.** $2x^2(3x + 2y)(3x - 2y)$ **15.** $(x^2 + 9y^2)(x + 3y)(x - 3y)$

16. $(x - 3)(x - 5)$ **17.** $(a + 3b)(a + 9b)$ **18.** $(xy + 2)(xy - 5)$ **19.** $(9 + m)(7 - m)$
20. $(ab - 8c)(ab + 7c)$ **21.** $(2a - 5b)(2a - 5b)$ **22.** $(3x - 7)(2x + 1)$
23. $-2(x - 3y)(x + y)$ **24.** $5(2m + 11)(m - 4)$ **25.** $(4x - 5y)(4x - 5y)$
26. $(13 + 2a)(13 - 2a)$ **27.** $2xy(5x + y)(3x - y)$ **28.** $(8x - 3)(x - 1)$
29. $(x + y)(4 + z)$ **30.** $(x - 4)(x + 1)(x - 1)$
31. $(x^2 - 5)(x + 3)(x - 3)$ **32.** $3(y^2 + 25)(y - 2)(y + 2)$

CHAPTER 6

Exercise 6.1 (page 243)
1. $x = 3, x = 2$ **3.** $x = -2, x = 6$ **5.** $a = -13, a = -5$ **7.** $x = 0, x = 8$
9. $x = 7, x = -9$ **11.** $y = 3, y = -1$ **13.** $x = 2, x = 4$ **15.** $x = 4, x = -4$
17. $x = -3, x = -6$ **19.** $m = 0, m = -15$ **21.** $x = -9, x = 6$ **23.** $x = 0, x = 5$
25. $x = 8, x = -7$ **27.** $x = 15$ **29.** $x = 10, x = -10$ **31.** $n = 2, n = 0$
33. $y = -11, y = 3$ **35.** $x = -1$ **37.** $a = 9, a = -9$ **39.** $z = -9, z = -11$

Checkup 6.1 (page 244)
1. $x = 13, x = 0$ **2.** $a = 9, a = -7$ **3.** $y = 12, y = -12$ **4.** $x = -10, x = 2$
5. $x = -9, x = 4$ **6.** $x = 0, x = 4$ **7.** $x = -3, x = -7$ **8.** $x = 8$
9. $x = 1, x = -1$ **10.** $x = -9, x = 2$

Exercise 6.2 (page 251)
1. $x = -6, x = 3$ **3.** $x = 9, x = -9$ **5.** $x = 0, x = 5$ **7.** $y = 4, y = 9$
9. $y = 9, y = -8$ **11.** $a = -5, a = 2$ **13.** $x = 8$ **15.** $x = 13, x = -4$
17. $a = 12, a = -12$ **19.** $y = 0, y = -17$ **21.** $x = 7, x = -3$ **23.** $y = 20, y = -5$
25. $x = 9, x = 5$ **27.** $x = 0, x = 9$ **29.** $x = -2, x = -3$ **31.** $y = 2, y = -2$
33. $y = 3$ **35.** $y = 0, y = 1$ **37.** $x = 9, x = -3$ **39.** $x = 6, x = -6$
41. $x = 5, x = -5$ **43.** $a = 0, a = 3$ **45.** $x = 8, x = -6$ **47.** $x = 5$ **49.** $x = -1$
51. $y = 0, y = 9$

Checkup 6.2 (page 252)
1. $x = 8, x = -5$ **2.** $x = 0, x = 7$ **3.** $a = -9, a = 7$ **4.** $x = 7$ **5.** $a = 3, a = -3$
6. $x = 10, x = -7$ **7.** $x = 3, x = 2$ **8.** $x = 8, x = -8$ **9.** $y = 0, y = 7$
10. $y = -5, y = -2$

Exercise 6.3 (page 256)
1. $-7, -2$ or $4, 9$ **3.** 9 ft, 8 ft   **5.** $-7, 5$ **7.** 9 cm   **9.** 15, 8   **11.** 8 by 16 ft
13. 150 by 200 ft   **15.** 36 ft   **17.** 5, 6 or $-6, -5$ **19.** 4, 6

Checkup 6.3 (page 258)
1. $-6, -2$ or 4, 8   **2.** 7 by 5 cm   **3.** 12 cm   **4.** 3 in., 4 in.   **5.** $-3, -2$ or 2, 3

Review Exercises (page 260)
1. $x = 7, x = -8$ **2.** $a = 0, a = 3$ **3.** $y = 9, y = -10$ **4.** $x = 8, x = 2$
5. $x = 6, x = -4$ **6.** $x = 13, x = -13$ **7.** $m = 0, m = -15$ **8.** $a = 9, a = -9$
9. $n = 0, n = 3$ **10.** $z = -13, z = -4$ **11.** $y = -8, y = -3$ **12.** $x = 11, x = -11$
13. $y = 0, y = -23$ **14.** $x = -10, x = 4$ **15.** $x = 21, x = 3$ **16.** $x = 9, x = -9$
17. $z = 0, z = 1$ **18.** $x = 10, x = -3$ **19.** $y = 0, y = 3$ **20.** $x = 8, x = -2$
21. 6 by 10 ft   **22.** 15 centimeters   **23.** 9, 13 or $-13, -9$ **24.** 12 feet
25. 5, 7 or $-3, -1$

CHAPTER 7

Exercise 7.1 (page 273)
1. $\frac{4}{1}$ **3.** $\frac{-9}{10}$ **5.** $\frac{16}{3}$ **7.** $\frac{-13}{1}$ **9.** $\frac{13}{100}$ **11.** $\frac{-31}{1000}$ **13.** $x = \frac{9}{2}$ **15.** $y = \frac{14}{5}$
17. $a = \frac{-5}{7}$ **19.** $z = \frac{-5}{4}$ **21.** $x = \frac{9}{5}$ **23.** $y = \frac{3}{10}$ **25.** $y = \frac{49}{6}$ **27.** $a = 0$
29. $x = \frac{2}{5}, x = -3$ **31.** $z = \frac{4}{3}, z = \frac{-5}{2}$ **33.** $x = \frac{1}{5}, x = 2$ **35.** $a = \frac{2}{7}, a = \frac{1}{3}$

37. $x = \frac{7}{4}$, $x = 2$ **39.** $y = 0$, $y = \frac{4}{5}$ **41.** $z = \frac{2}{3}$, $z = \frac{-2}{3}$ **43.** $x = \frac{-5}{4}$, $x = \frac{2}{3}$

45. $x \neq 0$ **47.** $x \neq -1$ **49.** $x \neq 0$ **51.** $y \neq 8$ **53.** $x \neq \frac{1}{2}$ **55.** None

Checkup 7.1 (page 275)

1. $\frac{-7}{1}$ **2.** $\frac{31}{100}$ **3.** $x = \frac{-1}{3}$ **4.** $y = \frac{11}{6}$ **5.** $x = 0$ **6.** $z = \frac{-1}{5}$, $z = \frac{7}{4}$

7. $x = \frac{5}{2}$, $x = \frac{-5}{2}$ **8.** $y = 0$, $y = \frac{3}{2}$ **9.** $x \neq 0$ **10.** $y \neq -7$

Exercise 7.2 (page 291)

1. Yes   **3.** No   **5.** Yes   **7.** No   **9.** Yes   **11.** $\frac{1}{5}$ **13.** $\frac{7}{5}$ **15.** $\frac{9}{16}$ **17.** $\frac{-5}{7}$

19. $\frac{3}{2}$ **21.** $\frac{-1}{7}$ **23.** $2x^3$ **25.** $\frac{-a}{4}$ **27.** $\frac{3x^2}{y^2}$ **29.** $\frac{-3}{4a^2b^3}$ **31.** $\frac{5}{4x^2z}$ **33.** $\frac{x+y}{3}$

35. $a + 5b$ **37.** $9y - 7$ **39.** $\frac{2}{16+x}$ **41.** $\frac{1}{3x-5y}$ **43.** $\frac{6(a^2+b^2)}{5}$ **45.** $5(x^2 + y^2)$

47. $\frac{7x^2}{x+1}$ **49.** $\frac{-(4a+5c)}{6}$ **51.** $\frac{2x^2+3x+1}{10}$ **53.** $\frac{-(3y^2-5x^2y+7)}{5y^2}$ **55.** $\frac{a+b}{x+y}$

57. $\frac{x-y}{2(x+y)}$ **59.** $\frac{1}{2}$ **61.** $\frac{-1}{3}$ **63.** $\frac{x-y}{3}$ **65.** $\frac{x-7}{4}$ **67.** $\frac{3}{x+y}$ **69.** $\frac{-7}{3(a-b)}$

71. $\frac{x-2y}{3}$ **73.** $\frac{x+5}{x-4}$ **75.** $\frac{x+4}{x+5}$ **77.** $\frac{5(x+3)}{x-7}$ **79.** $\frac{x+5y}{x+y}$ **81.** $\frac{2(a+b)}{a-3b}$

83. $\frac{(x+5)(x-5)}{x^2+25}$ **85.** $\frac{1}{2b-1}$ **87.** $\frac{(x+2y)(x^2+4y^2)}{x+4y}$ **89.** $\frac{-1}{9}$

91. $-1(x+5)$ or $-x - 5$

Checkup 7.2 (page 294)

1. $\frac{3}{5}$ **2.** $\frac{-7}{9}$ **3.** a^4 **4.** $\frac{-1}{4x}$ **5.** $4y - 5$ **6.** $\frac{1}{5x-4y}$ **7.** $\frac{a+b}{4(a-b)}$ **8.** $\frac{-1}{3(x+y)}$

9. $\frac{m+2n}{2}$ **10.** $\frac{x+4}{x+5}$

Exercise 7.3 (page 309)

1. $\frac{2}{21}$ **3.** -3 **5.** $\frac{-3}{5}$ **7.** $\frac{1}{15}$ **9.** 3 **11.** $\frac{14}{15}$ **13.** $\frac{7}{3}$ **15.** $\frac{-1}{15}$ **17.** $\frac{55}{4}$

19. $\frac{9}{10}$ **21.** $\frac{x}{y^2}$ **23.** x **25.** $\frac{105y}{x^2}$ **27.** b **29.** $\frac{-x}{9}$ **31.** $4xz$ **33.** $\frac{5b^5}{a^2}$ **35.** $\frac{75}{4y^2}$

37. $\frac{4}{9b^2}$ **39.** $\frac{5x^6}{6y^{11}}$ **41.** $\frac{2}{x+3}$ **43.** $\frac{3}{8x}$ **45.** $\frac{2b}{a}$ **47.** $\frac{7}{3}$ **49.** $x - 3y$ **51.** $\frac{2(x-2)}{3(x+2)}$

53. $\frac{5y(x+3)}{8(x-3)}$ **55.** $\frac{x+4}{x+6}$ **57.** $\frac{x+5y}{5}$ **59.** $y(x-7)$ **61.** $\frac{x+5}{49}$ **63.** $\frac{1}{2(x-9)}$

65. $\frac{3}{x-2}$ **67.** $\frac{1}{x+3}$ **69.** $\frac{5(a-b)}{a^2b^2}$ **71.** 1 **73.** $\frac{5}{2}$ **75.** $\frac{(2x+1)(3x+2)}{4(2x-1)(3x-2)}$

Checkup 7.3 (page 312)

1. $\frac{2}{21}$ **2.** $\frac{3}{5}$ **3.** -15 **4.** y **5.** $\frac{-2y^2}{x}$ **6.** $\frac{2}{y}$ **7.** $\frac{2(x-3)}{3(x+3)}$ **8.** 1 **9.** $\frac{x-8}{y}$

10. $\frac{1}{3(x+6)}$

Review Exercises (page 315)

1. (a) $\frac{25}{8}$ (b) $\frac{21}{1000}$ (c) $\frac{-7}{1}$ **2.** (a) $x \neq 0$ (b) $x \neq -4$ (c) $x = \frac{1}{2}$ **3.** $x = \frac{5}{2}$ **4.** $y = \frac{17}{5}$

5. $a = \frac{16}{11}$ **6.** $x = \frac{5}{3}$, $x = \frac{-1}{2}$ **7.** $z = \frac{4}{5}$, $z = \frac{-4}{5}$ **8.** $x = \frac{1}{5}$, $x = \frac{3}{2}$ **9.** $\frac{1}{3}$ **10.** $\frac{-5}{3}$

11. $2x^5$ **12.** $\frac{5x^2y^2}{4}$ **13.** $a + 3b$ **14.** $\frac{7}{9+x}$ **15.** $\frac{8x^4}{x^2+1}$ **16.** $\frac{-1}{5}$ **17.** $\frac{x-y}{5}$

18. $\frac{x+11}{x}$ **19.** $\frac{1}{5a-6b}$ **20.** $\frac{3}{x+y}$ **21.** $\frac{x-5}{2x-3}$ **22.** $x + y$ **23.** $\frac{-4}{7}$ **24.** $\frac{5}{2}$

25. -21 **26.** $\dfrac{7y^2}{6}$ **27.** $\dfrac{9bc}{a^2}$ **28.** $\dfrac{30}{x^3y^4}$ **29.** $\dfrac{4}{x+4}$ **30.** $10b^3$ **31.** $\dfrac{x+5y}{3x}$

32. $\dfrac{y^2(x+4)}{8z(x-4)}$ **33.** $\dfrac{x-5}{12x}$ **34.** $\dfrac{5x+2}{5x-2}$ **35.** y **36.** $\dfrac{1}{2}$

CHAPTER 8

Exercise 8.1 (page 325)
1. 42 **3.** 96 **5.** -36 **7.** -51 **9.** 63 **11.** 45 **13.** 93 **15.** $2x$ **17.** $-3x$
19. $8xy$ **21.** $9xy^2$ **23.** $-30a^2$ **25.** $3xy$ **27.** $-3x^2y$ **29.** $-8a^2b^4$ **31.** $8(x-y)$
33. $25x^2$ **35.** $-4(a+b)$ **37.** $3x^2$ **39.** $2a(b-1)$ **41.** $6x^2(x+2)$ **43.** $x^2(x+5)$
45. $(a-1)(a-4)$ **47.** $-3y(x-1)$ **49.** $(x-y)(x-3y)$

Checkup 8.1 (page 327)
1. 40 **2.** -100 **3.** $-3xy^2$ **4.** $16a^4b^2$ **5.** $-30a^2$ **6.** $28x^2$ **7.** $2a^2b$
8. $5x(x-6)$ **9.** $(x-2)(x+4)$ **10.** $(a-b)(a+b)$

Exercise 8.2 (page 337)
1. $\dfrac{4}{5}$ **3.** $\dfrac{1}{3}$ **5.** $\dfrac{-5}{13}$ **7.** $\dfrac{1}{11}$ **9.** 1 **11.** 0 **13.** $3x$ **15.** $\dfrac{2x+5}{3}$ **17.** $\dfrac{5a-6}{7}$

19. $\dfrac{5}{x}$ **21.** $\dfrac{1}{x}$ **23.** $\dfrac{19}{xy}$ **25.** $\dfrac{2x-1}{y}$ **27.** $\dfrac{-8}{3b}$ **29.** $\dfrac{4}{x+y}$ **31.** $\dfrac{9y}{x+3y}$ **33.** 1

35. 3 **37.** $\dfrac{2x}{2x+5}$ **39.** $\dfrac{1}{a+4}$ **41.** $\dfrac{x}{x-2}$ **43.** $\dfrac{x^2+4}{x+2}$ **45.** $3(x-5)$ **47.** $x+2$

49. $\dfrac{2a-3}{a}$ **51.** $\dfrac{3x+1}{2x+5}$ **53.** $\dfrac{1}{x-2y}$ **55.** $\dfrac{2}{x-3}$ **57.** $\dfrac{x+1}{x-4}$ **59.** $\dfrac{x-2}{x-4}$

Checkup 8.2 (page 339)
1. $\dfrac{1}{3}$ **2.** $\dfrac{-x}{2}$ **3.** $\dfrac{4x+1}{5}$ **4.** $\dfrac{1}{x}$ **5.** $\dfrac{2a-1}{b}$ **6.** 1 **7.** 3 **8.** $\dfrac{1}{2x+3}$ **9.** $\dfrac{5}{2x}$
10. $x-8$

Exercise 8.3 (page 349)
1. $\dfrac{8}{15}$ **3.** $\dfrac{2}{9}$ **5.** $\dfrac{4}{21}$ **7.** $\dfrac{-11}{5}$ **9.** 1 **11.** $\dfrac{-8}{45}$ **13.** $\dfrac{5x}{2}$ **15.** $\dfrac{40x+63}{70}$

17. $\dfrac{5x-3y}{30}$ **19.** $\dfrac{3(3a+1)}{a^2}$ **21.** $\dfrac{15x+23}{5x^2}$ **23.** $\dfrac{7y-1}{y}$ **25.** $\dfrac{2(2-x^2)}{x}$

27. $\dfrac{5y+2x}{xy}$ **29.** $\dfrac{-x-3}{6x}$ **31.** $\dfrac{2(18y-25x)}{15xy}$ **33.** $\dfrac{x-5}{24x^2}$ **35.** $\dfrac{4y-5x}{x^2y^2}$

37. $\dfrac{-(8y^2+3x^2)}{12x^2y^2}$ **39.** $\dfrac{8}{a+2}$ **41.** $\dfrac{7x-9}{(x+2)^2}$ **43.** $\dfrac{-2(x+16)}{(x+4)(x-4)}$ **45.** $\dfrac{(x+3)(x-2)}{3(x+1)}$

47. $\dfrac{-11}{3(x+1)}$ **49.** $\dfrac{x^2-x-5}{x(x+5)(x-5)}$ **51.** $\dfrac{1}{x^2}$ **53.** $\dfrac{x+4}{x^4}$ **55.** $\dfrac{3x+35}{(x-7)(x+7)^2}$

57. $\dfrac{4x}{(x+4)(x-2)}$ **59.** $\dfrac{2x-5}{x(2x-1)}$

Checkup 8.3 (page 351)
1. $\dfrac{10}{21}$ **2.** $\dfrac{22x}{15}$ **3.** $\dfrac{9x-2y}{6}$ **4.** $\dfrac{3y+2x}{xy}$ **5.** $\dfrac{11x-21}{x(x-3)}$ **6.** $\dfrac{7x-2}{(x+2)(x-2)}$

7. $\dfrac{15}{2(x+1)}$ **8.** $\dfrac{3x+25}{(x-5)(x+5)^2}$ **9.** $\dfrac{1}{x^2}$ **10.** $\dfrac{17}{(x-9)(x+7)}$

Exercise 8.4 (page 359)
1. $x=\dfrac{7}{2}$ **3.** $x=4$ **5.** $x=\dfrac{-14}{5}$ **7.** $y=-15$ **9.** $x=\dfrac{90}{11}$ **11.** $x=\dfrac{33}{2}$

13. $y=-1$ **15.** $x=3$ **17.** $x=-5$ **19.** $z=-3$ **21.** $x=3$ **23.** $x=\dfrac{13}{3}$

25. $x=-8$ **27.** $x=\dfrac{44}{9}$ **29.** No solution **31.** $x=\dfrac{33}{4}$ **33.** $x=\dfrac{-1}{10}$ **35.** $a=\dfrac{38}{7}$

37. $b=\dfrac{-1}{5}$ **39.** $x=\dfrac{-2}{5}$ **41.** $x=-41$ **43.** $x=\dfrac{-17}{3}$ **45.** $a=\dfrac{3}{5}$

47. No solution **49.** $x=0$

Checkup 8.4 (page 361)
1. $y = 18$ **2.** $x = -2$ **3.** $x = -11$ **4.** $y = -23$ **5.** $a = \dfrac{23}{2}$ **6.** $x = \dfrac{55}{17}$

7. $b = \dfrac{1}{13}$ **8.** No solution   **9.** $a = \dfrac{-12}{7}$ **10.** $x = \dfrac{17}{15}$

Exercise 8.5 (page 367)
1. $2800   **3.** 55 mph   **5.** 165 miles   **7.** 12 by 18 feet   **9.** 100 students

Review Exercises (page 371)
1. 48   **2.** $15x^2$ **3.** $15xy^4$ **4.** $2a^2$ **5.** $3y^2 + 6y$ **6.** $x^2 - 8x + 15$ **7.** $\dfrac{1}{x}$

8. $\dfrac{2x + 3}{y}$ **9.** 1   **10.** $\dfrac{1}{2x + 5}$ **11.** $\dfrac{2a + 1}{a}$ **12.** $\dfrac{4x + 5}{2x + 3}$ **13.** $\dfrac{-4}{15}$ **14.** $\dfrac{29x}{35}$

15. $\dfrac{32x + 1}{4x^2}$ **16.** $\dfrac{-7x + 24}{6x}$ **17.** $\dfrac{-4x^2 - 10xy + 3y^2}{12x^2y^2}$ **18.** $\dfrac{a + 9}{(a + 1)(a - 1)}$

19. $\dfrac{-2x}{x(x - 3)}$ **20.** $\dfrac{2a + 3}{(a + 2)(a + 1)}$ **21.** $\dfrac{-3x + 17}{(x + 6)(x - 3)}$ **22.** $\dfrac{3y + 4}{y(y + 1)(y + 2)}$

23. $x = 16$ **24.** $x = -15$ **25.** No solution   **26.** $x = 18$ **27.** $a = 1$
28. $x = -1$ **29.** 12 gallons   **30.** 25

CHAPTER 9

Exercise 9.1 (page 381)
1. $<$ **3.** $>$ **5.** $=$ **7.** $<$ **9.** $<$ **11.** $>$ **13.** $>$ **15.** $-11 < -5 < 0$

17. $-2 < 3 < 7$ **19.** $\dfrac{-4}{5} < \dfrac{-2}{5} < 0$

21.   **23.**

25.   **27.**

29.   **31.**

33.   **35.**

37.   **39.**

Checkup 9.1 (page 382)
1. $<$ **2.** $=$ **3.** $>$ **4.** $>$ **5.** $-5 < -2 < 0$ **6.** $\dfrac{3}{2} < 3 < \dfrac{7}{2}$

7.   **8.**

9.   **10.**

Exercise 9.2 (page 389)
1. $1 < 9$ **3.** $6 > -5$ **5.** $8 > -12$ **7.** $-6 < -2$ **9.** $-7 > -14$ **11.** $-2 < 1$
13. $13 > 1$ **15.** $4 > -8$

17. $x \le 4$

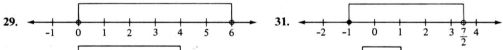
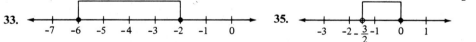
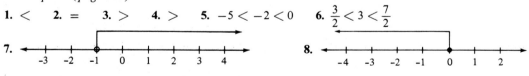
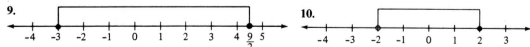

19. $x > -5$

21. $a < \dfrac{-7}{2}$

23. $a \geq 5$

25. $x < -3$

27. $z \geq -1$

29. $y < 2$

31. $x \geq 2$

33. $x \geq \dfrac{9}{2}$ **35.** $y < 0$ **37.** $x > 12$ **39.** $a < -1$ **41.** $x \leq \dfrac{1}{2}$ **43.** $x < \dfrac{-5}{4}$

45. $z > -3$ **47.** $x \leq 12$ **49.** $x \leq \dfrac{-1}{5}$

Checkup 9.2 (*page 390*)

1. $3 > -4$ **2.** $15 > 0$ **3.** $-2 < 3$ **4.** $x \geq -2$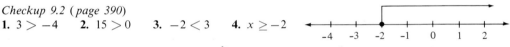

5. $a \geq -3$

6. $y < 1$

7. $x > 3$

8. $a \leq \dfrac{5}{3}$

9. $x > \dfrac{-5}{4}$

10. $y > \dfrac{-3}{2}$

Exercise 9.3 (*page 403*)

1. $-2 < x < 3$ **3.** $-4 \leq x \leq \dfrac{1}{2}$ **5.** $-3 \leq x \leq 4$ **7.** $-4 \leq x < 16$ **9.** $-1 \leq x \leq \dfrac{2}{3}$

11. $-2 \leq x \leq 3$ **13.** $x > 0$ **15.** $x < 5$ **17.** $x > 1$ or $x < -2$ **19.** No solution

21. $\dfrac{1}{2} < x < 4$ **23.** $x > 0$ **25.** $-2 < x < 2$ **27.** $2 \leq x < 3$ **29.** $x \geq -5$

31. $x < -1$ or $x > 0$ **33.** $x > 4$ **35.** $x < 3$ **37.** $x \leq 3$ **39.** $x > 9$
41. $5 \leq x < 11$ **43.** $x < -4$ or $x > 13$ **45.** $x \geq 3$ **47.** $x \geq -8$

Checkup 9.3 (*page 405*)

1. $-1 \leq y \leq 2$ **2.** $1 \leq x \leq 6$ **3.** $2 \leq x \leq 6$ **4.** $x > -2$ **5.** No solution

6. $x < -2$ or $x > 4$ **7.** $z > 9$ **8.** $-4 < x < 4$ **9.** $x > 5$ **10.** $-1 < x < \dfrac{3}{2}$

Exercise 9.4 (*page 409*)

1. 8 feet   **3.** 159   **5.** $n < 3$ **7.** $2 < x < 5$ **9.** 350   **11.** 24 feet   **13.** 2400

Review Exercises (*page 413*)

1. $<$ **2.** $>$ **3.** $-10 < -7 < 3$ **4.** $0 < \dfrac{5}{3} < 4$

5.

6.

7.

A number line from -5 to 8 with an open circle at -3 and a closed dot at 7, bracket connecting above from -3 to 7.

8.

A number line from -3 to 3 with closed dots at -2 and $2\frac{5}{2}$, bracket connecting above.

9. $4 < 13$ **10.** $-21 > -30$ **11.** $-5 < 2$ **12.** $-9 > -15$

13. $x \leq 4$ A number line from -3 to 7 with a closed dot at 4 and arrow pointing left.

14. $y < 5$ A number line from -2 to 6 with an open circle at 5 and arrow pointing left.

15. $a \geq -8$ A number line from -10 to 3 with a closed dot at -8 and arrow pointing right.

16. $x > 6$ A number line from 0 to 9 with an open circle at 6 and arrow pointing right.

17. $x > -5$ **18.** $x \geq 2$ **19.** $x > \dfrac{-9}{5}$ **20.** $y \leq -6$ **21.** $x \leq \dfrac{18}{5}$ **22.** $\dfrac{-2}{3} \leq y \leq 2$

23. $-2 < x \leq 6$ **24.** $\dfrac{-25}{3} \leq x \leq \dfrac{25}{3}$ **25.** $-9 < x \leq 3$ **26.** $x > 3$ **27.** $x \geq 6$

28. $x < -1$ or $x > 3$ **29.** 12   **30.** 12 by 24 feet   **31.** 300

Chapter 10
Exercise 10.1 (page 425)
1. Yes   **3.** No   **5.** Yes   **7.** No   **9.** No   **11.** Yes   **13.** Answers will vary.
15. Answers will vary.   **17.** Answers will vary.   **19.** Answers will vary.
21. Answers will vary.   **23.** Answers will vary.   **25.** Answers will vary.
27. Answers will vary.   **29.** $y = 4x + 8$ **31.** $x + y = 28$ **33.** $l = 2w - 5$
35. $x + 5y = 100$ **37.** $2l + 2w = 48$ **39.** $3x + 2y = 20$

Checkup 10.1 (page 427)
1. Yes   **2.** No   **3.** Yes   **4.** No   **5.** Answers will vary.   **6.** Answers will vary.
7. Answers will vary.   **8.** $y = 2x - 500$ **9.** $2x + 3y = 17$ **10.** $2l + 2w = 196$

Exercise 10.2 (page 441)
1.–7.

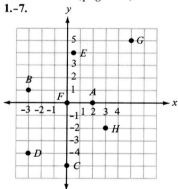

9. $A(2, 5)$ **11.** $C(2, -1)$ **13.** $E(0, 0)$ **15.** $G(-3, -3)$

17.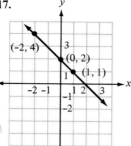

19.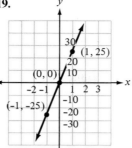

21.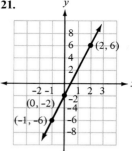

23.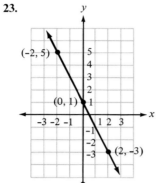

25.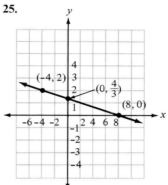

27.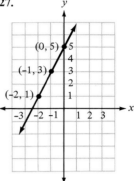

Checkup 10.2 (*page 442*)

1.–3.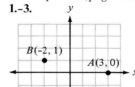

4. $A(-2, 0)$ **5.** $B(4, 2)$ **6.** $C(3, -2)$

7.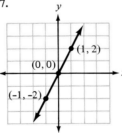

8.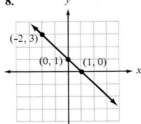

9.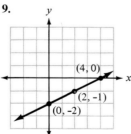

10.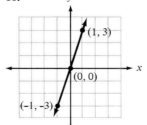

Exercise 10.3 (*page 457*)

1.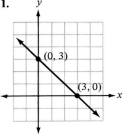

3.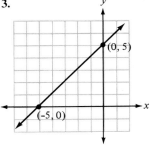

5.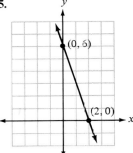

7.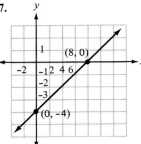

9.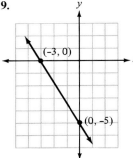

11.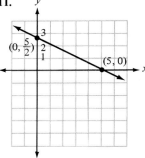

13.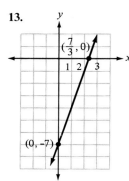

15.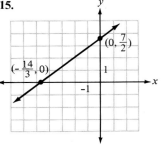

17.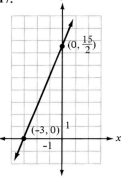

19.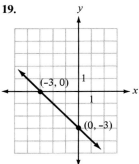

21.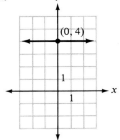

23.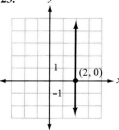

25.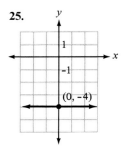

27.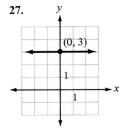

29.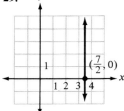

31.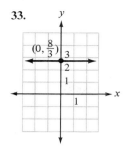

33.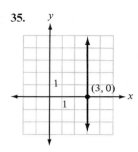

35.

Checkup 10.3 (page 458)

1.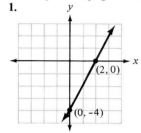

2.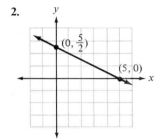

3.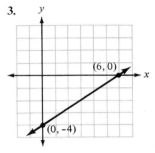

4.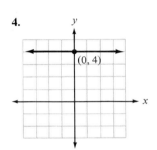

5.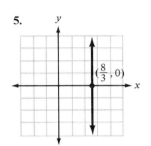

6.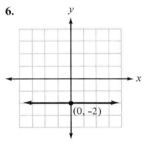

Review Exercises (page 461)
1. Yes   **2.** Yes   **3.** Answers will vary.   **4.** Answers will vary.   **5.** Answers will vary.
6. Answers will vary.   **7.** $l = 3w - 5$ **8.** $5x + 3y = 45$
9. 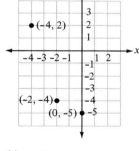   **10.** $A(-4, 1)$, $B(2, 0)$, $C(3, -4)$, $D(4, 3)$

11.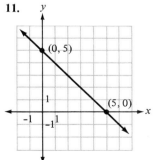

12.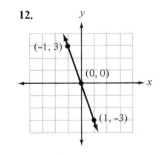

13.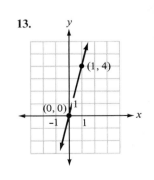

14.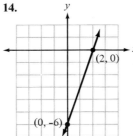

15.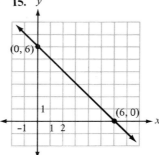

16.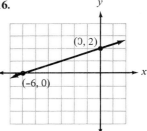

17.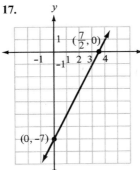

18.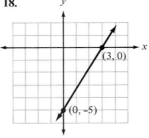

19.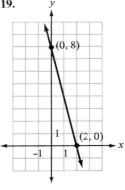

20.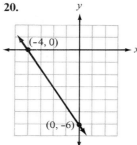

21.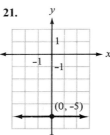

22.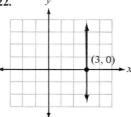

23.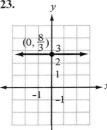

24.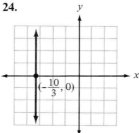

CHAPTER 11

Exercise 11.1 (page 469)
1. $(1, 3)$ **3.** $(-2, 6)$ **5.** $(-2, -12)$ **7.** Parallel lines   **9.** $(2, 0)$ **11.** Same line
13. $(2, 3)$ **15.** $(7, 4)$ **17.** $(4, 1)$ **19.** Parallel lines

Checkup 11.1 (page 470)
1. $(1, -1)$ **2.** Parallel lines   **3.** $(-3, 4)$ **4.** Same line

Exercise 11.2 (page 479)
1. $(2, 3)$ **3.** $(-1, -2)$ **5.** $(6, 19)$ **7.** $(17, 5)$ **9.** Ordered pairs satisfying $y = 2x - 4$
11. $(-5, -9)$ **13.** $\left(8, \dfrac{-9}{2}\right)$ **15.** $\left(\dfrac{24}{5}, \dfrac{6}{5}\right)$ **17.** $\left(3, \dfrac{7}{3}\right)$ **19.** No solution   **21.** $(1, 3)$
23. $(-23, -10)$ **25.** $\left(\dfrac{1}{10}, \dfrac{17}{10}\right)$

Checkup 11.2 (*page 480*)

1. $(3, 6)$ **2.** No solution   **3.** $\left(6, \dfrac{5}{2}\right)$ **4.** $(2, 2)$ **5.** $\left(\dfrac{5}{3}, \dfrac{4}{3}\right)$

Exercise 11.3 (*page 491*)

1. $(1, -1)$ **3.** $(-1, -1)$ **5.** $\left(4, \dfrac{7}{4}\right)$ **7.** $(2, -2)$ **9.** $\left(\dfrac{-1}{2}, -1\right)$ **11.** $\left(\dfrac{1}{2}, \dfrac{3}{4}\right)$
13. $(2, 1)$ **15.** $\left(\dfrac{-11}{3}, 4\right)$ **17.** No solution   **19.** $\left(0, \dfrac{-5}{2}\right)$ **21.** $\left(\dfrac{-2}{3}, \dfrac{11}{4}\right)$
23. $(1, 1)$ **25.** Ordered pairs satisfying $14x + 7y = 7$ **27.** $\left(1, \dfrac{-1}{3}\right)$ **29.** $(1, 1)$

Checkup 11.3 (*page 492*)

1. $(2, 0)$ **2.** $\left(\dfrac{-2}{3}, -2\right)$ **3.** No solution   **4.** $\left(\dfrac{8}{7}, -2\right)$ **5.** $\left(\dfrac{5}{2}, \dfrac{9}{10}\right)$

Exercise 11.4 (*page 497*)

1. -1 and 7 **3.** 8 by 21 ft   **5.** 13 five-dollar bills, 12 one-dollar bills
7. 9 books at \$2, 11 books at \$3   **9.** \$3000 invested in Abell's and \$5000 invested in Betti's
11. Jeans \$16, tops \$12   **13.** $24°$ and $66°$

Review Exercises (*page 501*)

1. $(2, 3)$ **2.** $(2, 1)$ **3.** $(-14, -7)$ **4.** $(-1, 2)$ **5.** $(-1, 3)$ **6.** $\left(\dfrac{1}{2}, \dfrac{-3}{2}\right)$
7. $(-2, -8)$ **8.** No solution   **9.** $(13, 3)$ **10.** $(-4, 1)$ **11.** $(5, 2)$ **12.** $(3, 2)$
13. No solution   **14.** $(11, -1)$ **15.** Ordered pairs satisfying $-6x + 5y = 9$ **16.** $(1, 2)$
17. $(0, 8)$ **18.** $(-10, -5)$ **19.** $(3, -2)$ **20.** $(5, 2)$ **21.** 10 and 12
22. 24 by 34 meters   **23.** \$450 at National, \$1250 at Union

CHAPTER 12

Exercise 12.1 (*page 519*)

1. $8x^6$ **3.** a^3 **5.** $\dfrac{5}{y^3}$ **7.** $\dfrac{-27a^6}{b^3}$ **9.** $\dfrac{1}{9x^2}$ **11.** $\dfrac{-1}{125}$ **13.** $\dfrac{25}{16}$ **15.** $\dfrac{1}{36x^2}$ **17.** $\dfrac{6}{x^2}$
19. 4 **21.** $\dfrac{1}{a^3}$ **23.** $-10x^4$ **25.** $\dfrac{-15}{x^3}$ **27.** $\dfrac{1}{a^{12}}$ **29.** y **31.** $\dfrac{9}{x^3}$ **33.** $\dfrac{1}{64x^3}$
35. $\dfrac{27}{a^{12}}$ **37.** $\dfrac{y^{10}}{64}$ **39.** $\dfrac{25a^6}{b^{10}}$ **41.** $\dfrac{x^{10}y^6}{81}$ **43.** $\dfrac{1}{7}$ **45.** x^{15} **47.** $\dfrac{3}{5a^{10}}$ **49.** $\dfrac{1}{3a^7b^{12}}$
51. $-2x^2$ **53.** $\dfrac{1}{x}$ **55.** $\dfrac{27}{8x^3}$ **57.** a^2b^4 **59.** $\dfrac{x^{12}}{y^6}$ **61.** $\dfrac{25b^8}{a^2}$ **63.** $\dfrac{1}{16y^6}$

Checkup 12.1 (*page 521*)

1. $\dfrac{1}{32x^5}$ **2.** a^4 **3.** $\dfrac{3}{y^2}$ **4.** 2 **5.** $-3x^7$ **6.** $\dfrac{a^2}{2}$ **7.** $\dfrac{25}{49x^2}$ **8.** $\dfrac{4x^2}{3y^4}$ **9.** $-5x^4$
10. $\dfrac{3}{x^8}$

Exercise 12.2 (*page 533*)

1. Rational   **3.** Irrational   **5.** Rational   **7.** Irrational   **9.** Rational   **11.** Irrational
13. Irrational   **15.** Rational   **17.** 12 **19.** a^2 **21.** 8 **23.** x^5 **25.** $5b$ **27.** $\dfrac{x}{6}$
29. $3\sqrt{5}$ **31.** $10\sqrt{6}$ **33.** $5\sqrt{3}$ **35.** $2\sqrt{11}$ **37.** x **39.** $y^2\sqrt{y}$ **41.** $xy\sqrt{y}$
43. $4a\sqrt{2a}$ **45.** $2x\sqrt{y}$ **47.** $9a^2b^2\sqrt{a}$ **49.** $8xy^2z\sqrt{3z}$ **51.** $6a^3b\sqrt{5abc}$ **53.** $\dfrac{2x}{5}$
55. $\dfrac{4a\sqrt{2a}}{b^2}$ **57.** $\dfrac{2x\sqrt{10x}}{3y}$ **59.** $\dfrac{ab\sqrt{6}}{10}$ **61.** $5x$ **63.** $x^2\sqrt{3}$ **65.** $\dfrac{\sqrt{y}}{11}$

Checkup 12.2 (*page 535*)

1. 7 **2.** $3a$ **3.** 12 **4.** $4\sqrt{5}$ **5.** $5y^2\sqrt{3}$ **6.** $3ab\sqrt{5a}$ **7.** $\dfrac{5x}{6y}$ **8.** $\dfrac{4x^2\sqrt{2x}}{y^3}$
9. $5a^2b$ **10.** $a^2\sqrt{3}$

Exercise 12.3 (page 545)
1. $3\sqrt{3}$ **3.** $11\sqrt{x}$ **5.** $-7\sqrt{3}$ **7.** $5\sqrt{6a}$ **9.** $9 - 6\sqrt{3}$ **11.** $23\sqrt{3x}$
13. $10\sqrt{7} - 20\sqrt{2}$ **15.** $11\sqrt{2}$ **17.** $-5x\sqrt{3}$ **19.** $\dfrac{-12\sqrt{a}}{5}$ **21.** $\sqrt{21}$ **23.** 12
25. $11a$ **27.** $x - 1$ **29.** $3a\sqrt{5}$ **31.** 30 **33.** $-120x$ **35.** $15x\sqrt{2}$ **37.** $2\sqrt{5} - 5$
39. $2\sqrt{6} + 12$ **41.** $-3\sqrt{x} + 6\sqrt{5} - 3\sqrt{6}$ **43.** $x - 16\sqrt{x}$ **45.** $31 - 10\sqrt{6}$
47. $x - 25$ **49.** $23 - 7\sqrt{11}$ **51.** $2x - 5\sqrt{x} - 3$ **53.** $5 - 5\sqrt{3} - \sqrt{2} + \sqrt{6}$
55. $11 + \sqrt{5}$ **57.** $8 - 2\sqrt{3x} - 3x$ **59.** -4 **61.** $x - 100$ **63.** $58 - 23\sqrt{14}$
65. $y - 2\sqrt{7y} + 7$ **67.** $2a^2 - 10a\sqrt{2} + 25$

Checkup 12.3 (page 547)
1. $5\sqrt{6x}$ **2.** $3\sqrt{3}$ **3.** $3\sqrt{3}$ **4.** $3a\sqrt{5}$ **5.** $-30\sqrt{14}$ **6.** $7\sqrt{5} - 5\sqrt{2}$
7. $2\sqrt{6} + 4\sqrt{3}$ **8.** $\sqrt{10} - 8\sqrt{2} - 14$ **9.** $-18 - \sqrt{5}$ **10.** $x - 3$

Exercise 12.4 (page 555)
1. $\sqrt{3}$ **3.** $\dfrac{2\sqrt{7}}{7}$ **5.** $2\sqrt{2}$ **7.** $\dfrac{6\sqrt{10}}{5}$ **9.** $\dfrac{9\sqrt{2}}{8}$ **11.** $\dfrac{\sqrt{10}}{4}$ **13.** $\sqrt{2}$ **15.** $\dfrac{\sqrt{2}}{2}$
17. $\dfrac{2\sqrt{5}}{5}$ **19.** $\dfrac{\sqrt{3x}}{x}$ **21.** $\dfrac{\sqrt{5} + \sqrt{10}}{5}$ **23.** $2 - 3\sqrt{2}$ **25.** $\dfrac{2\sqrt{5} + 5\sqrt{2}}{10}$
27. $2 - \sqrt{3}$ **29.** $-12 - 4\sqrt{6}$ **31.** $-\sqrt{2} - 2$ **33.** $\sqrt{6} - \sqrt{2}$ **35.** $2\sqrt{15} - 4\sqrt{3}$
37. $\sqrt{2}$ **39.** $5 + 2\sqrt{6}$

Checkup 12.4 (page 557)
1. $\sqrt{6}$ **2.** $2\sqrt{10}$ **3.** $\dfrac{7\sqrt{3}}{6}$ **4.** $\dfrac{\sqrt{10}}{5}$ **5.** $\dfrac{2\sqrt{6}}{3}$ **6.** $\sqrt{6} - 5\sqrt{2}$ **7.** $\dfrac{1 + \sqrt{3}}{2}$
8. $\dfrac{2\sqrt{3} + 1}{11}$ **9.** $\sqrt{11} - 2$ **10.** $\dfrac{11 - 4\sqrt{7}}{3}$

Exercise 12.5 (page 567)
1. $x = \pm 2\sqrt{3}$ **3.** $a = \pm\dfrac{\sqrt{5}}{3}$ **5.** $x = \pm\dfrac{\sqrt{5}}{5}$ **7.** $y = \pm\dfrac{\sqrt{10}}{2}$ **9.** No real solution
11. $y = \pm\sqrt{5}$ **13.** $a = \pm\sqrt{2}$ **15.** $x = 0$ **17.** $a = \pm 4\sqrt{5}$ **19.** $x = \dfrac{-1 \pm \sqrt{13}}{2}$
21. $a = \dfrac{5 \pm 3\sqrt{5}}{2}$ **23.** $x = 2 \pm \sqrt{7}$ **25.** $y = 1 \pm \sqrt{2}$ **27.** $a = \dfrac{-7 \pm \sqrt{53}}{2}$
29. $x = \dfrac{1 \pm \sqrt{7}}{3}$ **31.** No real solution **33.** $y = 3 \pm \sqrt{5}$ **35.** $z = \dfrac{-1 \pm \sqrt{31}}{2}$

Checkup 12.5 (page 568)
1. $x = \pm\sqrt{2}$ **2.** $a = \pm\sqrt{5}$ **3.** $y = \pm 4\sqrt{2}$ **4.** $x = \pm 2\sqrt{5}$ **5.** $x = 0$
6. $x = -2 \pm 2\sqrt{2}$ **7.** $x = \dfrac{3 \pm \sqrt{5}}{2}$ **8.** $x = \dfrac{3 \pm \sqrt{3}}{3}$ **9.** $x = \dfrac{5 \pm \sqrt{33}}{4}$
10. No real solution

Review Exercises (page 571)
1. $\dfrac{1}{25x^2}$ **2.** $\dfrac{4}{y^3}$ **3.** $-7x^9$ **4.** $\dfrac{a^3}{4}$ **5.** $\dfrac{16}{81x^4}$ **6.** $\dfrac{-5}{2y}$ **7.** $3x^4$ **8.** $\dfrac{5}{x^7}$ **9.** 12
10. $8a$ **11.** 13 **12.** $3y^2\sqrt{5}$ **13.** $4ab\sqrt{3a}$ **14.** $\dfrac{3x^2\sqrt{3x}}{y^2}$ **15.** $7a^3b$ **16.** $3a^3$
17. $3\sqrt{5x}$ **18.** $7a\sqrt{3}$ **19.** $-60\sqrt{6}$ **20.** $3\sqrt{2} - 2\sqrt{7}$ **21.** $15\sqrt{6}$
22. $\sqrt{15} - 6\sqrt{5} - 12$ **23.** $-32 + 5\sqrt{6}$ **24.** $a - 5$ **25.** $5\sqrt{3}$ **26.** $\dfrac{\sqrt{5}}{2}$ **27.** $\dfrac{\sqrt{35}}{5}$
28. $2 - 7\sqrt{3}$ **29.** $\dfrac{\sqrt{2} + \sqrt{6}}{2}$ **30.** $\dfrac{5\sqrt{11} + 1}{14}$ **31.** $-(11 + 6\sqrt{3})$ **32.** $\dfrac{17 - 4\sqrt{13}}{9}$
33. $y = \pm 4\sqrt{3}$ **34.** $a = \pm\sqrt{2}$ **35.** $x = 0$ **36.** $x = -3 \pm 2\sqrt{3}$ **37.** $x = 2 \pm \sqrt{3}$
38. $x = \dfrac{5 \pm \sqrt{15}}{5}$ **39.** $x = \dfrac{3 \pm \sqrt{17}}{4}$ **40.** $x = \dfrac{2 \pm \sqrt{10}}{3}$

Index